T0275827

Scaling, self-similarity, and intermediate asymptotics

Cambridge Texts in Applied Mathematics

MANAGING EDITOR
D.G. Crighton, University of Cambridge

EDITORIAL BOARD
M. Ablowitz, University of Colorado; J.-L. Lions, Collège de France;
A. Majda, New York University; J. Ockendon, University of Oxford;
E.B. Saff, University of South Florida

Maximum and Minimum Principles
M.J. Sewell

Solitons
P.G. Drazin and R.S. Johnson

The Kinematics of Mixing
J.M. Ottino

Introduction to Numerical Linear Algebra and Optimisation
Philippe G. Ciarlet

Integral Equations
David Porter and David S.G. Stirling

Perturbation Methods
E.J. Hinch

The Thermomechanics of Plasticity and Fracture
Gerard A. Maugin

Boundary Integral and Singularity Methods for Linearized Viscous Flow
C. Pozrikidis

Nonlinear Wave Processes in Acoustics
K. Naugolnykh and S. Ostrovsky

Nonlinear Systems
P.G. Drazin

Stability, Instability and Chaos
Paul Glendinning

Viscous Flow
H. Ockendon and J.R. Ockendon

Applied Analysis of the Navier-Stokes Equations
Charles R. Doering and J.D. Gibbon

Scaling, self-similarity, and intermediate asymptotics

Grigory Isaakovich Barenblatt

Emeritus G. I. Taylor Professor of Fluid Mechanics, University of Cambridge;

Fellow of Gonville and Caius College, Cambridge

CAMBRIDGE
UNIVERSITY PRESS

CAMBRIDGE UNIVERSITY PRESS
Cambridge, New York, Melbourne, Madrid, Cape Town, Singapore, São Paulo, Delhi

Cambridge University Press
32 Avenue of the Americas, New York, NY 10013-2473, USA

www.cambridge.org
Information on this title: www.cambridge.org/9780521435222

© Cambridge University Press 1996

This publication is in copyright. Subject to statutory exception
and to the provisions of relevant collective licensing agreements,
no reproduction of any part may take place without the written
permission of Cambridge University Press.

First published 1996
Reprinted 1997, 2002, 2004, 2005

A catalog record for this publication is available from the British Library

ISBN 978-0-521-43516-1 hardback
ISBN 978-0-521-43522-2 paperback

Transferred to digital printing 2009

Cambridge University Press has no responsibility for the persistence or
accuracy of URLs for external or third-party Internet websites referred to in
this publication, and does not guarantee that any content on such websites is,
or will remain, accurate or appropriate. Information regarding prices, travel
timetables and other factual information given in this work are correct at
the time of first printing but Cambridge University Press does not guarantee
the accuracy of such information thereafter.

To the glowing memory of his beloved parents,

Dr. Nadezhda Veniaminovna Kagan, physician–virologist,
heroically lost for the sake of the healthy future of humanity,

and

Dr. Isaak Grigorievich Barenblat, physician–endocrinologist,

the author dedicates his work.

Contents

Preface

Scaling (power-law) relationships have wide application in science and engineering. Well-known examples of scaling relations are the following (we will discuss them later in detail):

G.I. Taylor's scaling law for the shock-wave radius r_f after a nuclear explosion,[†]

$$r_f = \left(\frac{Et^2}{\rho_0} \right)^{1/5} ;$$

the scaling law for the velocity distribution u near a wall in a turbulent shear flow,[‡]

$$u = Ay^n;$$

the scaling law for the breathing rate R of animals,[§]

$$R = AW^n;$$

and many others.

A very common view is that these scaling or power-law relations are nothing more than the simplest approximations to the available experimental data, having no special advantages over other approximations. It is not so. Scaling laws give evidence of a very deep property of the phenomena under consideration – their *self-similarity*: such phenomena reproduce themselves, so to speak, in time and space. Self-similar

---

[†] E explosion energy; t, time after explosion; ρ_0, air density.

[‡] y distance from the wall; A, n, constants.

[§] W body mass of an animal; A, n constants.

phenomena entered mathematical physics rather early, perhaps with the famous memoir of Fourier (1822) on the analytical theory of heat conduction. In this he arrived at a 'source-type' solution[†]

$$\theta(x,t) = \frac{A}{\sqrt{t}} f\left(\frac{x}{\sqrt{t}}\right), \quad f(\xi) = e^{-\xi^2/4}, \quad A = \text{Const}$$

to the heat conduction equation

$$\partial_t \theta = \partial^2_{xx} \theta.$$

Subsequently the phenomena under consideration, and the equations entering their mathematical models, became more and more complicated and very often nonlinear. Obtaining self-similar solutions was considered as a success, particularly in the pre-computer era. Indeed, the construction of such solutions always reduces to solving the boundary-value problems for ordinary, not partial, differential equations, which was considered as a substantial simplification. Moreover, in 'self-similar' coordinates $\theta\sqrt{t}$, $x/\sqrt{t}$ (and analogous coordinates in other problems), self-similar phenomena become time independent. This gives important evidence of a certain type of stabilization. Thus very often obtaining a self-similar solution was the only way to understand the qualitative features of the phenomena.

The exponents of the independent variables x, t in self-similar variables such as $\theta\sqrt{t}$, $x/\sqrt{t}$ in the heat conduction problem mentioned above were obtained at the outset in some simple way giving no special trouble to the researcher, often *dimensional analysis*. Dimensional analysis is merely a simple sequence of rules based on the fundamental *covariance principle* of physics: *all physical laws can be represented in a form equally valid for all observers*. Such classical self-similarities were discussed and summarized in a book by Sedov (1959) and in a monographic review by Germain (1973), in which a general approach to problems leading to such solutions was also discussed.

In fact, the situation changed drastically after the paper by Guderley (1942), in which a solution to the problem of a very intense implosion (a converging spherical or cylindrical shock wave) was obtained, and the papers by von Weizsäcker (1954) and Zeldovich (1956) treating the plane analogue of the implosion wave problem, the problem of an *impulsive*

[†] Here θ, x, t are the temperature, the spatial coordinate, and time. This solution is remarkable for two reasons. Firstly, the temperature θ, a function of two variables x, t, is represented via a function of one variable $x/\sqrt{t}$. Furthermore, according to this solution the temperature distributions at various times can be obtained one from another by a similarity transformation: the solution remains similar to itself.

loading. In these problems a delicate analytical procedure, qualitative investigation of the portrait in a phase plane, was needed to obtain the power n in which the time enters the self-similar variable x/t^n. These powers appeared generally speaking to be certain transcendental numbers rather than simple fractions as for classical self-similarities. In fact solutions with such 'anomalous' dimensions had appeared for different variables even earlier. I am referring to the fundamental papers by Kolmogorov, Petrovsky, and Piskunov (1937), and by Fisher (1937), devoted to the propagation of an advantageous gene, and by Zeldovich and Frank–Kamenetsky (1938a, b), dealing with flame propagation in gases. In these papers the wave-type solutions $\theta(x - \lambda t)$ of the nonlinear parabolic equation

$$\partial_t \theta = \partial_{xx}^2 \theta + f(\theta)$$

were considered, and the wave phase speed λ has been calculated by a complicated analytical procedure: phase-plane portrait investigation. Transforming the variables $x = \ln \xi, t = \ln \tau$ one arrives at the same problem of determining the exponent of τ in the self-similar variable ξ/τ^λ.

An important question arose: what is the real nature of such a difference in self-similar solutions? To understand that, in the papers of Barenblatt and Sivashinsky (1969, 1970) two special problems were considered, containing a parameter that entered to the problem's formulation naturally. For a single value of this parameter a classical self-similar solution appeared, in which all powers were obtained from dimensional considerations. However, for all other values of the parameter anomalous dimensions appeared as continuous functions of the parameter; they are obtained from the solution to a nonlinear eigenvalue problem. These results allowed one to understand the fundamental nature of the difference between the two types of self-similar solution mentioned above. Indeed, self-similar solutions are always 'intermediate asymptotics' to the solutions of more general problems, valid for times, and distances from boundaries, large enough for the influence of the fine details of the initial and/or boundary conditions to disappear, but small enough that the system is far from the ultimate equilibrium state. So, the reason for the difference is the character of these intermediate asymptotics. If an asymptotics is represented by a function that tends to a finite limit when approaching the self-similar state, self-similarity of the first kind appears. If, however, a finite (different from zero) limit does not exist, but the asymptotics is a power-type (scaling) one, with the exponents depending on the fine details of the analytical properties of pre-self-similar behaviour, self-similarity of the second kind occurs. So, it became clear how anomalous, transcendental dimensions appear in self-similar solutions. It is also the case that only a power-type asymptotics preserves self-similarity.

Independently but later an activity started in theoretical physics, basically in quantum field theory and in the theory of phase transitions in statistical physics, related to the scaling and renormalization group. Anomalous dimensions entered the language of physicists. The names and works of Stückelberg and Peterman (1953), Gell-Mann and Low (1954), Bogolyubov and Shirkov (1955), Kadanoff (1966; see also Kadanoff *et al*, 1967), Patashinsky and Pokrovsky (1966), and Wilson (1971), as well as the books by Bogolubov and Shirkov (1959), Ma (1976), Amit (1989), and Goldenfeld (1992) should be mentioned. It is essential to emphasize, however, that in contrast with the researchers in applied mechanics mentioned above, researchers in theoretical physics considered problems where rigorous mathematical formulations such as initial and/or boundary value problems for partial differential equations were lacking.

Rather early it became clear that the concepts of intermediate asymptotics developed in applied mechanics and the concepts of scaling and renormalization group developed in theoretical physics are closely related. This relationship was emphasized in the author's first book concerning this subject (Barenblatt, 1979), the Foreword to which, by Acad. Ya.B. Zeldovich, follows this Preface; in that book, theoretical physicists were invited to look at how the approach of intermediate asymptotics can work in problems previously considered by the renormalization group approach.

In a remarkable series of works by N. Goldenfeld, Y. Oono, O. Martin, and their students (see the book by Goldenfeld, 1992) several problems in continuum mechanics (filtration, elasticity, turbulence, etc.) which had been solved previously by the method of intermediate asymptotics were solved by the traditional renormalization group method. Moreover, on the one hand using the singular expansion method widely applied in theoretical physics (ϵ-expansion) Goldenfeld, Oono and their colleagues were able to obtain some instructive and useful approximate solutions to these problems. On the other hand, they obtained by the method of intermediate asymptotics the solutions to several problems of statistical physics, solved previously by the renormalization group approach.

These important works helped to represent in final form the renormalization group approach from the viewpoint of intermediate asymptotics. In particular it appeared useful to give a proper definition of the renormalization group using the concept of intermediate asymptotics. Ultimately the works by Goldenfeld and his colleagues were among the basic stimuli for me to write this book. Of course, in this writing I have used essential materials from my previous books devoted to this subject (Barenblatt, 1979, 1987), so the continuity is completely preserved.

I want to express in conclusion my deep gratitude to the memory of my great mentors, A.N. Kolmogorov and Ya.B. Zeldovich whose approach in particular to self-similarities and intermediate asymptotics greatly influenced my views.

I want to thank Professor D.G. Crighton, FRS, for his kind offer to publish this book in the series under his editorship at Cambridge University Press. I am pleased to express my deep gratitude to him, to Professor G.K. Batchelor, FRS, and to Professor H.K. Moffatt, FRS, for the honour and pleasure of writing this book here at the Department of Applied Mathematics and Theoretical Physics in Cambridge. I am grateful to Professor M.D. van Dyke for his valuable advice. I thank Miss Sarah Kirkup for her help in preparing the manuscript.

Foreword

Professor Grigorii Isaakovich Barenblatt has written an outstanding book that contains an attempt to answer the very important question of how to *understand* complex physical processes and how to *interpret* results obtained by numerical calculations.

Progress in numerical calculation brings not only great good but also notoriously awkward questions about the role of the human mind. The human partner in the interaction of a man and a computer often turns out to be the weak spot in the relationship. The problem of formulating rules and extracting ideas from vast masses of computational or experimental results remains a matter for our brains, our minds.

This problem is closely connected with the recognition of patterns. It is not just a coincidence that in both the Russian and English languages the word 'obvious' has two meanings – not only something easily and clearly understood, but also something immediately evident to our eyes. The identification of forms and the search for invariant relations constitute the foundation of pattern recognition; thus, we identify the similarity of large and small triangles, and so on.

Let us assume now that we are studying a certain process, for example a chemical reaction in which heat is released and whose rate depends on temperature. For a wide range of parameters and initial conditions, a completely definite type of solution is obtained – flame propagation. The chemical reaction occurs in a relatively narrow region separating the cold combustible substance from the hot combustion products; this region moves relative to the combustible substance with a velocity that is independent of the initial conditions. (Of course, the very occurrence of combustion depends on the initial conditions.)

This result can be obtained by direct numerical integration of the partial differential equations that describe the heat transfer, diffusion, chemical reaction, and (in some cases) hydrodynamics. Such a computational approach is difficult; the result is obtained in the form of a listing of quantities such as temperature and concentration as functions of temporal and spatial coordinates. To make manifest the flame propagation, i.e., to extract from the mass of numerical material the regime of uniform temperature propagation, $T(x - ut)$, is a difficult problem! It is necessary to know the type of the solution in advance in order to find it; anyone who has made a practical attempt to apply mathematics to the study of nature knows this truth.

The term 'self-similarity' was coined and is by now widespread: a solution $T(x, t_1)$ at a certain moment t_1 is similar to the solution $T(x, t_0)$ at a certain earlier moment. In the case of uniform propagation considered above, similarity is replaced by simple translation. Similarity is connected with a change of scales:

$$T = \left(\frac{t_1}{t_0}\right)^n T\left(x\left(\frac{t_1}{t_0}\right)^m, t_0\right)$$

or

$$T = \varphi(t)\Psi(x/\xi(t)).$$

In geometry, this type of transformation is called an affine transformation. The existence of a function Ψ that does not change with time allows us to find a similarity of the distributions at different moments.

Barenblatt's book contains many examples of analytic solutions of various problems. The list includes heat propagation from a source in the linear case (for constant thermal conductivity) and in the nonlinear case, and also in the presence of heat loss. The problem of the hydrodynamic propagation of energy from a localized explosion is also considered. In both cases, the problem in its ordinary formulation – without loss – was solved many years ago; in these problems the dimensions of the constants that characterize the medium (its density, equations of state, and thermal conductivity) and the dimensions of energy uniquely dictate the exponents of self-similar solutions.

However, with properly introduced losses the problems turn out to be essentially different. If $dE/dt = -\alpha E^{3/2}/R^{5/2}, dR/dt = \beta E^{1/2}/R^{3/2}$ (E being the total energy referred to the initial density of the gas, R the radius of the perturbed domain and t the time) so that $dE/dR =$

$-\gamma E/R$, then the conservation of energy does not hold:

$$E \sim R^{-\gamma}, \; E = E_0 R_0^{\gamma} R^{-\gamma} \neq \text{const};$$

however, self-similarity remains.

The dimensionless numbers α, β, and γ depend on the functions describing the solution, but the equations that determine these solutions contain indeterminate exponents. Mathematically we have to deal with the determination, from nonlinear ordinary differential equations and their boundary conditions, of certain numbers that can be called eigenvalues.

The new exponents in the problem are not necessarily integers or rational fractions; as a rule they are transcendental numbers that depend continuously on the parameters of the problem, including the parameters of energy loss. Thus arises a new type of self-similar solution, which we shall call the second type, reserving the title of first type for the case where naive dimensional analysis succeeds.

An important point arises here. The solution does not describe the point source asymptotically: if R_0 (the value of R at $t = 0$) is taken to be equal to zero, then it must necessarily be that $E_0 = \infty$ for $t = 0$, which is physically inconsistent. Hence the new solution is considered as an intermediate asymptotics. We assume that up to a certain finite time t_0 there is no loss. At this moment, when the radius of the perturbed domain reaches the finite value R_0, we switch on the loss. Or, to be more general, we can start with a finite energy E created by some other means, that has already spread out to the finite radius R_0. It is assumed that asymptotically, for sufficiently large time, the solution assumes a self-similar form corresponding to the given loss.

We want to emphasize the asymptotic character of the self-similar solution for $t \gg t_0$. In nonlinear problems, exact special solutions sometimes appear to be useless: since there is no principle of superposition, one cannot immediately find a solution of the problem for arbitrary initial conditions.

Here asymptotic behaviour is the key that partially plays the role of the lost principle of superposition. However, for arbitrarily given initial conditions this asymptotic behaviour must be proved. The problem is difficult, and in many cases numerical computations give only a substitute for rigorous analytic proof.

The preceding arguments may seem unusual in a Foreword: but I wanted, using the simplest examples, to introduce the reader as quickly as possible to the advantages and difficulties of the new world of solutions of the second kind.

There are also other types of solutions, among which convergent spherical shock waves are the most important. In this case there is no external loss, but the region in which self-similarity holds is contracting; it it therefore impossible to assume that the entire energy is always concentrated in the shrinking region, and this energy in fact decreases according to a power law, since part of the energy remains in the exterior regions of the gas. Again it is necessary to find the exponents as eigenvalues of a nonlinear operator.

The specific character of this class of equations is connected with the finiteness of the speed of sound; the point where the phase velocity of propagation of a self-similar variable is equal to the velocity of sound plays a decisive role in the construction of the solution.

Barenblatt also discusses in his book another problem of analogous type: the problem of a strong impulsive load in a half-space filled with gas. This problem abounds in paradoxes. In particular, why do the laws of conservation of energy and momentum not make it possible to determine the exponents? The answer to this question is contained in chapter 4, and it would be against the rules to give it here in the Foreword.

Problems involving the nonlinear propagation of waves on the surface of a heavy fluid, described by the Korteweg–de Vries equation, give a remarkable example. Here there are long-established and well-known solutions describing solitary waves (called 'solitons'), propagating with a velocity dependent on the amplitude. This example is remarkable in that there exist theorems proving the stability of solitons even after their collisions, and theorems determining the asymptotic behaviour of initial distributions of general type, which are transformed into a sequence of solitons. At first suggested by numerical computations, these properties are now rigorously proved by analytic methods of extraordinary beauty. In these solutions all the properties of ideal self-similar solutions of the second kind appear.

In some sense the problems of turbulence, considered at the end of the book, differ from those mentioned above. These are farther from my interests and I will not dwell on them here. A complete outline of all that is contained in the book can be found in the Table of Contents and should not be sought in the Foreword.

We shall now return to the nature of the book as a whole; we shall not hesitate to repeat for the general situation some considerations that have already been presented above in connection with simple examples.

The problems are chosen carefully. Each of them taken separately is a pearl, important and cleverly presented. In the solution of many of the problems the role of the author was essential, and this gives to the presentation the flavour of something lived. But I must emphasize that the importance of this book far exceeds its value as a collection of interesting special examples; from the special problems considered, very general ideas develop.

Most of the problems are nonlinear. What is the use of special solutions if there is no principle of superposition? The fact is that as a rule these special solutions represent the asymptotics of a wide class of other more general solutions that correspond to various initial conditions. Under these circumstances the value of exact special solutions increases immensely. This aspect of the question is reflected in the title of the book in the words 'intermediate asymptotics'. The value of solutions as asymptotics depends on their stability. The questions of the stability of a solution and of its behavior under small perturbations are also considered in this book; in particular, there is presented a rather general approach to the stability of invariant solutions developed in a paper by Barenblatt and myself.

The very idea of self-similarity is connected with the group of transformations of solutions. As a rule, these groups are already represented in the differential (or integro-differential) equations of the process. The groups of transformations of equations are determined by the dimensions of the variables appearing in them; the transformations of the units of time, length, mass, etc. are the simplest examples. This type of self-similarity is characterized by power laws with exponents that are simple fractions defined in an elementary way from dimensional considerations.

Such a course of argument has led to results of immense and permanent importance. It is sufficient to recall the theory of turbulence and the Reynolds number, of linear and nonlinear heat propagation from a point source, and of a point explosion. Nevertheless, we shall see that dimensional analysis determines only a part of the problem, the tip of the iceberg; we shall call the corresponding solutions *solutions of the first kind*, as mentioned above. We shall reserve the name *solutions of the second kind*, for the large and ever growing class of solutions for which the exponents are found in the process of solving the problem, analogously to the determination of eigenvalues for linear equations. For this case, conservation laws and dimensional considerations prove to be insufficient.

The establishment of an intrinsic connection between nonlinear propagation problems with solutions of the type $f(x - ut)$ and self-similar problems with solutions of the form $t^n f(x/t^m)$ has turned out to be a very important step. The general procedures for determining the speed parameter u and the powers n, m have, as it turns out, many points of contact. By the same token, self-similarity touches on a new stream of problems arising from the theory of combustion and from applications to chemical technology. Barenblatt's book contains concrete, detailed consideration of certain problems, giving a wealth of information. It also contains brilliant generalizations, foresights touching on developments of the future, and hints about discoveries not yet made.

You can read this book and study it, but you can also use it as a source of inspiration. Possibly this is the best compliment for a book with a title that sounds so special.

Ya. B. Zeldovich
*Member, Academy of Sciences
of the USSR*

0

Introduction

0.1 Dimensional analysis and physical similarity

The starting point of this book is dimensional analysis and it is used throughout. Like unhappy families, every unfortunate scientific idea is unfortunate in its own way. Many of those who have taught dimensional analysis (or have merely thought about how it should be taught) have realized that it has suffered an unfortunate fate.

In fact, the idea on which dimensional analysis is based is very simple, and can be understood by everybody: physical laws do not depend on arbitrarly chosen basic units of measurement. An important conclusion can be drawn from this simple idea, using a simple argument: the functions that express physical laws must possess a certain fundamental property, which in mathematics is called generalized homogeneity or symmetry. This property allows the number of arguments in these functions to be reduced, thereby making it simpler to obtain them (by calculating them or determining them experimentally). This is, in fact, the entire content of dimensional analysis – there is nothing more to it.

Nevertheless, using dimensional analysis, researchers have been able to obtain remarkably deep results that have sometimes changed entire branches of science. The mathematical techniques required to derive these results turn out to be simple and accessible to all. The list of great names involved runs from Newton and Fourier to Maxwell, Rayleigh and Kolmogorov. Among recent developments, it is sufficient to recall the triumph of the Kolmogorov–Obukhov theory in turbulence.

Everyone would like to score a classical triumph. Many people therefore attacked what one would think were almost identical problems using the same simple dimensional analysis approach. Alas, they almost always failed. Dimensional analysis was cursed and reproached for being untrustworthy and unfounded, even mystical. Paradoxically, the reason for this lack of success was that only a few people understood the content and real abilities of dimensional analysis.

It was like the old Deanna Durbin film: a girl with a small suitcase arrives in New York and, in no time, charms the son of a millionaire. Films like this are pleasant to watch. However, if they are treated as a guide to what provincial girls should do, disillusionment is inevitable.

Let us describe here what dimensional analysis is using several simple examples.

From elementary physics, the reader knows that the period θ for small oscillations of a simple pendulum of length l (Figure 0.1) is

$$\theta = 2\pi\sqrt{\frac{l}{g}} \approx 6.28\sqrt{\frac{l}{g}}, \tag{0.1}$$

where g is the gravitational acceleration.

Figure 0.1. A pendulum performs small oscillations. Experiment shows that the period of the small oscillations is independent of the maximum deviation of the pendulum.

Equation (0.1) is usually obtained by deriving and solving a differential equation for the oscillations of the pendulum. We shall now obtain it from completely different considerations without any use of calculus. First of all, we ask ourselves; on what can the period of oscillation of the pendulum depend? It is clear that in principle, it can depend only on (a) the length of the pendulum, (b) the mass of the bob, and (c) the gravitational acceleration – if there were no gravitational force (i.e., under weightless conditions), the pendulum would not oscillate. The length of the pendulum, mass of the bob, period of oscillation, and gravitational acceleration can be written in terms of the numbers l, m, θ, and g, which are obtained in the following way. Definite objects representing units of length, mass, and time are chosen; these are agreed standards, which are either carefully preserved or reproducible. Then, the number l is obtained by measuring the length of the pendulum, i.e., comparing the length of the pendulum with the unit of length. The number m is obtained by comparing the mass of the bob with the unit of mass, and the number θ is obtained by comparing the period of oscillation with the unit of time. The situation is slightly more complicated for the gravitational acceleration:

First, we recall that velocity by its very definition is the ratio of the distance travelled in an infinitesimal time interval to the magnitude of that time interval. We therefore adopt the velocity of uniform motion in which one unit of length is traversed per unit of time as the unit for velocity. Analogously, the acceleration is the variation in velocity over an infinitesimal time interval divided by the magnitude of that time interval. We therefore adopt the acceleration of uniformly accelerated motion in which the velocity increases by one velocity unit per unit time as the unit for acceleration.

Let us now decrease the unit of length by a factor L, the unit of mass by a factor M, and the unit of time by a factor T. We are justified in doing so, and in selecting the abstract positive numbers L, M, and T as we like: the choice of units for mass, length, and time – the fundamental units – is arbitrary. In doing so, since the units have been decreased in magnitude, the numerical values of all lengths increase by a factor L, all masses increase by a factor M, and all times increase by a factor T. The velocity increases by a factor LT^{-1} with respect to its original magnitude under this transformation. Indeed, in uniform motion at a velocity assumed to be equal to the new unit of velocity, the new unit of length (which is a factor L smaller than the original unit) is now traversed in one new unit of time (which is a factor T smaller than the original unit). Because of this, the numerical values of all velocities increase by a factor LT^{-1}. Analogously, the unit of acceleration decreases by a factor LT^{-2} under this transformation of fundamental units. Thus, the numerical values of all accelerations (and, in particular, the gravitational acceleration) increase by a factor LT^{-2}.

Therefore, in general, when the magnitudes of the fundamental units – those in which length, mass, and time are measured – are changed, the numerical value of a physical quantity also changes. The factor which gives the magnitude of this change is determined by the *dimension* of the quantity in question. For example, if the unit of length is decreased by a factor L, the numerical values of all lengths are increased by a factor L. We say that length has dimension L. Analogously, mass has dimension M, time has dimension T, velocity has dimension LT^{-1}, and acceleration has dimension LT^{-2}. We emphasize once again: L, M, and T are nothing more than *abstract positive numbers*. As is evident from these examples, the dimensions of velocity and acceleration are functions of these numbers, in fact *very special* functions of these numbers, power-law monomials. (In chapter 1 the reason why is explained in detail.)

Consider the quantity l/g. This quantity is a ratio of two numbers that depend on the units defined for length and acceleration. If the unit of length is decreased by a factor L, and the unit of time by a factor T, the numerical value of the length in the numerator increases by a factor L, and the numerical value of the acceleration in the denominator increases by a factor LT^{-2}. Consequently, the ratio l/g increases by a factor T^2, and the quantity $(l/g)^{1/2}$ increases by a factor T, i.e., exactly the same factor by which the period of oscillation θ increases in this case. Therefore, the ratio

$$\Pi = \frac{\theta}{\sqrt{l/g}}$$

is invariant under a change of fundamental units. Quantities like Π, which do not change when the fundamental units of measurement are changed, are called *dimensionless*: their dimensions are equal to unity. All other physical quantities are called *dimensional*. A correspondence is set up between each dimensional physical quantity and its dimension, which differs from unity and indicates the factor by which the numerical value of this quantity increases for a given decrease in the magnitude of the fundamental units of measurement.

We shall now proceed from the fact that, just like the period of oscillation θ, the quantity Π may, in principle, depend on these same quantities l, m and g; thus, $\Pi = \Pi(l, m, g)$. Once again we recall that l, m, and g are numbers that hold for one particular system of fundamental units of measurement. We now decrease the unit of mass by some factor M, and leave all the other units unchanged. In this case, the number m increases by an arbitrary factor M, but the numbers Π, l, and g remain unchanged. But this means that the function $\Pi(l, m, g)$ remains unchanged for any change in the argument m while the other two arguments l and g remain unchanged, i.e., that this function is *independent* of m. Next, we decrease the unit of time by some factor T, leaving the unit of length unchanged. Then, in accordance with the above, the numerical value of g increases by a factor T^{-2}. (The dimension of acceleration is actually LT^{-2}, but L is equal to unity in this case.) The quantities Π and l remain unchanged (recall that Π is dimensionless). But this means that the function $\Pi(l, m, g)$ is also independent of g. Finally, we decrease the unit of length by some factor L. The numerical value of the last remaining argument l then increases by a factor L, and the dimensionless quantity Π once again remains unchanged. This means that Π is also independent of l, and is therefore completely independent of all the parameters. Therefore, it is in fact a constant:

$$\Pi = \frac{\theta}{\sqrt{l/g}} = \text{const}, \qquad (0.2)$$

from which we obtain

$$\theta = \text{const} \sqrt{\frac{l}{g}}, \qquad (0.3)$$

which is the same as (0.1) up to a constant. The constant in (0.3) can be determined fairly accurately from a single experiment that the reader may carry out him- or herself by measuring the period of oscillation of a weight hung on a thread. With this step, the derivation of the relation (0.1) for the period of oscillation of a pendulum is complete. The example just presented (which is due to the French mathematician P. Appell) is instructive. It would seem that we have succeeded in obtaining an answer to an interesting problem from nothing – or, more precisely, only from a list of the quantities on which the period of oscillation of the pendulum is expected to depend, and a comparison (analysis) of their dimensions.

The following example concerns the steady uniform motion of a body in a gas at high velocity. To be specific, we shall discuss the simplest such case: the motion of a sphere (Figure 0.2(a)). At high velocities, it intuitively seems possible to neglect the internal friction in the gas (the viscosity), since the resistance to the motion of the body is mainly due to the inertia of the gas as it is pushed apart by the body. Therefore, the drag force that the gas exerts against the motion of the sphere in it depends on the static gas density ρ, the static gas pressure p, the velocity with which the body moves, U, and the diameter of the sphere, D. Let us now determine the dimensions of density, force, and pressure, since we already know the dimensions of the remaining quantities.

Density is by definition the ratio of a mass to the volume occupied by that mass. Consequently, the density of a homogeneous body in which a unit mass occupies a unit volume can be adopted as the unit for density. In decreasing the unit of mass by a factor M and the unit of length by a factor L, we decrease the unit of density by a factor ML^{-3}. Therefore, all the density values increase by this same factor. This means that density has dimension ML^{-3}.

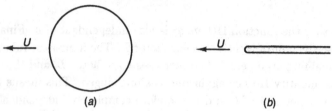

Figure 0.2. (a) A sphere moving in a gas at high velocity. (b) A plate moving in a fluid at high velocity.

Force is related to mass and acceleration via Newton's second law, force equals mass times acceleration. Now it must be true that the dimensions of both sides of any equation having physical sense must be identical. Otherwise, the equation would no longer hold under a change of fundamental units of measurement[†]. Thus, the dimension of force must be identical to the dimension of the product of mass and acceleration. When the unit of mass is decreased by a factor M, the unit of length is decreased by a factor L, and the unit of time is decreased by a factor T, the value of the product of mass and acceleration increases by a factor of MLT^{-2}. Force therefore has dimension MLT^{-2}. Pressure is normal force per unit area; area has dimension L^2, so pressure has dimension $ML^{-1}T^{-2}$.

We shall now turn to the quantity p/ρ. From the above discussion, when the fundamental units of measurement are decreased by factors M, L, and T, respectively, the numerator of this quantity increases by a factor $ML^{-1}T^{-2}$, and the denominator increases by a factor ML^{-3}, so that the quantity p/ρ increases by a factor L^2T^{-2}, i.e., as the velocity squared. The quantity $(p/\rho)^{1/2}$ therefore has the same dimension as velocity. Indeed, this is not surprising, since the quantity $c = (\gamma p/\rho)^{1/2}$, where γ is a dimensionless constant that is a characteristic property of a given gas,[‡] is the speed of sound in the gas. Thus, it may be assumed that the drag force against the motion of the sphere, f, depends on the density of the gas ρ, the velocity of the sphere U, the diameter of the

---

[†] Such equations which hold only in one system of fundamental units do exist and may be very useful. For instance, my colleague Professor A. Yu. Ishlinsky proposed a formula for the time taken to drive a given distance in Moscow: the time (in minutes) is equal to the distance (in kilometres) plus the number of traffic lights. Of course, the formula

time = distance + number of traffic lights

does not work in other units.

[‡] γ is the ratio of the specific heat at constant pressure to the specific heat at constant volume ($\gamma = 1.4$ for air at room temperature).

sphere D, and the speed of sound in the gas at rest, c, which may be conveniently introduced in place of the static gas pressure:

$$f = f(\rho, U, D, c). \tag{0.4}$$

We now form the combination $f/\rho\, U^2 D^2$. The reader may easily verify that it is dimensionless. Equation (0.4) may obviously be rewritten in the form

$$\Pi = \frac{f}{\rho\, U^2 D^2} = \Pi(\rho, U, D, \Pi_1), \tag{0.5}$$

where $\Pi_1 = U/c$ is also a dimensionless quantity – the ratio of the velocity of the body to the sound speed. This quantity is called the *Mach number*, in honour of the Austrian scientist who performed pioneering experiments with shock waves in a gas. We now decrease the unit of mass by an arbitrary factor M, and leave all other units unchanged. The numerical value of the density then increases by a factor M, while the numerical values of the dimensional quantities U and D and the dimensionless quantities Π and Π_1 remain unchanged. But this means that the quantity Π is independent of the density ρ. Further, upon changing the unit of time alone by an arbitrary factor, we find that the numerical value of the velocity U changes by the same arbitrary factor, while the quantities D, Π, and Π_1 remain unchanged. This means that Π is also independent of the velocity. Finally, upon changing the unit of length, we find that Π is independent of D as well, so that

$$\Pi = \Pi(\Pi_1). \tag{0.6}$$

However, further simplification is now impossible! The quantity Π_1 is dimensionless, and does not change when the fundamental units are changed. But even without further simplification, the result is impressive: restoring dimensional variables in (0.6), we find that

$$f = f(\rho, U, D, c) = \rho\, U^2 D^2 \Pi\left(\frac{U}{c}\right); \tag{0.7}$$

thus, the problem has been reduced to determining a function of *one* (!) variable instead of a function of *four* variables.

In order to complete the analysis, the function of one variable, $\Pi(U/c)$, must be determined either experimentally or by calculation. For example, it can be obtained from experiments on a small model in a wind tunnel (Figure 0.3). A graph of the function $\Pi(U/c)$ obtained in this way is shown in Figure 0.4.

The above examples illustrate a complete recipe for applying dimensional analysis. Outwardly, it appears very simple.

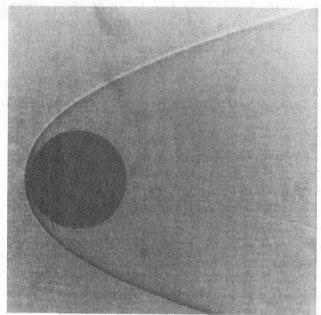

Figure 0.3. A nylon sphere moves in air at Mach number 7.6. A detached shock wave is visible ahead of the sphere. (From van Dyke (1982)).

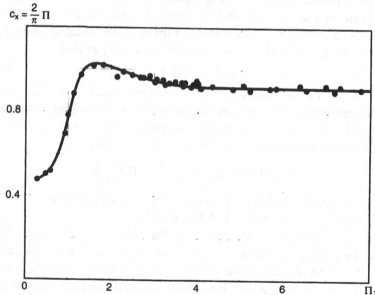

Figure 0.4. The dimensionless drag on a sphere, $c_x = (2/\pi)\Pi$, as a function of the dimensionless governing parameter $\Pi_1 = U/c$ (the Mach number) (Chernyi, 1961). The quantity Π approaches a constant for large values of Π_1.

(1) The parameters on which the quantity to be determined depends are ascertained;

(2) Those parameters whose dimensions are independent (so that their numerical values can change independently of one another when the fundamental units of measurement are changed) are selected; and then

(3) The relations being studied are transformed into relations between dimensionless quantities.

The advantage of using dimensional analysis is that the number of dimensionless quantities is smaller than the total number of dimensional quantities between which we are searching for a relationship. Once again, we find that the difference between the total number of dimensional parameters and the number of dimensionless parameters is equal to the number of dimensional parameters with independent dimensions.

0.2 Assumptions underlying dimensional analysis

However, the apparent simplicity of the above procedure is illusory. In fact, it is most effective when the problem is in the end reduced to determining a constant or a function of one dimensionless variable. This is why it is important to restrict oneself to the minimum necessary number of parameters when finding out on which of them the quantity to be determined depends. Moreover, none of the essential parameters may be left out. How should we proceed here, especially in those cases where we do not have a mathematical formulation for the problem for instance, in turbulence?

We shall now illustrate the real conceptual difficulties that arise in doing this, using the following apparently very similar example. Namely, consider the steady uniform motion of a thin plate in a fluid (Figure 0.2(b)). In this case the effects of compressibility can be neglected for small enough values of the Mach number whereas viscosity effects are essential. Then the drag force f (per unit width) will depend on the velocity U with which the plate moves, the plate length l, and the fluid's properties, its density ρ and its kinematic viscosity ν; so

$$f = f(\rho, U, l, \nu). \qquad (0.8)$$

Repeating the discussion from the preceding example, this relation can also be reduced to the form (0.6), where now

$$\Pi = \frac{f}{\rho U^2 l}, \qquad \Pi_1 = \frac{Ul}{\nu}. \qquad (0.9)$$

Thus, the problem has again been reduced to one of determining the function of one variable $\Pi(\Pi_1)$. If the dependence on kinematic viscosity were not essential, the problem would be reduced to one of determining a single constant, as in the first example: a clear gain. How can one determine that a certain governing parameter is not essential? The reasoning very often (it could even be said, usually) goes like this. If the dimensionless parameter Π_1 corresponding to some dimensional parameter is either very small or very large compared to unity, it may be assumed to be not essential, and the function $\Pi(\Pi_1)$ can be assumed to be a constant (or, in general, when there are several dimensionless parameters Π_1, Π_2, ..., a function of one fewer arguments).

A very strong assumption, which is usually *not* mentioned, and which is satisfied in some cases and not in others, is in fact being made when reasoning along these lines. Namely, the function $\Pi(\Pi_1)$ is assumed to approach a *finite, non-zero limit* at either large or small Π_1 (i.e., as Π_1 goes to either zero or infinity): $\Pi(0) = C$ or $\Pi(\infty) = C$. If it is in fact the case and the parameter Π_1 is sufficiently large or sufficiently small, the equation $\Pi = \Pi(\Pi_1)$ can, to the required accuracy, be written as a simpler relation similar to (0.2):

$$\Pi = C. \tag{0.10}$$

This is precisely the situation that obtains in the motion of a sphere at high velocity (see Figure 0.4): to sufficient accuracy, the function $\Pi(U/c)$ is constant for ratios U/c greater than four, and so for large velocities $f = C\rho\,U^2 D^2$.

However, it is obvious that this is far from always the case. If, for example, for a certain process

$$\Pi(\Pi_1) = \ln \Pi_1 \quad \text{or} \quad \Pi(\Pi_1) = \sin \Pi_1 \tag{0.11}$$

(which obviously cannot be excluded in a general discussion), it is not permissible to replace the function by a constant, no matter how large or small the parameter Π_1 is. That is, no matter how large or small the parameter Π_1, the assumption that a particular dimensional parameter may be neglected is a strong hypothesis that must be supported by experiment, numerical calculation, or (at least) the intuition of the investigator. However, since dimensional analysis is normally used only when we cannot obtain a more complete solution to the problem, this means that we can rarely answer in advance the subtle question of whether the function $\Pi(\Pi_1)$ has a non-zero limit as Π_1 goes to zero or infinity. Moreover, there is yet another, rather insidious situation that may arise here. It is illustrated by the example of plate motion considered above.

Let us make the same assumption of a finite non-zero limit at large Π_1, i.e. at large velocities or small viscosities. We then obtain a limiting relation in the form (0.10), i.e.

$$f = \text{const}\,\rho\,U^2 l. \tag{0.12}$$

In fact, however (see chapter 6), the function $\Pi(\Pi_1)$ in this case approaches zero at large Π_1, according to

$$\Pi = \text{const}/\sqrt{\Pi_1}, \tag{0.13}$$

so, for large Π_1, the relation (0.8) can be represented in the form

$$\Pi^* = C, \text{ where } \Pi^* = f/\rho U^{3/2}(l\nu)^{1/2} \tag{0.14}$$

whence

$$f = \text{const}\,\rho\,U^{3/2}(l\nu)^{1/2}. \tag{0.15}$$

So, we have obtained scaling laws identical in form to (0.2) and (0.10). However, although the relation for Π^* has in principle the same monomial form as Π, it differs from Π in two important respects. First, the powers in (0.14) cannot be determined from simple considerations, that is, an analysis of the dimensions of the quantities in the problem. Second, in contrast with the sound velocity c in the first example, the parameter ν remains in (0.14) and (0.15). Thus the simplification that has occurred here is no longer due to dimensional analysis alone but to a special property of the problem being studied: the existence of the power-law representation (0.13) of the function Π for large Π_1.

We shall now give one more example (this time, geometric) of this type of situation. Consider two continuous curves. One of them (Figure 0.5) is a normal circle. We inscribe a regular n-gon with side length η in it. The perimeter of this inscribed polygon, L_η obviously depends only on the diameter of the circle d and side length η:

$$L_\eta = f(d, \eta). \tag{0.16}$$

Proceeding in much the same way as in the previous examples, we transform this relation, using dimensional analysis, to the form (0.6),

$$\Pi = \Pi(\Pi_1), \tag{0.17}$$

where this time $\Pi = L_\eta/d$, $\Pi_1 = \eta/d$, whence

$$L_\eta = d\Pi(\eta/d).$$

Let the number of sides of the polygon, n, approach infinity, i.e., let the side length η approach zero. From elementary geometry, it is known that the perimeter of the inscribed polygon approaches the finite limit $L_0 = \pi d$ (which is, in fact, adopted as the circumference of a circle). Thus, as $\eta/d \to 0$, the function $\Pi(\eta/d)$ approaches a finite limit equal

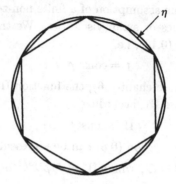

Figure 0.5. A circle with inscribed regular polygons. As the number of sides in the polygon approaches infinity, and the side length approaches zero, the perimeter of the polygon approaches a finite limit.

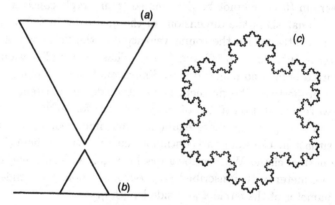

Figure 0.6. A fractal curve – the Koch triad. (*a*) The original triangle, (*b*) the elementary operation, and (*c*) the broken line that approximates the fractal curve for a large number of sides. As the number of sides increases, the perimeter of the broken line approaches infinity according to a power law.

to π. Therefore, for sufficiently small η/d, it is possible to neglect the influence of the parameter η and to assume that the following relation is satisfied to the required accuracy for polygons with a large number of sides:

$$\Pi = \text{const} = \pi, \tag{0.18}$$

i.e., $L_\eta = \pi d$.

The second curve is obtained in the following way[†] (Figure 0.6). An equilateral triangle of side d is taken, and each of its three sides is subjected to the following *elementary operation*: the side is divided into three sections, and the middle section is replaced by two sides of an equilateral triangle constructed using it as a base. The sides of the polygon obtained are once again subjected to the same elementary operation, and so on to infinity. Obviously, the side length of this polygon at the nth stage, η is equal to $d/3^n$, and the perimeter of the entire polygon, L_η, is equal to $3d(4/3)^n$. Equations (0.16) and (0.17) clearly also hold in this case. However, it can easily be shown that (since it is obvious that $n = \log(d/\eta)/\log 3$)

$$L_\eta = 3d[10^{n(\log 4 - \log 3)}] = 3d[10^{\alpha \log(d/\eta)}] = 3d(d/\eta)^\alpha, \qquad (0.19)$$

where

$$\alpha = (\log 4 - \log 3)/\log 3 \simeq 0.26\ldots.$$

Comparing (0.19) and (0.17), we find that

$$\Pi(\eta/d) = 3(\eta/d)^{-\alpha} \qquad (0.20)$$

(i.e., the length of the curve L_0 is infinite in this case, so that only the empty relation $\Pi = \infty$ is obtained in going to the limit $\eta/d \to 0$). Thus, if one is interested in the perimeter of the polygon for large n, it is not possible to pass to the limit and use a limiting relationship such as (0.18). At the same time, equation (0.19) can be rewritten in the form $\Pi^* = C$, setting

$$\Pi^* = \frac{L_\eta}{d^{1+\alpha}\eta^{-\alpha}}, \quad C = 3. \qquad (0.21)$$

The parameter Π^* is (like Π) a power-law combination of the parameters that determine it. However, the structure of (0.21) is not determined by dimensional considerations alone; we do not know the number α beforehand as we did in the case of a circle. Furthermore, unlike the case of a circle (equation (0.18)), the parameter η remains in the resulting equation no matter how small η/d is. Therefore, the length of the inscribed broken line, $L_\eta = 3d^{1+\alpha}/\eta^\alpha$, turns out to be proportional to

---

[†] Curves and, in general, geometric objects of this type are called fractals, as suggested by Mandelbrot (1975). They were intensively studied by mathematicians at the beginning of this century. Mandelbrot in his papers, and especially in his illuminating collections of essays (Mandelbrot, 1975, 1977, 1982) has revived interest in such geometric objects by showing that they provide adequate descriptions of important objects in nature. The curve shown in Figure 0.6 was constructed by von Koch (1904) and is called the Koch triad. We will discuss such objects and the whole concept at some length in chapter 12.

$d^{1+\alpha}$ rather than d, and the length of a segment of this broken line, η remains in the constant of proportionality.

The discussion presented above shows that, for a given quantity of interest, correctly choosing the parameters on which it depends and correctly evaluating the nature of this dependence are what really come first, rather than the formal procedure behind dimensional analysis. We should not assume that too many of the parameters are essential, since dimensional analysis then becomes ineffective. However, only with extreme caution should one discard particular parameters as nonessential merely because corresponding dimensionless parameters are either large or small.

Dimensional analysis and the general concepts of dynamical similarity are presented in chapter 1. Our exposition is essentially different from those available in the literature, although it follows in its general ideas the excellent book of P.W. Bridgman (1931), undeservedly forgotten in recent years.

0.3 Self-similar phenomena

A time-developing phenomenon is called *self-similar* if the spatial distributions of its properties at various different moments of time can be obtained from one another by a similarity transformation.[†] Establishing self-similarity has always represented progress for a researcher: self-similarity has simplified computations and the representation of the properties of phenomena under investigation. In handling experimental data, self-similarity has reduced what would seem to be a random cloud of empirical points so as to lie on a single curve or surface, constructed using self-similar variables chosen in some special way. The self-similarity of the solutions of partial differential equations has allowed their reduction to ordinary differential equations, which often simplifies the investigation. Therefore, with the help of self-similar solutions researchers have attempted to envisage the characteristic properties of new phenomena. Self-similar solutions have also served as standards in evaluating approximate methods for solving more complicated problems.

The appearance of computers changed the general attitude toward self-similar solutions but did not decrease the interest in them. Previously it had been considered that the reduction of partial to ordinary

---

[†] The fact that we identify one of the independent variables with time is of no significance.

differential equations simplified matters, and hence self-similar solutions had attracted attention, first of all, because of the simplicity of obtaining and analyzing them. Gradually, however, the situation grew more complicated, and in many cases it turned out that the simplest method of numerically solving boundary-value problems for the systems of ordinary equations that resulted from the construction of self-similar solutions was computation by the method of stabilization of the solutions of the original partial differential equations. Nevertheless, self-similarity continued as before to attract attention as a profound physical fact indicating the presence of a certain type of stabilization of the processes under investigation, valid for a rather wide range of conditions. Moreover, self-similar solutions were used as a first step in starting numerical calculations on computers. For all these reasons the search for self-similarity was undertaken at the outset, as soon as a new domain of investigation was opened up.

Instructive examples of self-similarities are given by several highly idealized problems in the mathematical theory of filtration – slow ground–water motion in porous media.

Suppose that, in a porous stratum over an underlying horizontal impermeable bed, at an initial instant $t = 0$ a finite volume of water V is supplied instantaneously by a well of very small, let us say infinitesimally small, radius (Figure 0.7). Then at time t the local height h of the ground water mound formed in such a way will be given (Barenblatt, 1952, see the details below in chapter 12) by

$$h = \frac{Q^{1/2}}{(\kappa t)^{1/2}} \left[8 - \frac{r^2}{(Q\kappa t)^{1/2}} \right] \qquad (0.22)$$

for $r < r_f = \sqrt{8}(Q\kappa t)^{1/4}$, and by $h = 0$ for $r \geq r_f$. Here

$$\kappa = \frac{k\rho g}{2m\mu}, \qquad Q = \frac{V}{2\pi m} \qquad (0.23)$$

where k, m are the permeability and porosity (the relative volume occupied by the pores) of the stratum which are statistical geometric properties of the porous medium; ρ and μ are the water density and dynamic viscosity, g is the gravitational acceleration, and r is the distance from the well of the point at which the observation is made.

The form of the relation (0.22) is instructive: there exist a mound height scale $h_0(t)$ and a linear scale $r_0(t)$, both depending on time,

$$h_0(t) = \frac{Q^{1/2}}{(\kappa t)^{1/2}}, \qquad r_0 = (Q\kappa t)^{1/4}, \qquad (0.24)$$

such that the spatial distribution of the ground-water mound height,

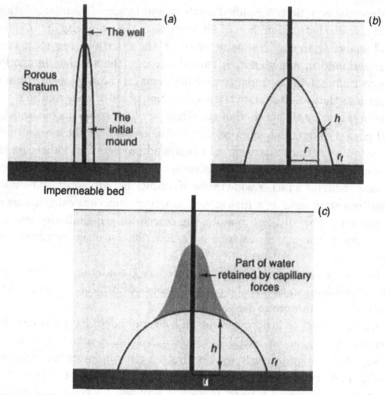

Figure 0.7. (a) A ground-water mound is formed rapidly through a well of small radius. (b) If capillary forces are negligible the mound volume stays constant as the mound height decreases. (c) When part of the water is retained by capillary forces the mound volume decreases with time.

when expressed in these scales, ceases to depend on time:

$$\frac{h}{h_0(t)} = \Phi\left(\frac{r}{r_0(t)}\right) \tag{0.25}$$

where

$$\Phi(\xi) = \begin{cases} (8 - \xi^2)/16 & \text{for } \xi \le \xi_f = \sqrt{8} \\ 0 & \text{for } \xi \ge \xi_f \end{cases}$$

and

$$\xi = \frac{r}{r_0(t)}$$

The example just considered is typical. Suppose that we are faced with a mathematical physics problem in two independent variables r and t, requiring the solution of a system of partial differential equations. In this problem, self-similarity means that we can choose variable scales $\mathbf{u}_0(t)$

and $r_0(t)$ such that in the new scales the properties of the phenomenon can be expressed by functions of one variable:

$$\mathbf{u} = \mathbf{u}_0(t)\mathbf{U}(\xi), \qquad \xi = r/r_0(t). \tag{0.26}$$

The solution of the problem thus reduces to the solution of a system of ordinary differential equations for the vector–function $\mathbf{U}(\xi)$.

It was natural to attempt to clarify the nature of self-similarity. Here at a certain stage the attraction of dimensional analysis played an essential role. Indeed, let us apply dimensional analysis to the idealized ground-water motion problem considered above. The total ground-water mound head $u = \rho g h$ satisfies the Boussinesq equation (Polubarinova–Kochina, 1962, see also further discussion, in chapter 12),

$$\partial_t u = a\frac{1}{r}\partial_r\left(r\partial_r u^2\right) \qquad a = \frac{k}{2m\mu} = \frac{\kappa}{\rho g}, \tag{0.27}$$

under the initial condition that the groundwater is concentrated initially in a well of infinitesimally small radius

$$u(r,0) \equiv 0 \qquad (r \neq 0); \qquad u(\infty, t) \equiv 0,$$

$$2\pi m \int_0^\infty ru(r,0)dr = \rho g V \tag{0.28}$$

Of course, such an initial ground-water head distribution $u(r,0)$ is actually given by a generalized function. It is evident from the mathematical statement of the problem, (0.27), (0.28), that the total head u depends on the time t, on the quantities a and $\rho g V/2\pi m$, and on the distance r of the observation point from the well. All these quantities are dimensional and their numerical values depend on the choice of units of mass, length, and time. Dimensional analysis gives us as before the relation (0.6), where this time

$$\Pi = \frac{u(at)^{1/2}}{(\rho g Q)^{1/2}} = \frac{h(\kappa t)^{1/2}}{Q^{1/2}}, \qquad \Pi_1 = \frac{r}{(\rho g Q a t)^{1/4}} = \frac{r}{(Q\kappa t)^{1/4}}, \tag{0.29}$$

$Q = V/2\pi m$ and $\kappa = a\rho g$, and from this we get the relation (0.25), i.e.,

$$u = \rho g h = \frac{\rho g Q^{1/2}}{(\kappa t)^{1/2}}\Phi(\xi), \qquad \xi = \Pi_1 = \frac{r}{(Q\kappa t)^{1/4}}. \tag{0.30}$$

Thus, we have in this case succeeded in establishing the self-similarity of the solution, and in determining the scales $h_0(t)$ and $r_0(t)$ using only dimensional analysis.

When we substitute the relations (0.30) into the partial differential equation (0.27) we obtain for the function Φ an ordinary differential equation

$$\frac{d^2\Phi^2}{d\xi^2} + \frac{1}{\xi}\frac{d\Phi^2}{d\xi} + \frac{\xi}{4}\frac{d\Phi}{d\xi} + \frac{\Phi}{2} = 0. \tag{0.31}$$

Furthermore, from (0.28) at $\xi = \infty$ we have

$$u(\infty, t) = \rho g Q^{1/2}(\kappa t)^{-1/2}\Phi(\infty) = 0,$$

whence $\Phi(\infty) = 0$. It is easy to find a bounded solution to (0.31) under the condition $\Phi(\infty) = 0$ to within a constant:

$$\Phi = \begin{cases} (\xi_f^2 - \xi^2)/16 & (\xi \leq \xi_f) \\ 0 & (\xi \geq \xi_f). \end{cases} \tag{0.32}$$

The constant ξ_f is determined as follows. The solution (0.30) to the problem considered must satisfy the law of conservation in time of the total volume of water:

$$2\pi m \int_0^\infty rh(r,t)dr = \frac{Q^{1/2}[Q^{1/2}(\kappa t)^{1/2}]2\pi m}{(\kappa t)^{1/2}} \int_0^{\xi_f} \xi\Phi(\xi)d\xi \tag{0.33}$$
$$= \text{const } V$$

where, according to (0.32),

$$\text{const} = \int_0^{\xi_f} \xi\Phi(\xi)d\xi = \frac{\xi_f^4}{64}. \tag{0.34}$$

However, the constant must be equal to unity, because the initial amount of water was equal to V. Therefore we obtain $\xi_f = \sqrt{8}$, and this completes the solution to the problem.

The analysis of the idealized problem presented here is typical. In many other cases too, considerations of dimensional analysis turn out to be quite sufficient for proving the self-similarity of the solution starting from the formulation of the mathematical problem, and for obtaining expressions for the scales and the self-similar variables. We shall see later, however, that self-similar solutions for which dimensional analysis is sufficient to establish the self-similarity are relatively rare; as a rule the situation is more complicated.

0.4 Self-similar solutions as intermediate asymptotics. The solutions of the first and second kind. Renormalization group.

Self-similarities, even as understood from the point of view of dimensional analysis, have been regarded by the majority of researchers merely as isolated exact solutions of special problems – elegant, sometimes rather useful, but all the same limited in their significance as properties of physical theories. It has only gradually been realized that the significance of these solutions is much broader. In fact they do not merely describe the behaviour of physical systems under certain special conditions. They also describe the 'intermediate-asymptotic' behaviour of solutions of wider classes of problem in the range where these solutions no longer depend on the details of the initial and/or boundary conditions, yet the system is still far from being in an ultimate equilibrium state.

The concept of intermediate asymptotics has an important general significance, and not only in mathematical physics. For instance, this concept is always used in our perception of visual art. We have to look at paintings at distances great enough not to see the brush-strokes, but at the same time small enough to enjoy not only the painting as a whole but also its important details; think of van Gogh's work, for example. (Remember also *Gulliver's Travels* by Jonathan Swift (1992). Gulliver's impressions of the fine details of the skin of a giant Brobdingnag beauty, who had the custom of putting him upon her breast, are especially instructive from this viewpoint. It is clear from his description that her admirers restricted themselves to an intermediate-asymptotic perception of her!) As far as I know the concept of intermediate asymptotics was formally introduced into mathematical physics by Ya.B. Zeldovich and myself (Barenblatt and Zeldovich, 1971, 1972; see also Barenblatt, 1959b and Zeldovich and Raizer, 1966, 1967), although it was used implicitly long before this.

It is precisely the consideration of self-similar solutions as intermediate asymptotics that allows one to understand properly the role of dimensional analysis in establishing self-similarity and determining self-similar variables. As it turns out, dimensional considerations are far from always sufficient to establish self-similarity. What is more, one can even assert that as a rule they are not.

Zeldovich (1956) first explicitly distinguished a particular class of self-similar solutions for which dimensional analysis is insufficient for establishing self-similarity and determining self-similar variables. He called these *self-similar solutions of the second kind*. He had in mind also the solutions considered by Guderley (1942), Landau and Staniukovich (see Staniukovich, 1960), and von Weizsäcker (1954). In the book of Zeldovich and Raizer (1967)[‡], and also in the important paper by Brushlinsky and Kazhdan (1963), a detailed analysis was given of the solutions of this type known at that time.

To understand what constitutes the intrinsic nature of the classification of self-similarities in this, the simplest, case, one can modify somewhat the problem of ground-water motion considered above. Thus, we

---

[‡] The term 'self-similarity of the second kind' was used by Ya.B. Zeldovich in his earlier papers and in Zeldovich and Raizer (1967) in a narrower sense than that in which we use it here. Zeldovich himself agreed with such an extension of the definition; see Barenblatt and Zeldovich (1971, 1972).

assume that as the mound spreads from the initially occupied space some
fixed part of the groundwater remains there. (The solution to this problem discussed below and in chapter 12, is given in Kochina, Mikhailov,
and Filinov (1983); the work was performed under the author's supervision.)

In this case the following equation for the mound head $u = \rho g h$ is
obtained instead of (0.27):

$$\partial_t u = a_1 \frac{1}{r} \partial_r (r \partial_r u^2) \qquad (\partial_t u \leq 0),$$

$$\partial_t u = a \frac{1}{r} \partial_r (r \partial_r u^2) \qquad (\partial_t u \geq 0) \tag{0.35}$$

where

$$a = \frac{\kappa}{\rho g} = \frac{k}{2m\mu}, \qquad a_1 = \frac{\kappa_1}{\rho g} = \frac{k}{2m(1 - \sigma_0)\mu},$$

and σ_0 is the fraction of the porous volume assumed to remain water-
filled, by assumption this fraction is constant. At first sight it appears
that on applying dimensional analysis one can construct a solution to
equation (0.35) for an instantaneous line-concentrated source with $\kappa_1 \neq
\kappa$ in virtually the same way as for the classical Boussinesq equation
(0.27) with $\kappa_1 = \kappa$. In fact, to the parameters t, κ, Q ,r governing the
solution of (0.27) there is added in this case only an additional constant
parameter κ_1 of the same dimension as κ. Hence it seems at first glance
that the solution can be represented in the same form as (0.30) and that
the additional constant dimensionless parameter κ_1/κ does not change
the situation. The reader will see in chapter 12, however, that for $\kappa_1 \neq \kappa$
such a solution simply does not exist!

The resolution of this apparent paradox from a general viewpoint is
also instructive, and we shall present this a little further on in the text;
a detailed treatment can be found in chapter 12. We are actually not
interested in the solution of the idealized prolem of a line-concentrated
instantaneous water source, but rather in the asymptotic behaviour for
large times of the solution corresponding to the discharge of a quantity
of water at the initial instant in a cylinder of small but finite radius r_*.
The problem then involves this new parameter r_*, having the dimension
of length, and a new dimensionless parameter

$$\Pi_2 = \frac{r_*}{(Q\kappa t)^{1/4}} = \eta. \tag{0.36}$$

The parameter r_* immediately spoils the self-similarity of the solution,
since the solution can no longer be expressed as a function of one vari-

able, but has the form

$$u = \rho g \frac{Q^{1/2}}{(\kappa t)^{1/2}} \Phi \left(\xi, \eta, \frac{\kappa_1}{\kappa} \right) . \tag{0.37}$$

Just as for $\kappa_1 = \kappa$, we are interested in the behaviour of the solution for small η. However, for $\kappa_1 \neq \kappa$ it is impossible simply to pass to the limiting form of the solution corresponding to $\eta = 0$ in (0.37). The reason for this is trivial: it turns out that for small η the function Φ does not tend to a finite non-zero limit; instead, we have

$$\Phi \simeq \eta^{\alpha} \varphi \left(\frac{\xi}{\eta^{\alpha/2}}, \frac{\kappa_1}{\kappa} \right) \tag{0.38}$$

where φ is a finite quantity and α is a constant that depends only on κ_1/κ; it is non-zero for $\kappa_1 \neq \kappa$, but equal to zero for $\kappa_1 = \kappa$. If one tries to pass to the limit $\eta \to 0$ with $\kappa_1 \neq \kappa$, then on the right-hand side of (0.37) and (0.38) one gets either zero or infinity depending on the sign of α, i.e., one gets an empty relation. However, we do not in fact need the limit, so, without passing to the limit, we substitute the asymptotics (0.38) into the general representation (0.37) of the solution, while regarding η as small, i.e., the time as large or r_* as small. Thus we obtain a self-similar asymptotic representation of the solution of the original non-self-similar problem, valid for large t or small r_*:

$$u = \rho g \frac{A}{(\kappa t)^{(2+\alpha)/4}} \varphi \left(\frac{r}{B t^{(2-\alpha)/8}}, \frac{\kappa_1}{\kappa} \right) ,$$

$$A \sim \left(Q r_*^{\frac{4\alpha}{2-\alpha}} \right)^{\frac{2-\alpha}{4}} , \quad B \sim \sqrt{A} . \tag{0.39}$$

Equation (0.39) shows that the representation for large times is given not by a solution of linear-source type but by another self-similar solution. Indeed, let us now decrease the size r_* of the region of the initial water discharge. It is evident that the solution is such that we cannot preserve the magnitude of Q as was the case for $\sigma_0 = 0$, $\kappa_1 = \kappa$. On the contrary, we must vary the output Q of the source in such a way that the product $Q r_*^{4\alpha/(2-\alpha)}$ preserves its magnitude. Physically it is clear: when reducing r_* we have to leave some water in the pores when reaching the previous size of the initial discharge region.

If we now substitute the relation (0.39) for the solution into equation (0.35) we obtain for the function φ an ordinary differential equation containing α as a parameter. It turns out that for arbitrary α this equation has no solution with the necessary properties. However, for each value of the parameter κ_1/κ there exists one value of α for which the required solution of the ordinary differential equation does exist.

Thus, to determine φ and the parameter α one obtains a *nonlinear eigenvalue problem*. Under such a direct construction of the self-similar intermediate asymptotics the constant A remains undetermined. It is impossible to find it from an integral conservation law when $\kappa_1 \neq \kappa$ since in this case the equation for bulk water balance assumes the non-integrable form

$$\frac{d}{dt} \int_0^\infty r u(r,t) dr = (\kappa_1 - \kappa)[r\partial_r u^2]_{r=r_0}, \qquad (0.40)$$

$r_0(t)$ being the coordinate of the point at which, at the given moment, the derivative $\partial_t u$ vanishes.

Thus, the behaviour of the solution for large t turns out to be self-similar also for $\kappa_1 \neq \kappa$, but the self-similarity here is not the same as for $\kappa_1 = \kappa$. First of all, the parameter r_* with the dimension of length, which spoils the self-similarity of the original problem, does not disappear from the limiting solution. Further, dimensional analysis does not in this case allow us, starting from the mathematical formulation of the problem to find the self-similar variables or to establish the self-similarity of the limiting solution; the fact is that the dimension of the constant A is unknown in advance and must be found in the course of the solution. At the end, the constant A turns out to be undefined. To find it one must 'match' the constructed solution with the solution to the original non-self-similar problem, for example by means of numerical computation.

A peculiar situation arises also with similarity laws. Dimensional analysis allows one to obtain to within a constant factor the law of attenuation of the ground-water mound's maximum height in the classical case $\kappa_1 = \kappa$. In fact the height at the centre depends only on the quantities $\rho g Q$, κ, and t, from which one can construct only one combination with the dimension of height, $Q^{1/2}(\kappa t)^{-1/2}$. It is impossible to form a dimensionless combination from these quantities, since their dimensions are independent. It is therefore clear that

$$u_{\max} = \rho g h_{\max} = \text{const}\, \rho g \frac{Q^{1/2}}{(\kappa t)^{1/2}}. \qquad (0.41)$$

According to the solution given above, the constant here equals $1/2$. A naive application of dimensional analysis, i.e., the assumption of water discharge at a line, would lead to the same similarity law also for $\kappa_1 \neq \kappa$, although in this case, as (0.39) shows, such similarity law does not hold. In fact,

$$u_{\max} = \rho g h_{\max} = \text{const}\, \rho g \frac{Q^{(2-\alpha)/4} r_*^\alpha}{(\kappa t)^{(2+\alpha)/4}}, \qquad (0.42)$$

so that although the law of attenuation is also a power law, one cannot now obtain the exponent by dimensional analysis. Besides, the radius of the initial discharge region, r_*, remains in the resulting relation (0.42). The situation is that until a non-self-similar problem has been solved in its entirety, it is impossible, generally speaking, to say in advance, whether one can use dimensional analysis to analyse the similarity laws.

It is now easy to understand what happens in the general case. Consider a physical relationship

$$a = f(a_1, \ldots, a_k, b_1, b_2). \qquad (0.43)$$

Here the arguments $a_1, \ldots a_k$, have independent dimensions. That means that by properly choosing the fundamental units it is possible to vary arbitrarily and independently the values of $a_1, \ldots, a_k$ so that

$$a_1' = A_1 a_1, \quad \ldots, \quad a_k' = A_k a_k. \qquad (0.44)$$

Here $A_1, \ldots, A_k$ are arbitrary positive numbers. At the same time the dimensions of a, b_1, b_2 can be represented as power monomials in the dimensions of $a_1, \ldots, a_k$:

$$[b_1] = [a_1]^{p_1} \ldots [a_k]^{r_1},$$
$$[b_2] = [a_1]^{p_2} \ldots [a_k]^{r_2}, \qquad (0.45)$$
$$[a] = [a_1]^{p} \ldots [a_k]^{r}.$$

(We have restricted ourselves to two arguments with dependent dimensions, b_1, b_2, to show the main idea.) The relations (0.45) mean that, after the transformation (0.44), the values of a, b_1, b_2 will transform as

$$b_1' = A_1^{p_1} \ldots A_k^{r_1} b_1,$$
$$b_2' = A_1^{p_2} \ldots A_k^{r_2} b_2, \qquad (0.46)$$
$$a' = A_1^{p} \ldots A_k^{r} a.$$

The transformations (0.44), (0.46) form a *group of continuous transformations*, and the positive numbers $A_1, \ldots, A_k$ are the *parameters* of this group. The fundamental physical covariance principle claims that all physical laws can be represented in a form equally valid for all observers. This principle is valid for observers using different magnitudes of basic units. Therefore the relationship (0.43) can be represented as a relation between the invariants of the group (0.44), (0.46):

$$\Pi = \Phi(\Pi_1, \Pi_2) \qquad (0.47)$$

where the dimensionless invariants Π, Π_1, Π_2 are

$$\Pi_1 = \frac{b_1}{a_1^{p_1} \cdots a_k^{p_1}}, \ \Pi_2 = \frac{b_2}{a_1^{p_2} \cdots a_k^{r_2}}, \ \Pi = \frac{a}{a_1^{p} \cdots a_k^{r}}. \qquad (0.48)$$

Returning in (0.47) to dimensional variables we obtain that the function f in (0.43) possesses the fundamental property of 'generalized homogeneity':

$$f = a_1^p \ldots a_k^r \Phi \left(\frac{b_1}{a_1^{p_1} \ldots a_k^{p_1}}, \; \frac{b_2}{a_1^{p_2} \ldots a_k^{r_2}} \right). \qquad (0.49)$$

Self-similar solutions always correspond to idealized problems in which the parameters of the problem that have the dimensions of the independent variables (a characteristic length, a characteristic time, etc.) are equal to zero or infinity. In the non-idealized case, the arguments would include the ratios of the independent variables to these parameters, and there would be no self-similarity. This means that upon passage from the non-idealized problem statement, corresponding to finite values of the parameters, to the idealized one, the dimensionless parameter Π_2 can tend to zero (or infinity). But what will occur if the dimensionless parameter Π_2, corresponding to the dimensional parameter b_2, is small but finite (or large but finite)? This question is important, not only for self-similar solutions but in every physical investigation, because in mathematical models we always drop certain factors considered as inessential.

In principle three possibilities are available.

(1)   As $\Pi_2 \to 0$ the function Φ tends to a finite limit different from zero.

(2)   As $\Pi_2 \to 0$ such a finite non-zero limit of the function Φ does not exist; however the function Φ has a power-law asymptotics possessing the property of generalized homogeneity:

$$\Phi = \Pi_2^{\alpha_1} \Phi_1 \left(\frac{\Pi_1}{\Pi_2^{\alpha_2}} \right). \qquad (0.50)$$

(3)   Neither (1) nor (2) holds: as $\Pi_2 \to 0$ the function Φ has no finite limit different from zero and no power-type asymptotics.

In case (1), for sufficiently large (or small) Π_2 one can simply replace the function Φ by its limiting expression, corresponding to Π_2 equal to zero (or infinity). Here the number of its arguments is diminished, and the values of the corresponding dimensional parameters (for example, the size r_* of the domain of the initial discharge of water for $\kappa_1 = \kappa$) turn out to be immaterial and drop out of consideration. This case is called *complete similarity* or *similarity of the first kind in the parameter* Π_2.

In case (2) the relation (0.47) can for small (or large) Π_2 be rewritten using (0.50) in the form

$$\Pi^* = \Phi_1(\Pi_1^*)$$

where

$$\Pi^* = \frac{\Pi}{\Pi_2^{\alpha_1}} = \frac{a}{a_1^{p-\alpha_1 p_1} \cdots a_k^{r-\alpha_1 r_1} b_2^{\alpha_1}} \,,$$

$$\Pi_1^* = \frac{\Pi_1}{\Pi_2^{\alpha_2}} = \frac{b_1}{a_1^{p_1-\alpha_2 p_2} \cdots a_k^{r_1-\alpha_2 r_2} b_2^{\alpha_2}} \qquad (0.51)$$

Thus in this case too the number of arguments in (0.47) is reduced. The parameters Π^*, Π_1^* are completely analogous in their structure to common similarity parameters, being dimensionless combinations of powers. The difference, however, consists in two facts: first, the quantities Π^*, Π_1^* retain the dimensional parameter that originally spoiled the self-similarity, and second, they cannot be obtained by dimensional analysis. In such cases one speaks of *incomplete similarity* or *similarity of the second kind in the parameter* Π_2.

Finally, in case (3) the parameter Π_2 remains essential no matter how large or small it is and no similarity in it ensues. The nature of the classification of self-similar solutions becomes transparent now. If the passage from the solution of the non-self-similar original problem to a self-similar intermediate asymptotics corresponds to complete similarity in a dimensionless parameter that spoils self-similarity in the original problem, the self-similar solution is a solution of the first kind. If the passage corresponds to incomplete similarity, the self-similar solution is a solution of the second kind. The real difficulty is that similarity methods usually apply when the solution of the complete non-self-similar problem is unknown. Hence, *a priori* it is impossible to say which type of self-similarity we are dealing with.

Complete similarity in the parameter Π_2 means that in addition to the transformation group (0.44), (0.46) that forms the basis of the dimensional analysis, the mathematical model possesses at small Π_2 an additional property of invariance with respect to the additional group

$$a_1' = a_1, \ \ldots, \ a_k' = a_k; \ b_1' = b_1, \ b_2' = Bb_2, \ a' = a \qquad (0.52)$$

where $0 < B < 1$ is the group parameter. Incomplete similarity (case (2)) is equivalent to the statement that at small Π_2 the mathematical model possesses an additional invariance with respect to the more complicated group

$$a_1' = a_1, \ \ldots, \ a_k' = a_k, \ b_1' = B^{\alpha_2} b_1, \ b_2' = Bb_2, \ a' = B^{\alpha_1} a \qquad (0.53)$$

which is a simple example of the *renormalization group*. Indeed, the quantities b_1, a do not remain invariable as in (0.52); they are renormalized.

In a previous book by the author (Barenblatt, 1979) the identity of the concepts of scaling, widely used by physicists in quantum-field theory and the theory of phase transitions, and of self-similarity of the second kind (incomplete similarity) was noted. In a series of remarkable papers by N. Goldenfeld, O. Martin and Y. Oono and their students (Goldenfeld, Martin, and Oono, 1989, 1991; Goldenfeld, Martin, Oono and Liu, 1990; Chen, Goldenfeld, and Oono, 1991; Chen and Goldenfeld 1992), summarized in a book by Goldenfeld (1992), the connection between the two different approaches to scaling, that is, the renormalisation group used by physicists and the approach using intermediate asymptotics was traced. Some basic problems solved previously by the intermediate-asymptotic approach were solved by the renormalization group technique and vice versa. The renormalization group technique is presented in chapter 6 both from the viewpoint of intermediate asymptotics, developed in this book, and from the more traditional viewpoint.

0.5 Self-similarities and travelling waves

Clarification of the nature of self-similarity of the second kind was aided by the establishment of a close connection between the classification of self-similarities and of nonlinear travelling waves (Barenblatt and Zeldovich, 1971, 1972). Travelling waves are solutions of the form

$$u = f(\zeta - \lambda\tau + c) \qquad (0.54)$$

(ζ being the spatial and τ the temporal variable, λ the constant speed of propagation of the wave, and c a constant), for which the distributions of properties at different moments of time are obtained from one another by means of a simple translation. It is well known (see, e.g., Sedov 1971) that travelling waves are divided into two types. For waves of the first type the speed of propagation λ is found from the conservation laws alone and is independent of the internal structure of the wave. Examples of such waves are shock waves in gas dynamics and detonation waves. For waves of the other type the speed of propagation λ is found from the condition for the existence of a global solution describing the internal structure of the wave, and is completely determined by that structure. An example of a wave of this type is a flame wave or the propagation wave of a gene having an advantage in the struggle for existence. It should be noted that consideration of the problem of the propagation of an advantageous gene in the classical papers by Kolmogorov, Petrovsky, and Piskunov (1937) and by Fisher (1937) was in fact the first example of analysis of the intermediate asymptotics of nonlinear problems.

We now set $\zeta = \ln x$, $\tau = \ln t$ and $c = -\ln A$, in (0.54). Then this relation is transformed into a self-similar form:

$$u = f\left(\ln \frac{x}{At^\lambda}\right) = F\left(\frac{x}{At^\lambda}\right) . \qquad (0.55)$$

By this transformation the classification of self-similar solutions into solutions of the first and second kind is put into one-to-one correspondence with the classification of travelling waves mentioned above.

Self-similarity is connected with a generally nonlinear eigenvalue problem, the existence of a solution of which guarantees the existence of a global self-similar intermediate asymptotics. A non-trivial question appears concerning the set of eigenvalues in this problem – the spectrum determined by the possible values of the exponents in the self-similar variables. Everything is simple if the spectrum consists of one point, as in the ground-water mound problem considered above. But if the spectrum consists of more than one point, in particular if it is continuous, the exponents in the self-similar variables can depend on the initial conditions of the original non-self-similar problem. A remarkable example here is provided by the self-similar interpretation of the well-known Korteweg–de Vries equation (for details see chapter 7).

Recently, ideas connected with the concepts of incomplete self-similarity and self-similar solutions of the second kind have been used to solve many important problems, which are of independent, non-illustrative interest. Some of these problems are considered below in chapters 9–12. Of notable importance is the analysis of incomplete self-similarity in the theory of turbulence and geophysical fluid dynamics, where a complete mathematical formulation of the problem is lacking at present and the comparison of similarity laws with experimental data is therefore of decisive importance in estimating the character of the self-similarity. For the convenience of readers we sometimes present a brief discussion of the basic models involved. Also, the book has many entries: after reading this Introduction the reader may turn immediately to specialized chapters.

We have outlined here the underlying ideas in this book; in what follows the reader is invited to step further into the fascinating world of self-similarities and intermediate asymptotics.

1

Dimensions, Dimensional Analysis and Similarity

1.1 Dimensions

1.1.1 Measurement of physical quantities, units of measurement. Systems of units

We say without thinking that the mass of water in a glass is 200 grams, the length of a ruler is 0.25 metres, the half-life of radium is 1 600 years, the speed of a car is 60 miles per hour. In general, we express all physical quantities in terms of numbers; these numbers are obtained by *measuring* the physical quantities. Measurement is the direct or indirect comparison of a certain quantity with an appropriate standard, or, to put it another way, with an appropriate *unit of measurement*. Thus, in the examples discussed above, the mass of the water is compared with a *standard* – a unit for the measurement of mass – the gram; the length of the ruler is compared with a unit for the measurement of length – the metre; the half-life of radium is compared with a unit for the measurement of time – the year; and the velocity of the car is compared with a unit for the measurement of velocity – the velocity of uniform motion in which a distance of one mile is traversed in a time equal to one hour.

The units for measuring physical quantities are divided into two categories: *fundamental units* and *derived units*. This means the following.

A class of phenomena (for example, mechanics, i.e., the motion and equilibrium of bodies) is singled out for study. Certain quantities are listed, and standard reference values for these quantities – either natural or artificial – are adopted as fundamental units; there is a certain amount of arbitrariness here. For example, when describing mechanical phenomena, we may adopt mass, length, and time standards as the fundamental units, though it is also possible to adopt force, length, and time standards. However, these standards are insufficient for the description of, for example, heat transfer. Additional standards also become necessary when studying electromagnetic phenomena, economic phenomena[†] etc.

Once the fundamental units have been decided upon, derived units are obtained from the fundamental units using the definitions of the physical quantities involved. These definitions always involve describing at least a conceptual method for measuring the physical quantity in question. For example, velocity is by definition the ratio of the distance traversed during some interval of time to the size of that time interval. Therefore, the velocity of uniform motion in which one unit of length is traversed in one unit of time can be adopted as a unit of velocity. In exactly the same way, density is by definition the ratio of some mass to the volume occupied by that mass. Thus, the density of a homogeneous body that contains one unit of mass per unit of volume – a cube with a side equal to one unit of length – can be adopted as a unit of density, and so on. We see that it is precisely the class of phenomena under discussion, i.e., the complete set of physical quantities in which we are interested that ultimately determines whether or not a given set of fundamental units is sufficient for its measurement. For example, it is impossible to define a unit for the measurement of density using only the fundamental units of length and time. It becomes possible to define such a unit by adding a unit of mass.

A set of fundamental units that is *sufficient* for measuring the properties of the class of phenomena under consideration is called a *system of units*. Until recently, the cgs (centimetre-gram-second) system, in which units for mass, length, and time are used as the basic units, and one gram[‡] (g) is adopted as the unit of mass, one centimetre[§] (cm) is

---

[†] Recently the analysis of economic and, especially, financial phenomena using the traditional approaches of applied mathematics has attracted serious attention. For such applications the correct measurement of the quantities involved is of prime importance.

[‡] The gram is one-thousandth of the mass of a specially fabricated standard mass which is carefully preserved at the Bureau of Weights and Measures in Paris.

[§] The centimeter is one-hundredth of the length of a specially fabricated, carefully preserved standard length – the metre. There is another, more precise definition of this standard based on a natural process: 1 650 736.73 wavelengths in vacuo of the radiation corresponding to the transition between the $2p^{10}$ and $5d^5$ levels of the krypton 86 atom.

adopted as the unit of length, and one second[**] (s) is adopted as the unit of time, has customarily been used.

The unit of velocity in the system is the velocity of uniform motion in which a distance of one centimetre is traversed in one second. This unit is written in the following way: cm/s. The unit of density in the system is the density of a homogeneous body in which one cubic centimetre contains a mass of one gram. This unit is written in the following way: g/cm^3. This method of writing units is, to a certain extent, a matter of convention: for example, the ratio cm/s cannot be thought of as a quotient of the length standard – the centimetre – and the time standard – the second. Such a quotient would be totally meaningless: one may divide one number by another, but not an interval of length by an interval of time!

A system of units consisting of two units (a unit for the measurement of length and a unit for the measurement of time, for example the centimetre and the second) is sufficient for measuring the properties of *kinematic* phenomena, while a system of units consisting of only one length unit (for example, the centimetre) is sufficient for measuring the properties of *geometric* objects.

On the other hand, in order to be able to measure the properties of *heat transfer* in a flowing liquid or gas, the system of units for the measurement of mechanical quantities must be supplemented by an independent standard (the degree Kelvin (kelvin), a temperature standard, is convenient for this purpose). We would require an additional standard, for example, a unit for the measurement of current (the ampere) in order to be able to measure electromagnetic phenomena, and so forth.

However, a system of units need not be *minimal*. For example, one can use a system of units in which the unit of length is 1 cm, the unit of time is 1 s, and the unit of velocity is 1 knot (approximately 50 cm/s). However, in this case, the velocity will not be numerically equal to the ratio of the distance traversed to the magnitude of the time interval in which the distance was traversed. We shall discuss this important point in greater detail below.

1.1.2 Classes of systems of units

Let us now consider, in addition to the cgs system, a second system, in

---

[**]   The second is, by definition, 1/86 400 of a mean solar day. A more precise definition of the second is 9 192 621 770 periods of the radiation corresponding to the transition between two hyperfine levels in the ground state of the caesium-133 atom.

which 1 kilometre ($= 10^5$ cm) is used as the unit of length, 1 metric tonne ($= 10^6$ g) is used as the unit of mass, and 1 hour ($= 3600$ s) is used as the unit of time. These two systems of units have the following property in common: standard quantities of the same physical nature (mass, length and time) are used as the fundamental units. Consequently, we say that these systems belong to the same *class*. To generalize, a set of systems of units that differ only in the magnitude (but not in the physical nature) of the fundamental units is called a *class of systems of units*. The system just mentioned and the cgs system are members of the class in which standard lengths, masses and times are used as the fundamental units. The corresponding units for an arbitrary system in this class are as follows:

$$\text{unit of length} = \text{cm}/L \, ,$$
$$\text{unit of mass} = g/M \, , \tag{1.1}$$
$$\text{unit of time} = s/T \, ,$$

where L, M and T are abstract positive numbers that indicate the factors by which the fundamental units of length, mass and time decrease in passing from the original system (in this case, the cgs system) to another system in the same class. This class is called the LMT class[†]. The SI system has recently come into widespread use. This system, in which one metre ($= 100$ cm), is adopted as the unit of length, one kilogram ($= 1000$ g) is adopted as the unit of mass, and one second is adopted as the unit of time, also belongs to the LMT class. Thus, when passing from the original system to the SI system, $M = 0.001, L = 0.01$, and $T = 1$.

Systems in the LFT class, where units for length, force, and time are chosen as the fundamental units are also frequently used; the fundamental units for this class are as follows:

$$\text{unit of length} = \text{cm}/L \, ,$$
$$\text{unit of force} = \text{kgf}/F \, , \tag{1.2}$$
$$\text{unit of time} = s/T \, .$$

The unit of force in the original system, the kilogram-force (kgf), is the force that imparts an acceleration of 9.80665 m/s^2 to a mass equal to that of the standard kilogram. We emphasize that a change in the

---

[†] The designation of a class of systems of units is obtained by writing down, in consecutive order, the symbols for the quantities whose units are adopted as the fundamental units. These symbols simultaneously denote the factor by which the corresponding fundamental unit decreases upon passage from the original system to another system in the same class.

magnitudes of the fundamental units in the original system of units does not change the class of systems of units. For example, the class where the units of length, mass and time are given by

$$\frac{m}{L}, \frac{kg}{M}, \frac{hr}{T}$$

is the same as that defined in (1.1), LMT. The only difference is that the numbers L, M and T for a given system of units (for example, the SI system) will be different in the two representations of the LMT class: in the second representation, we obviously have $L = 1$, $M = 1$ and $T = 3600$.

1.1.3 Dimensions

As we mentioned in the Introduction, if the unit of length is decreased by a factor L, and the unit of time is decreased by a factor T, the new unit of velocity is a factor LT^{-1} smaller than the original unit, so that the numerical values of all velocities are increased by a factor LT^{-1}. Upon decreasing the unit of mass by a factor M and the unit of length by a factor L, we find that the new unit of density is a factor ML^{-3} smaller than the original unit, so that the numerical values of all densities are thus increased by a factor ML^{-3}. Other quantities may be treated similarly. The changes in the numerical values of physical quantities upon passage from one system of units to another system within the same class are determined by their *dimensions*. *The function that determines the factor by which the numerical value of a physical quantity changes upon passage from the original system of units to another system within a given class is called the dimension function, or dimension[†], of that quantity.* It is customary (following a suggestion of Maxwell) to denote the dimension of a quantity ϕ by $[\phi]$. We emphasize that the dimension of a given physical quantity is different in different classes of systems of units. For example, the dimension of density ρ in the LMT class is $[\rho] = ML^{-3}$; in the LFT class, it is $[\rho] = L^{-4}FT^2$. *Quantities whose numerical values are identical in all systems of units within a given class are called dimensionless*; clearly, the dimension function is equal to unity for a dimensionless quantity. All other quantities are called *dimensional*.

The following definition can also be found in the literature: the relation that describes a derived unit for a certain quantity in terms of the fundamental units is called the dimension of that quantity. The definition introduced above is more precise, and removes certain teaching

---

[†] Our use of the singular should be noted.

difficulties which arise when using it. For example, according to this definition, the dimension of work in the kgf-m-s system is kgf m. The natural question of the precise meaning of the product of a force standard (the kilogram force) and a length standard (the metre) arises in the student's mind. However, according to the definition in the previous paragraph, the dimension of work W in the LFT class is given by the function $[W] = LF$, where L and F, the arguments of the dimension function, are abstract positive numbers whose multiplication and division raise no doubts. The question of the definition of the dimension of a quantity was precisely what made it necessary to first define the concept of a class of systems of units.

Furthermore, if the relations for the derived units mentioned above were actually to make sense as products or quotients of the fundamental units, they would have to be independent of what we mean when carrying out the multiplication or division. For example, according to the above definition, kgf m is the derived unit for the moment of a force as well as for work; m^2/s is the derived unit for the stream function as well as the kinematic viscosity, etc. But it is not implied that the stream function is measured in multiples of a basic amount of kinematic viscosity or that the moment of a force is measured in multiples of a basic amount of work! In contrast, using our definition, the fact that the dimensions of two physical quantities of different nature are identical does not seem unnatural.

We shall now cite a few additional examples. If (in the LMT class) the unit of length is decreased by a factor L, the unit of mass is decreased by a factor M, and the unit of time is decreased by a factor T, the numerical values of all forces are increased by a factor of MLT^{-2}, since, according to Newton's second law, the force f is the product of the mass m and the acceleration a:

$$f = ma\,.$$

For the decreases in the fundamental units mentioned above, the numerical values of all masses are increased by a factor M, and the numerical values of all accelerations are increased by a factor LT^{-2}. We have already mentioned that the dimensions of both sides of any equation with physical sense must be identical: otherwise, an equation in one system of units would become an inequality in another system, and this is not permissible for equations with physical sense. Thus, we find that the dimension of force in the LMT class is

$$[f] = [m][a] = MLT^{-2}\,. \tag{1.3}$$

Analogously, the dimension of mass in the LMT class is M, while it is $[M] = L^{-1}FT^2$ in the LFT class; the dimension of energy, $[E]$, is L^2MT^{-2} in the LMT class and LF in the LFT class. In the LMT class, the ratio of the velocity of uniform motion to the distance divided by the time required to traverse it is dimensionless. However, if we use the $LMTV$ class, in which the unit of velocity (knot/V) is independent, this ratio has a dimension different from unity: $L^{-1}TV$.

Dimension functions possess two important properties that we shall now discuss.

1.1.4 The dimension function is always a power-law monomial

We have seen that the dimension function is a power-law monomial in all the cases discussed above. This brings up the following question: are there physical quantities for which this is not so, and for which the dimensions in the LMT class are given, for example, by dimension functions of the form Me^L or $\sin M \log T$? In fact, there are no such physical quantities, and *the dimension function for any physical quantity is always a power-law monomial*. This follows from a simple, naturally formulated (but actually very deep) principle: all systems within a given class are equivalent, i.e., there are no distinguished, somehow preferred, systems among them.

We shall show this using the LMT class of systems; the reader may easily make the generalization to the general case. By virtue of the fact that the systems within a given class are equivalent, the dimension of any mechanical quantity a depends only on the quantities L, M and T:

$$[a] = \phi(L, M, T). \tag{1.4}$$

If there existed some distinguished system within the LMT class, it would be necessary to include the relationship between the system of units we were working in and the distinguished system. In this case, the dimension function ϕ would depend on three additional arguments: l_0/l_d, m_0/m_d, and t_0/t_d, the ratios of the units of length, mass, and time l_0, m_0, and t_0, in the original system of a given class to the corresponding units, l_d, m_d and t_d, in the distinguished system. According to the principle formulated above, this is not so, and the dimension function ϕ depends on the arguments L, M and T in the LMT class, independently of which system is adopted as the original system.

We shall now choose two systems of units within the LMT class, system 1, which is obtained from the original system by decreasing the fundamental units by factors of L_1, M_1 and T_1, and system 2, which is obtained from the original system by decreasing the fundamental units by factors of L_2, M_2 and T_2.

By the definition of the dimension function, the numerical value of the quantity under discussion (equal to a in the original system) is $a_1 = a\phi(L_1, M_1, T_1)$ in the first system, and $a_2 = a\phi(L_2, M_2, T_2)$ in the second system. Thus, we have

$$\frac{a_2}{a_1} = \frac{\phi(L_2, M_2, T_2)}{\phi(L_1, M_1, T_1)}. \tag{1.5}$$

We now note that by virtue of the equivalence of systems within a given class, we may assume that system 1 is the original system of the class, without altering the class of the systems of units. In this case, system 2 can be obtained from the new original system (system 1) by decreasing the fundamental units by factors of L_2/L_1, M_2/M_1 and T_2/T_1, respectively. Consequently, the numerical value a_2 of the quantity under discussion in the second system of units is, by the definition of the dimension function,

$$a_2 = a_1\phi(L_2/L_1, M_2/M_1, T_2/T_1).$$

We emphasize that a_1, the numerical value of the quantity a in system 1, remains unchanged under the change in the original system made above. Thus, $a_2/a_1 = \phi(L_2/L_1, M_2/M_1, T_2/T_1)$. Setting this expression equal to that in (1.5), we obtain the following equation for the dimension function ϕ:

$$\frac{\phi(L_2, M_2, T_2)}{\phi(L_1, M_1, T_1)} = \phi(L_2/L_1, M_2/M_1, T_2/T_1). \tag{1.6}$$

Equations of this type are called functional equations. We shall now show that only power-law monomials satisfy this equation.

To solve (1.6), we differentiate[†] both sides of this equation with respect to L_2, and *then* set $L_2 = L_1 = L$, $M_2 = M_1 = M$, and $T_2 = T_1 = T$. We find that

$$\frac{\partial_L \phi(L, M, T)}{\phi(L, M, T)} = \frac{1}{L}\partial_L \phi(1, 1, 1) = \frac{\alpha}{L}, \tag{1.7}$$

where the quantity $\alpha = \partial_L \phi(1, 1, 1)$ is a constant independent of L, M and T. Integrating (1.7), we find that

$$\phi(L, M, T) = L^\alpha C_1(M, T). \tag{1.8}$$

Substituting this expression into (1.6), we obtain an equation for the function C_1 of the same form as (1.6), but with one fewer argument:

$$\frac{C_1(M_2, T_2)}{C_1(M_1, T_1)} = C_1(M_2/M_1, T_2/T_1). \tag{1.9}$$

Once again, we proceed in the same way: we differentiate both sides of

---

[†] It is natural to assume that the dimension function is smooth.

(1.9) with respect to M_2 and set $M_2 = M_1 = M$ and $T_2 = T_1 = T$:

$$\frac{\partial_M C_1(M,T)}{C_1(M,T)} = \frac{1}{M}\partial_M C_1(1,1) = \frac{\beta}{M},$$

from which

$$C_1 = M^\beta C_2(T),\tag{1.10}$$

where $\beta = \partial_M C_1(1,1)$ is a constant similar to α. Following the same line of reasoning again, we find that

$$C_2(T) = C_3 T^\gamma,$$

so that

$$\phi = C_3 L^\alpha M^\beta T^\gamma.\tag{1.11}$$

The constant C_3 is obviously equal to unity, since $L = M = T = 1$ means that the fundamental units remain unchanged, so that the value of the quantity a remains unchanged, and $\phi(1,1,1) = 1$.

And so, we have shown that the solution to the functional equation (1.6) is the power-law monomial $L^\alpha M^\beta T^\gamma$, where α, β and γ are constants, so that the dimensions of any physical quantity can be expressed in terms of a power-law monomial.

If there existed a distinguished system of units within a given class, (1.6) would have been of the form

$$\frac{\phi(L_2, M_2, T_2, l_0/l_d, m_0/m_d, t_0/t_d)}{\phi(L_1, M_1, T_1, l_0/l_d, m_0/m_d, t_0/t_d)}$$

$$= \phi\left(\frac{L_2}{L_1}, \frac{M_2}{M_1}, \frac{T_2}{T_1}, \frac{l_0/l_d}{L_1}, \frac{m_0/m_d}{M_1}, \frac{t_0/t_d}{T_1}\right).$$

Indeed, we decreased the fundamental units of mass, length and time by factors of L_1, M_1 and T_1 in passing to the new original system of units (system 1), so that the ratios of the fundamental units in the original and distinguished systems were also changed. It thus turns out (if we return to the line of reasoning just presented above) that the quantities α, β and γ are no longer constants, but depend on L, M and T. For example, α turns out to be equal to $\partial_L\phi\,(1,1,1,l_0/Ll_d, m_0/Mm_d, t_0/Tt_d)$. Thus, if we give up the principle that all systems of units within a given class are equivalent, the main result – that dimension functions are power-law monomials – does not hold.

The following statement can be found in the literature:

The fact that the dimensions of all physical quantities are power-law monomials is a result of the following physical condition: the ratio of two numerical values (denoted by a and A) of any derived quantity must be independent of the scales chosen for the fundamental units.

This statement is incorrect, and this condition is not sufficient. In actual fact, as was just shown, if there exists a distinguished system of units within a given class, the dimension is not required to be a power-law monomial. However, even in this case, the numerical values of the derived quantity will be equal to a and A multiplied by the identical factors ϕ $(L, M, T, l_0/l_d, m_0/m_d, t_0/d_d)$, so that the ratio of the numerical values of the derived quantity remains unchanged.

It should be noted that systems of units convenient for use with some special classes of problem have frequently been proposed. For example, Kapitza (1966), has proposed a natural system of units for classical electrodynamics. Kapitza's system uses the classical radius of the electron as the unit of length, the rest-mass energy of the electron as the unit of energy, and the mass of the electron as the unit of mass. This system is convenient in classical electrodynamics problems, since it allows one to avoid very large or very small numerical values for all quantities of practical interest. It is important to note that this system is not 'distinguished' in the sense described above: the dimensions of physical quantities in the LEM class (E is the symbol for energy) do not depend on the ratios of the units of length, energy and mass in an original system for the class to the units in this natural system.

1.1.5 Quantities with independent dimensions

The quantities $a_1, \ldots, a_k$ are said to have *independent dimensions* if none of these quantities has a dimension function that can be represented in terms of a product of powers of the dimensions of the remaining quantities.

For example, density ($[\rho] = ML^{-3}$), velocity ($[U] = LT^{-1}$), and force ($[f] = MLT^{-2}$) have independent dimensions. Indeed, let us assume the converse. Then, since the dimension functions for both density and force contain M, and the dimension function for velocity does not, there must exist numbers x and y such that $[f] = [\rho]^x[U]^y$. Substituting the expressions for the dimensions $[f]$, $[\rho]$ and $[U]$ in terms of L, M and T into this relation, we find that

$$LMT^{-2} = (ML^{-3})^x(LT^{-1})^y. \tag{1.12}$$

Equating the exponents of L, M and T on the two sides of the equation, we obtain a system of three equations for the two unknowns x and y:

$$-3x + y = 1, \quad x = 1, \quad y = 2, \tag{1.13}$$

which obviously has no solution: $x = 1$ and $y = 2$, do not satisfy the first equation.

On the other hand, it is easy to show that the dimensions of density, velocity and pressure are dependent: the dimension of pressure, $L^{-1}MT^{-2}$, is equal to the product of the dimension of density and the square of the dimension of velocity. Furthermore, it is clear that none of the quantities $a_1, \ldots, a_k$ having independent dimensions may be dimensionless: the dimension of a dimensionless quantity (which is equal to unity) is equal to the product of the dimensions of the remaining quantities (whatever they are) raised to the power zero.

Another fact which will be important below is that *it is always possible to pass from the original system to some system within the same class such that any quantity, say a_1, in the set of quantities with independent dimensions $a_1, \ldots, a_k$ changes its numerical value by an arbitrary factor A, and while other quantities remain unchanged.*

In fact, within the adopted class of systems of units $P, Q, \ldots$ (P and Q denote the symbols L, M and T, as well as other similar quantities), the dimensions of the quantities $a_1, \ldots, a_k$ are given by

$$[a_1] = P^{\alpha_1}Q^{\beta_1}\ldots, \qquad \ldots, \qquad [a_k] = P^{\alpha_k}Q^{\beta_k}\ldots, \qquad (1.14)$$

where at least one of the quantities $\alpha_m, \beta_m, \ldots$ is non-zero for every m, $1 \leq m \leq k$. Therefore, according to the definition of the dimension function, the numbers $P, Q, \ldots$ must satisfy the following relations when passing from the original system of units to the desired system of units:

$$P^{\alpha_1}Q^{\beta_1}\ldots = A, \quad P^{\alpha_2}Q^{\beta_2}\ldots = 1, \quad \ldots, \quad P^{\alpha_k}Q^{\beta_k}\ldots = 1. \quad (1.15)$$

Taking the logarithm of the equations in (1.15), we obtain a system of linear algebraic equations for the logarithms of the unknown transformation coefficients, $\ln P, \ln Q, \ldots$; this system always has at least one solution:

$$
\begin{aligned}
\alpha_1 \ln P &+ \beta_1 \ln Q &+ \ldots &= \ln A, \\
\alpha_2 \ln P &+ \beta_2 \ln Q &+ \ldots &= 0, \\
&\;\vdots &\vdots\;\; & \\
\alpha_k \ln P &+ \beta_k \ln Q &+ \ldots &= 0.
\end{aligned}
\qquad (1.16)
$$

In fact, if the number of unknowns $\ln P, \ln Q, \ldots$ is greater than the number of equations, system (1.16) has infinitely many solutions. Indeed, it is clear that this system is only insoluble if the left-hand side of the first equation is a linear combination of the left-hand sides of the remaining equations:

$$\alpha_1 \ln P + \beta_1 \ln Q + \ldots = N_2(\alpha_2 \ln P + \beta_2 \ln Q + \ldots) + \ldots$$
$$+ N_k(\alpha_k \ln P + \beta_k \ln Q + \ldots), \qquad (1.17)$$

where $N_2, \ldots, N_k$ are certain constants. However, if we return from the logarithms to the exponents, this implies that

$$P^{\alpha_1}Q^{\beta_1}\ldots = \left(P^{\alpha_2}Q^{\beta_2}\ldots\right)^{N_2}\ldots\left(P^{\alpha_k}Q^{\beta_k}\ldots\right)^{N_k}. \qquad (1.18)$$

This would mean that the dimension of a_1 was equal to the product of powers of the dimensions of $a_2, \ldots, a_k$, which contradicts the assumption that the dimensions of the quantities $a_1, \ldots, a_k$ are independent.

If the number of unknowns is equal to the number of equations in system (1.16), its determinant must be non-zero, so that system (1.16) has a solution. In fact, if the determinant were zero, the left-hand side of any one of the equations in system (1.16) would be a linear combination of the others, so that (cf. the argument above), the dimension of any one of the quantities $a_1, \ldots a_k$ could be written as a product of the powers of the dimensions of the others, and this would contradict the assumption. Finally, the number of unknowns in system (1.16) cannot be less than the number of equations k. Indeed, if the number of unknowns m were less than the number of equations, i.e., the number of quantities $a_1, \ldots, a_k$, it would be possible to take the portion of system (1.16) corresponding to the quantities $a_1, \ldots, a_m$, and obtain a system where the number of equations is equal to the number of unknowns. Its determinant cannot be equal to zero; otherwise, the dimensions of the quantities $a_1, \ldots, a_m$ would not be independent (see above). Thus, the partial system has a solution. But it is then possible to express the quantities $P, Q, \ldots$ as a product of powers of the dimensions of $a_1, \ldots, a_m$. Substituting this expression into, for example, the last equation in (1.14), we find that the dimension of the quantity a_k can be expressed as a product of the powers of the dimensions of the first m quantities, which contradicts the assumption we made above that the dimensions of the quantities $a_1, \ldots, a_m$ were independent.

Thus, we have proved the statement made above. The properties derived above for the dimension function will now be used to construct the theory of dimensional analysis.

1.2 Dimensional Analysis

1.2.1 Governing parameters

In any concrete physical study (theoretical or experimental), we attempt to obtain relationships among the quantities that characterize the phenomenon being studied. Thus, the problem always reduces to determining (one or several) relationships of the form

$$a = f(a_1, \ldots, a_k, b_1, \ldots b_m), \tag{1.19}$$

where a is the quantity being determined in the study, and the $a_1, \ldots, a_k, b_1, \ldots, b_m$ are quantities that are assumed to be given; they are called *governing parameters.* The governing parameters in (1.19) are divided up in such a way that parameters $a_1, \ldots, a_k$ have independent dimensions, while the dimensions of parameters $b_1, \ldots, b_m$ can be expressed as products of powers of the dimensions of the parameters $a_1, \ldots, a_k$:

$$b_1 = [a_1]^{p_1} \ldots [a_k]^{r_1},$$

$$\cdots\cdots\cdots$$

$$b_i = [a_1]^{p_i} \ldots [a_k]^{r_i}, \qquad\qquad (1.20)$$

$$\cdots\cdots\cdots$$

$$b_m = [a_1]^{p_m} \ldots [a_k]^{r_m}.$$

Such a division may always be made. In some special cases, we may have $m = 0$ (if the dimensions of all of the governing parameters are independent) or $k = 0$ (if all the governing parameters are dimensionless). In general $k > 0$, $m > 0$.

For example, in the flow of a fluid through a long cylindrical pipe, the pressure drop per unit length of pipe (assumed constant over the length of the pipe) dp/dx, which corresponds to the quantity a in (1.19), is completely determined by the mean fluid velocity (averaged over the cross-section of the pipe) U, the diameter of the pipe D, the fluid density ρ, and the fluid viscosity μ:

$$dp/dx = f(U, D, \rho, \mu).$$

As may easily be shown, the dimensions of all involved quantities can be written in the following form:

$$[dp/dx] = ML^{-2}T^{-2},$$
$$[U] = LT^{-1},$$
$$[D] = L,$$
$$[\rho] = ML^{-3},$$
$$[\mu] = ML^{-1}T^{-1}.$$

Thus $k + m = n = 4$ in the case at hand. The dimensions of the first three governing parameters U, D, ρ are obviously independent, since the dimension function for the first parameter U contains T, while the other two do not and the dimension function for the third parameter ρ contains M, while the other two do not. The dimension of the fourth parameter μ can be obviously expressed in terms of the dimensions of the first three:

$$[\mu] = [\rho][U][D].$$

Thus, $k = 3$, $m = 1$.

The dimension of the quantity a to be determined must be expressible in terms of the dimensions of the governing parameters in the first group $a_1, \ldots, a_k$:

$$[a] = [a_1]^p \ldots [a_k]^r . \qquad (1.21)$$

If this were not so, the dimensions of the quantities $a, a_1, \ldots, a_k$ would be independent. Then, by the property proved in Section 1.5, it would be possible to change the value of the quantity a by an arbitrary factor (via a change in the system of units within the class in question) and leave the quantities $a_1, \ldots, a_k$ unchanged. In doing so, the quantities $b_1, \ldots, b_m$ (whose dimensions can be expressed in terms of the dimensions of the quantities $a_1, \ldots, a_k$) would likewise remain unchanged. Thus, the quantity to be determined, a, can be changed by any amount, with the values of all the governing parameters remaining unchanged; this is impossible as long as the list of governing parameters is complete. Thus, there always exist numbers $p, \ldots, r$ such that (1.21) holds.

In the example discussed above (the flow of a fluid in a pipe), we find (as is not difficult to verify) that

$$[dp/dx] = [U]^2 [D]^{-1} [\rho] .$$

1.2.2 Transformation to dimensionless parameters

We shall now introduce the parameters

$$\Pi_1 = \frac{b_1}{a_1^{p_1} \ldots a_k^{r_1}} ,$$

$$\ldots \ldots \ldots$$

$$\Pi_i = \frac{b_i}{a_1^{p_i} \ldots a_k^{r_i}} ,$$

$$\ldots \ldots \ldots \qquad (1.22)$$

$$\Pi_m = \frac{b_m}{a_1^{p_m} \ldots a_k^{r_m}} ,$$

$$\Pi = \frac{a}{a_1^{p} \ldots a_k^{r}}$$

where the exponents of the governing parameters with independent dimensions are chosen such that all the parameters $\Pi, \Pi_1, \ldots, \Pi_m$ are dimensionless. Relation (1.19) may be rewritten, replacing the parameters $a, b_1, \ldots, b_m$ (whose dimensions depend on those of the parameters $a_1, \ldots, a_k$) via the dimensionless quantities $\Pi, \Pi_1, \ldots, \Pi_m$ defined in

(1.22) and the parameters $a_1, \ldots, a_k$. We find that

$$\Pi = \frac{f(a_1, \ldots, a, b_1, \ldots, b_m)}{a_1^p \ldots a_k^r}$$

$$= \frac{1}{a_1^p \ldots a_k^r} f(a_1, \ldots, a_k, \Pi_1 a_1^{p_1} \ldots a_k^{r_1}, \ldots, \Pi_m a_1^{p_m} \ldots a_k^{r_m}).$$

Thus, we find that

$$\Pi = \mathcal{F}(a_1, \ldots, a_k, \Pi_1, \ldots, \Pi_m), \qquad (1.23)$$

where $\mathcal{F}$ is a certain function.

Now, the most important point to be discussed here is as follows.

It was shown in the previous chapter that it is always possible to pass to a system of units within the class in question such that any one of the parameters with independent dimensions $a_1, \ldots, a_k$ (for example, a_1) can be changed by an arbitrary factor, with all of the the remaining parameters $a_1, \ldots, a_k$ remaining unchanged. Obviously, the dimensionless arguments $\Pi_1, \ldots, \Pi_m$ of the function $\mathcal{F}$ and the value of the function, Π, also remain unchanged under such a transformation. However, it follows from this that the function $\mathcal{F}$ is in fact independent of the argument a_1. In exactly the same way, it can be shown that it is also independent of the arguments $a_2, \ldots, a_k$, so that $\mathcal{F} = \Phi(\Pi_1, \ldots, \Pi_m)$. Equation (1.23) can therefore in fact be written in terms of a function of m rather than $n = k + m$ arguments:

$$\Pi = \Phi(\Pi_1, \ldots, \Pi_m). \qquad (1.24)$$

However, since $\Pi = f/a_1^p \ldots a_k^r$, it follows that any function f that defines some physical relationship possesses the property of *generalized homogeneity* or symmetry, i.e., it can be written in terms of a function of a smaller number of variables, and is of the following special form:

$$f(a_1, \ldots, a_k, b_1, \ldots, b_m) = a_1^p \ldots a_k^r \Phi \left(\frac{b_1}{a_1^{p_1} \ldots a_k^{r_1}}, \ldots, \frac{b_m}{a_1^{p_m} \ldots a_k^{r_m}} \right). \qquad (1.25)$$

These results lead to the central theorem in dimensional analysis, the so-called Π-theorem: *A physical relationship between some dimensional (generally speaking) quantity and several dimensional governing parameters can be rewritten as a relationship between some dimensionless parameter and several dimensionless products of the governing parameters; the number of dimensionless products is equal to the total number of governing parameters minus the number of governing parameters with independent dimensions.* The term 'physical relationship' is used to emphasize that it should obey the covariance principle. In the example discussed above (fluid flow in a pipe), (1.25) obviously takes the form

$$\frac{dp}{dx} = f(U, D, \rho, \mu) = U^2 D^{-1} \rho \, \Phi \left(\frac{\mu}{U D \rho} \right), \qquad (1.26)$$

i.e., a function of four variables has been expressed in terms of a function of one variable.

Note that the Π-theorem is, in fact, completely obvious at an intuitive level. Indeed, it is clear that physical laws should not depend on the choice of units. They must therefore be expressed using relationships between quantities that do not depend on this arbitrary chance, i.e., dimensionless combinations of the variables. This was realized long ago, and concepts from dimensional analysis were in use long before the Π-theorem had been explicitly recognized, formulated and proved formally. The outstanding names that should be mentioned here are Galileo, Newton, Fourier, Maxwell, Reynolds and Rayleigh.

Dimensional analysis may be successfully applied (see below) in theoretical studies where a mathematical model of the problem is available, in the processing of experimental data, and also in the preliminary analysis of physical phenomena. The point that we are trying to make here is the following.

In order to determine the functional dependence of some quantity a on each of the governing parameters, it is necessary to either measure or calculate the function f for, let us say, 10 values of each governing parameter. Of course, the number 10 is somewhat arbitrary; a smaller number of measurements or calculations may suffice for some smooth functions, while even a hundred measurements are insufficient for other functions. Thus, it is necessary to carry out a total of 10^{k+m} measurements or calculations. After applying dimensional analysis, the problem is reduced to one of determining a function Φ of m dimensionless arguments $\Pi_1, \ldots, \Pi_m$, and only 10^m (i.e., a factor of 10^k fewer) experiments or calculations are required to determine this function. As a result, we reach the following important conclusion: *the amount of work required to determine the desired function is reduced by as many orders of magnitude as there are governing parameters with independent dimensions.*

For the example of flow in a pipe discussed above, the amount of work required is decreased by a factor of a thousand! Indeed, at the end of the last century, when the great English hydrodynamicist Reynolds examined the experimental data on flow in pipes available at the time and derived the relationship between the dimensionless quantity

$$\Pi = \frac{dp/dx}{U^2 D^{-1} \rho}$$

and the dimensionless quantity $\rho\, UD/\mu$ which was later named in his honour, he found that, to quite high accuracy, the experimental data lay on a single curve (see Figure 1.1, where this relationship is illustrated using more recent data).

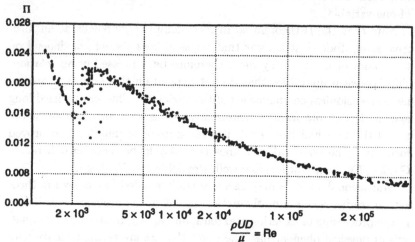

Figure 1.1. The dimensionless pressure drop (per unit length of pipe) of fluid flowing through a pipe as a function of Reynolds number Re = $\rho UD/\mu$. With the exception of the transition region between laminar and turbulent flow, the data from different experiments all lie on a single curve. The complicated nature of the curve indicates that the flow regime changes as a function of Re which is the only parameter that determines the global structure of the flow.

The following question naturally arises. If such substantial advantages are obtained for $n = k$, $m = 0$, why not make a transformation to a system of units in which the dimensions of all the quantities $a_1, \ldots a_k$, $b_1, \ldots, b_m$ would be independent?

Actually, nothing is gained in general by transforming to such a system. We shall show this using as an example a problem where quantities with dimensions of length l, time τ and velocity v are included among the governing parameters. We then change to the LTV class of systems, where the unit of velocity is independent. However, the formula $v = s/t$ (where s is the distance travelled, and t is the time of travel) is not valid in this class, and must be replaced by the formula $v = As/t$, where A is a constant having dimension $L^{-1}TV$ (see subsection 1.1.4). In general, therefore, the quantity A must be included among the governing parameters, thereby increasing the number of governing parameters by one. And, in general, the difference $m = n - k$ between the total number of governing parameters and the number of governing parameters with independent dimensions remains unchanged; thus, generally speaking, there is no advantage in transforming to a new class of systems of units. However, in some special cases, it may turn out that the additional parameters (like A) happen to be non-essential. In such cases, transforming to a new class increases the number of parameters with independent dimensions, and is useful. We shall see this later in some of the examples given.

1.2.3 Examples

We shall now present several illustrative examples.

1. At the beginning of this century – already many years after the work of Reynolds mentioned above – the physico-chemists E. Bose, D. Rauert and M. Bose published a series of experimental studies of the internal turbulent friction in various fluids (Bose and Rauert, 1909; Bose and Bose, 1911). The experiments were carried out in the following way (Figure 1.2).

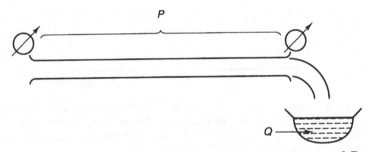

Figure 1.2. A schematic diagram showing the experiments of Bose, Rauert and Bose. The time τ required to fill a vessel of volume Q and the pressure drop between the ends of the pipe P were measured for the steady flow of various fluids through the pipe.

Various fluids (water, chloroform, bromoform, mercury, ethyl alcohol, etc.) were allowed to flow through a pipe in a regime of steady turbulence. The time τ required to fill a vessel with a certain fixed volume Q and the pressure drop P between the ends of the pipe were measured. As was customary, the results of the measurements were represented in the form of a series of tables and curves (similar to those in Figure 1.3) showing the pressure drop P as a function of the filling time τ.

At this time, the work of Bose and Rauert attracted the attention of T. von Kármán, at the time a young researcher, who later became one of the greatest men of the century in applied mechanics. He subjected their results to a processing procedure using (in modern terminology)

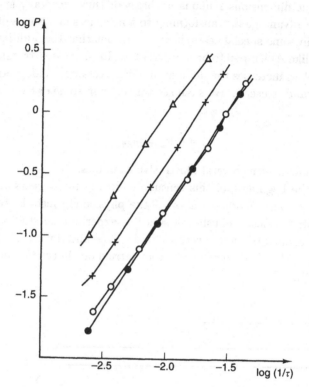

Figure 1.3. The experimental results of Bose, Rauert and Bose in their original form: $\circ$, water; $\bullet$, chloroform; $+$, bromoform; $\triangle$, mercury (P in kgf/cm^2, and τ in seconds). The curves are different for the different fluids. From von Kármán (1957).

dimensional analysis. Von Kármán's analysis can be presented in the following way. The pressure drop between the ends of the pipe, P, depends on the time τ required for the vessel to be filled and its volume Q, as well as on the properties of the fluid, its viscosity μ and density ρ. Obviously, $n = k + m = 4$ in the present case. It is instructive that von Kármán retained the none-too-promising original parameters Q and τ chosen by the experimenters. The dimensions of the quantities in question (for definiteness, in the LMT class) are given by the following expressions:

$$[P] = \frac{M}{LT^2}, \quad [\tau] = T, \quad [Q] = L^3, \quad [\mu] = \frac{M}{LT}, \quad [\rho] = \frac{M}{L^3}.$$

As may be seen, the three governing parameters τ, Q, and μ have independent dimensions. Indeed, the dimensional formula for μ contains the mass, while those for the other two governing parameters do not.

The dimension of viscosity therefore may not appear in the dimensions of the other governing parameters with any exponent other than zero. Furthermore, the dimensional formula for Q contains L alone, and the dimensional formula for τ contains T alone. It is therefore impossible to obtain the dimension of any one of these quantities in terms of the dimensions of the other two. On the other hand, the dimension of the governing parameter ρ can be expressed as a product of the dimensions of τ, Q, and μ raised to various powers:

$$[\rho] = [\tau][Q]^{-2/3}[\mu] . \tag{1.27}$$

The dimension of the determined parameter P can also be expressed in terms of the dimensions of the governing parameters τ, Q and μ:

$$[P] = [\tau]^{-1}[Q]^{0}[\mu] . \tag{1.28}$$

Thus $k = 3$, so that $m = n - k = 1$. Dimensional analysis yields

$$\Pi = \Phi(\Pi_1) , \tag{1.29}$$

where

$$\Pi = \frac{P}{\mu \tau^{-1}} , \tag{1.30a}$$

and

$$\Pi_1 = \frac{\rho}{\mu \tau Q^{-2/3}} . \tag{1.30b}$$

Thus, according to (1.29), the search for the desired relationship between the pressure drop and the four parameters that govern it, $P = f(\tau, Q, \mu, \rho)$, reduces to determining a single function Φ of one composite parameter – the function $\Phi(\Pi_1)$, since equation (1.29) can be written in the form

$$P = \frac{\mu}{\tau} \Phi \left(\frac{\rho}{\mu \tau Q^{-2/3}} \right) . \tag{1.31}$$

This means that all the experimental points should lie along a single curve in the coordinates $\rho/\mu \tau Q^{-2/3}$, $P/\mu \tau^{-1}$. Von Kármán's processing of the measured data of E. Bose, Rauert and M. Bose confirmed this (Figure 1.4). It is clear that if dimensional analysis had been carried out beforehand, the amount of experimental work required of the physico-chemists would have been reduced by a large factor.

2. The following example is famous. In an atomic explosion, a rapid (one might say, instantaneous) release of a significant amount of energy E occurs within a small region (one might say, at a point). A strong spherical shock wave (Figure 1.5) develops at the point of detonation; in the early stages, the pressure behind the wave front is several thousand times the initial air pressure, whose influence may be neglected in the early stages of the explosion. Thus, the radius of the shock wave front,

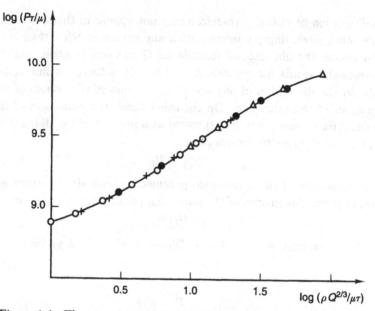

Figure 1.4. The experimental results of Bose, Rauert and Bose as represented by von Kármán (1911), who used dimensional analysis. All the experimental points lie on a single curve. From von Kármán (1957).

Figure 1.5. A photograph of a fireball 15 ms after an atomic explosion on the ground illustrates the spherical symmetry of the phenomenon and the sharp boundary of the perturbed region (Taylor, 1950a, b, 1963).

r_f, at an interval of time t after the explosion depends on the quantities E and t, and on the initial air density ρ_0; thus, $n = 3$. The dimensions of the governing parameters in the LMT class are

$$[E] = ML^2T^{-2}, \quad [t] = T, \quad [\rho_0] = ML^{-3}, \tag{1.32}$$

respectively. It may easily be shown that the dimensions of all three governing parameters are independent, so that k is also equal to three, and $m = n - k = 0$. Therefore, in this case, the function Φ in (1.24) has no arguments, i.e., it is simply a constant: $\Phi = \text{const}$. Furthermore, the dimension of the quantity to be determined r_f can be written in terms of the dimensions of the governing parameters raised to various powers: as may easily be shown,

$$[r_f] = [E]^{1/5}[t]^{2/5}[\rho_0]^{-1/5}.$$

Therefore,

$$\Pi = \frac{r_f}{E^{1/5}t^{2/5}\rho_0^{-1/5}}, \tag{1.33}$$

from which we find that

$$\Pi = \text{const}, \quad r_f = \text{const}\, E^{1/5}t^{2/5}\rho_0^{-1/5}. \tag{1.34}$$

The equation just obtained shows that if one measures the radius of the shock wave at various instants of time, the experimental points should lie on a straight line with a slope of unity if the logarithmic coordinates $\log t$, $(5/2)\log r_f$ are used:

$$\frac{5}{2}\log r_f = \frac{5}{2}\log(\text{const}) + \frac{1}{2}\log\frac{E}{\rho_0} + \log t. \tag{1.35}$$

The discussion presented above is due to Taylor (1941, 1950a,b), who processed data from a series of the high-speed photographs of the expansion of a fireball taken during an American nuclear test by J. Mack (Figure 1.6). The solution to the appropriate problem in gas dynamics (Taylor, 1941, 1950a; von Neumann, 1941, 1963; Sedov, 1946) showed that the constant has a value close to unity. Knowing this, it was possible to determine the energy of the explosion from the experimental dependence of the radius of the front on the time elapsed, i.e., from the y-intercept of the straight line constructed from the experimental points (Figure 1.6). At the time, Taylor's publication of this value (which turned out to be approximately 10^{21} erg) caused, in his words, 'much embarrassment' in American government circles: this figure was considered top secret, even though Mack's film was not classified.

3. We shall now present one more example (this time, rather amusing) of the application of dimensional analysis: we shall 'prove' Pythagoras' theorem using it[†]. The area of a right triangle, S_c, is completely determined by its hypotenuse c and, for definiteness, the smaller of its acute

---

[†] This example is also discussed in Migdal's book (1977). The word 'prove' has been placed in quotation marks for the following reason: this proof is based on similarity, which is presented after Pythagoras' theorem in rigorous geometry courses. In any case, the author does not recommend that this proof replace those used in geometry classes.

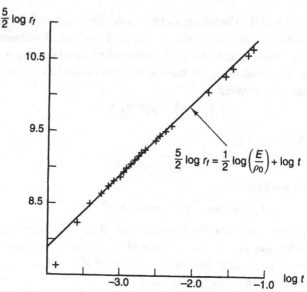

Figure 1.6. The experimental points determined by Taylor from Mack's movie film lay on a single straight line with slope unity in the coordinates $\log t$, $(5/2)\log r_f$. Taylor was thus able to determine the energy of the explosion from Mack's series of photographs (Taylor, 1950,a,b).

angles ϕ. Thus, $S_c = f(c, \phi)$, and $n = 2$. Clearly, $k = 1$, since ϕ is dimensionless. Dimensional analysis yields

$$\Pi = S_c/c^2 = \Phi(\Pi_1), \quad \Pi_1 = \phi, \quad S_c = c^2\Phi(\phi). \tag{1.36}$$

The altitude perpendicular to the hypotenuse of this triangle divides it into two similar right triangles (Figure 1.7) with hypotenuses equal to the sides a and b of the larger triangle. Equation (1.36) yields the following result for the areas of these triangles:

$$S_a = a^2\Phi(\phi), \quad S_b = b^2\Phi(\phi), \tag{1.37}$$

where $\Phi(\phi)$ is the same function as for the larger triangle.

But the sum of the areas of the two smaller triangles, S_a and S_b, is equal to the area of the larger triangle, S_c:

$$S_c = S_a + S_b, \quad c^2\Phi(\phi) = a^2\Phi(\phi) + b^2\Phi(\phi). \tag{1.38}$$

Cancelling out $\Phi(\phi)$ in the latter equation, we find that

$$c^2 = a^2 + b^2,$$

which is the desired result.

Note that the theorem essentially rests on the Euclidean nature of the geometry. In the Riemann and Lobachevskii geometries, there is an intrinsic parameter λ with dimension of length. This means that the

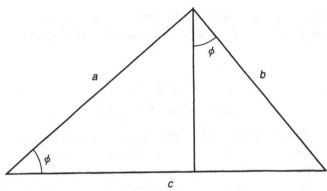

Figure 1.7. A proof of Pythagoras' theorem using dimensional analysis.

function Φ depends on not one, but two dimensionless parameters: the angle ϕ and the ratio of the length of the hypotenuse to this intrinsic parameter. The second argument is equal to c/λ for the basic triangle and to a/λ and b/λ for the first and second auxiliary triangles, respectively, so that it is impossible to cancel out Φ in (1.38), and the proof presented above no longer holds.

The examples just discussed support what we said earlier about the use of dimensional analysis. They demonstrate that the seemingly trivial concepts of dimensional analysis are capable of producing results with a great deal of content, especially when the difference between the number of governing parameters with independent dimensions and the total number of governing parameters is not large. Thus, correctly choosing the set of governing parameters becomes the most important factor: it is not only important to take all essential variables into account, but also not to include superfluous ones! The set of governing parameters may be determined relatively easily if a mathematical formulation of the problem is available.[‡] This must include the governing variables and constant parameters of the problem, which appear in the equations, boundary conditions, initial conditions, and so forth, and which determine the unique solution to the problem. Correctly choosing the set of governing parameters for problems that do not have an explicit mathematical formulation depends primarily on the intuition of the researcher. In such problems, success in applying dimensional analysis involves a correct understanding of which governing parameters are essential and which may be neglected. Remember that each governing parameter in

---

[‡] As we shall see later, there are many subtle points even here.

the problem that can be neglected reduces the amount of work involved in investigating the problem by roughly an order of magnitude!

1.3 Similarity

1.3.1 Similar phenomena

Before a large, expensive object (for example, a ship or aeroplane) is built, experimentation on models – *modelling* – is used to determine the best properties under future operating conditions. Many different kinds of measurement are carried out on models: for example, the lift and drag of an aeroplane model as air flows past it can be measured in a wind tunnel, as can the aerodynamic loading that causes a model of a television tower to collapse, etc. Clearly, one must know how to scale the results of the experiment carried out on the model up to the full-scale object being modelled. If one does not know how to do this, modelling is a useless pursuit. The concept of *physically similar phenomena* is central to correct modelling.

The concept of physical similarity is a natural generalization of the concept of similarity in geometry. For example, two triangles are similar if they differ only in the numerical values of the dimensional parameters, i.e. the lengths of the sides, while the dimensionless parameters, the angles at the vertices, are identical for the two triangles. Analogously, *physical phenomena are called similar if they differ only in respect of numerical values of the dimensional governing parameters; the values of the corresponding dimensionless parameters $\Pi_1, \ldots, \Pi_m$ being identical.*

In accordance with the definition that we have adopted for similar phenomena, the quantities $\Pi_1, \ldots, \Pi_m$ are called *similarity parameters*.

We shall now imagine that we propose to model a certain phenomenon; we shall call this the *prototype*. We require that the *model* that we wish to use to determine the desired properties of the prototype be a phenomenon *similar* to the prototype. Therefore, we have the following relationship between the parameter a to be determined and the governing parameters $a_1, \ldots, a_k, b_1, \ldots, b_m$:

$$a = f(a_1, \ldots, a_k, b_1, \ldots, b_m) \tag{1.39}$$

The function f is the same for both phenomena because they are similar, even though the numerical values of the governing parameters $a_1, \ldots, a_k, b_1, \ldots, b_m$ and the determined parameter a may differ. Thus, relationship (1.39) for the prototype takes the form

$$a^{(P)} = f\left(a_1^{(P)}, \ldots, a_k^{(P)}, b_1^{(P)}, \ldots, b_m^{(P)}\right). \tag{1.40}$$

The superscript P will hereinafter be used to refer to the properties of the prototype. Relation (1.39) for the model is similar in form, but the numerical values of the governing and determined parameters are different:

$$a^{(M)} = f\left(a_1^{(M)}, \ldots, a_k^{(M)}, b_1^{(M)}, \ldots, b_m^{(M)}\right). \qquad (1.41)$$

The superscript M will hereinafter be used to refer to the properties of the model. Via dimensional analysis, we obtain

$$\Pi^{(P)} = \Phi\left(\Pi_1^{(P)}, \ldots, \Pi_m^{(P)}\right), \qquad (1.42a)$$

$$\Pi^{(M)} = \Phi\left(\Pi_1^{(M)}, \ldots, \Pi_m^{(M)}\right), \qquad (1.42b)$$

where the function Φ is the same in both cases, since (see the preceding chapter) it can be expressed in terms of the function f in the same way in each case; $\Pi^{(P)}$, $\Pi^{M)}$, $\Pi_i^{(P)}$, $\Pi_i^{(M)}$ are dimensionless parameters.

1.3.2 The rule for scaling the results on a similar model up to the prototype

Since the model and prototype are similar, the following conditions must be satisfied, according to the definition of similar phenomena given above:

$$\Pi_1^{(M)} = \Pi_1^{(P)}, \quad \ldots, \quad \Pi_m^{(M)} = \Pi_m^{(P)}. \qquad (1.43)$$

Conditions (1.43) are sometimes called the *similarity criteria*.

Hence, as stated above,

$$\Phi\left(\Pi_1^{(M)}, \ldots, \Pi_m^{(M)}\right) = \Phi\left(\Pi_1^{(P)}, \ldots, \Pi_m^{(P)}\right),$$

and, in accordance with (1.42), the dimensionless parameters to be determined for the model and for the prototype are equal:

$$\Pi^{(P)} = \Pi^{(M)}. \qquad (1.44)$$

Returning to the dimensional parameters a, a_1, $\ldots$, a_k using (1.22), we find that

$$a^{(P)} = a^{(M)} \left(\frac{a_1^{(P)}}{a_1^{(M)}}\right)^p \cdots \left(\frac{a_k^{(P)}}{a_k^{(M)}}\right)^r, \qquad (1.45)$$

which is a simple rule for scaling the results of measurements on a similar model up to the prototype. It was precisely in order to be able to use this relation that it was necessary to require that the model be similar to the prototype.

1.3.3 Choosing the governing parameters in the model

The model parameters $a_1^{(M)}$, $\ldots$, $a_k^{(M)}$ may be selected arbitrarily, keeping in mind maximum simplicity and convenience in modelling. The conditions for similarity between the model and prototype – equality of the similarity parameters Π_1, ..., Π_m for both model and prototype, (1.43) – show how the remaining governing model parameters $b_1^{(M)}$, ..., $b_m^{(M)}$ must be chosen in order to ensure similarity between model and prototype. These conditions are as follows.

$$\Pi_1^{(M)} = \Pi_1^{(P)}: \qquad b_1^{(M)} = b_1^{(P)} \left(\frac{a_1^{(M)}}{a_1^{(P)}} \right)^{p_1} \quad \cdots \quad \left(\frac{a_k^{(M)}}{a_k^{(P)}} \right)^{r_1};$$

$$\cdots \qquad \qquad \cdots \cdots \qquad \cdots \qquad \cdots$$

$$\tag{1.46}$$

$$\cdots \qquad \qquad \cdots \cdots \qquad \cdots \qquad \cdots$$

$$\Pi_m^{(M)} = \Pi_m^{(P)}: \qquad b_m^{(M)} = b_m^{(P)} \left(\frac{a_1^{(M)}}{a_1^{(P)}} \right)^{p_m} \quad \cdots \quad \left(\frac{a_k^{(M)}}{a_k^{(P)}} \right)^{r_m}.$$

The simple definitions and statements presented above describe the entire content of the theory of similarity: we emphasize that there is nothing more to this theory. The simple examples below will demonstrate how to use the theory. Along the way, the reader will become familiar with the most important classical similarity parameters.

1.3.4 Examples

1. *The steady motion of a body in a fluid that fills a very large vessel* (we shall assume that it is infinite). The velocity of the body is assumed to be small relative to the velocity of sound, so that the compressibility of the fluid may be neglected, and its density ρ is assumed to be constant. The model's body must be geometrically similar to the prototype's body, and the direction of the velocity vector with respect to the principal axes of the body must be identical in both the model's and prototype's motion. Furthermore, the dimensional governing parameters of the motion are the characteristic length scale of the body, for example the maximum cross-sectional diameter D, the magnitude U of the body's velocity, the density ρ of the fluid, and its viscosity μ. The dimensions of the governing parameters in the LMT class are as follows:

$$[D] = L, \quad [U] = LT^{-1}, \quad [\rho] = ML^{-3}, \quad [\mu] = ML^{-1}T^{-1}. \tag{1.47}$$

Clearly, $n = 4$, $m = 1$ and $k = 3$, so that there is only one dynamical similarity parameter in addition to the geometric similarity parameters (expressing the similarity of the model's and prototype's body) and the kinematic similarity parameters (expressing the identical orientation of the velocity with respect to the principal axes of the model and of the prototype)[†]; this parameter can be written in the following form:

$$\Pi_1 = \frac{\rho\,UD}{\mu}. \tag{1.48}$$

As proposed by A. Sommerfeld, this parameter (as we saw, an analogous parameter is obtained for the flow of fluids in pipes) is called the *Reynolds number* or *Reynolds parameter* (the conventional symbol for this parameter is Re) in honour of the English scientist Osborne Reynolds who was one of the first to apply the ideas of similarity to hydrodynamics, with remarkable success. The dimensionless drag force on the body can be naturally defined in the following way:

$$\Pi = \mathcal{F}/\tfrac{1}{2}\rho\,U^2 S, \tag{1.49}$$

where $\mathcal{F}$ is the drag force on the body, $S \sim D^2$ is the cross-sectional area of the body, and the factor of $1/2$ is introduced by tradition.

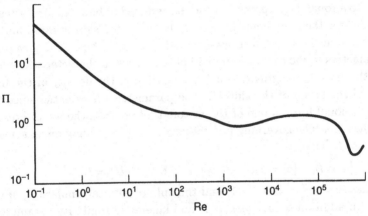

Figure 1.8. Dimensionless drag force on a sphere as a function of Reynolds number. The data from the various experiments shown here turn out to lie on a single curve. The complicated nature of the curve indicates that the flow regime changes with Reynolds number, and that the Reynolds number is the only parameter that governs the global structure of the flow.

The function $\Pi(\mathrm{Re})$ for the flow past a sphere is shown in Figure 1.8; to high accuracy, the data from a large number of experiments lie on a single curve. This curve appears to be very complicated: regions in which $\Pi(\mathrm{Re})$ varies smoothly alternate with sharp decreases and increases, and

---

[†] This follows from the identity of the corresponding geometric and kinematic similarity parameters for the model and prototype.

there are regions where Π is almost independent of Re. This all indicates that the flow regimes vary with Reynolds number (which is the only parameter that governs the structure of the flow as a whole as the fluid flows past the sphere).

The motion of a model is usually implemented in the same fluid as that in which the prototype moves. The similarity condition that parameter (1.48) be the same for the model and prototype motions indicates that in this case, the product UD must be identical for model and prototype; from this, we see that the velocity at which a dynamically similar model moves is inversely proportional to the size of the model compared with that of the prototype. From this, it follows that the drag forces are identical for the model and the prototype, so that the scaling coefficient for the drag force is equal to unity in this case.

2. *The motion of a streamlined surface ship at high speeds.* Here, we shall discuss the idealized case where the contribution from viscous drag can, to a rough first approximation, be assumed to be small for a stream-lined ship: the main contribution to the drag on a ship in rapid motion is that due to the surface waves created by the ship. The governing parameters in the case at hand will be as follows: a characteristic length for the ship, l, the gravitational acceleration g, the density of the fluid ρ, and the speed of the ship U. The parameter g is essential since the gravitational force is one of the factors that controls the waves created by the ship. The governing parameters have the following dimensions in the LMT class:

$$[l] = L, \quad [g] = LT^{-2}, \quad [\rho] = ML^{-3}, \quad [U] = LT^{-1}, \qquad (1.50)$$

so that $n = 4$, $k = 3$, $m = 1$, and the only dynamical similarity parameter (in addition to the geometric and kinematic similarity parameters, see above) is of the form

$$\Pi_1 = \frac{U}{l^{1/2}g^{1/2}}. \qquad (1.51)$$

This parameter is called the *Froude number* or *Froude parameter* (the conventional symbol is Fr) after the famous English engineer and ship-builder Wm. Froude.

Furthermore, the dimension of the drag force $\mathcal{F}$ in the same class LMT is $[\mathcal{F}] = LMT^{-2}$, so that $[\mathcal{F}] = [\rho][g][l]^3$. Thus, since the parameter g can be varied only with a great deal of effort by means of some subtle tricks, which are not normally used, the law for scaling the drag force from the model up to the prototype in the same fluid is of the form

$$\mathcal{F}^{(P)} = \mathcal{F}^{(M)} \left(\frac{l^{(P)}}{l^{(M)}} \right)^3, \qquad (1.52)$$

so that the drag force is proportional to the cube of the modelling scale. Relation (1.51) indicates that in order to ensure dynamical similarity, the ratio of the model velocity to the prototype velocity must be proportional to the square root of the modelling scale:

$$U^{(M)} = U^{(P)} \left(\frac{l^{(M)}}{l^{(P)}} \right)^{1/2}. \qquad (1.53)$$

If one does not neglect the role of viscosity, a second dynamical parameter appears – the Reynolds number Re, which is equal to $\rho Ul/\mu$. Modelling in which both similarity parameters – the Froude number and Reynolds number – are taken into account turns out to be impossible in the same fluid. Indeed, to do this, the products Ul (see example (1) above) and U^2/l would have to be identical for both model and prototype; this is only possible when modelling to full scale, which makes no sense. This is precisely why, for illustration, we have restricted ourselves to the case where the viscous drag is small compared with the wave drag. As a matter of fact, in ship-building practice, the viscous drag contribution is modelled separately from the wave drag using specially developed techniques.

3.   *Rowing* (McMahon, 1971) provides us with our next example of similarity analysis. The boats accommodating various numbers N of oarsmen; one, two, four and eight, are compared, and the following assumptions are made.

(1)   *There is a geometric similarity between boats.*

(2)   *the volume of a loaded boat per oarsman G is a constant, characteristic for boats of all classes*; this follows from assuming that the bulk weight of the boat per oarsman, including the oarsman's own weight, is constant. This means that the oarsmen are considered as indistinguishable in weight.

(3)   *The power per oarsman A is a constant, characteristic for all classes*, so the oarsmen are considered as indistinguishable also in power.

The principal force that hinders the motion of the boat through the water is, unlike the previous example, skin friction drag. Indeed, full-scale rowing-tank tests have shown that the resistance due to leeway and wave-making constitute together only a tiny part of total drag. In the range of Reynolds numbers characteristic for racing the drag coefficient λ can be considered as a constant. Therefore the bulk power supporting the motion is

$$P = \lambda \rho v^3 l^2 \qquad (1.54)$$

where ρ is the water density, v is the velocity of the motion, assumed to

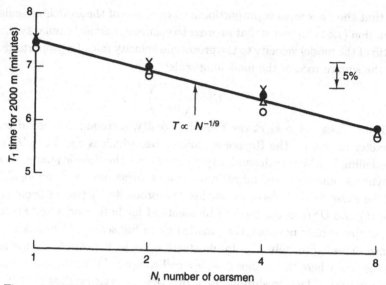

Figure 1.9. The -1/9 power-law dependence of rowing time on the number of oarsmen (solid line) compared with racing times for 2000 m, all at calm or near calm conditions. $\triangle$, 1964 Olympics, Tokyo; $\bullet$, 1968 Olympics, Mexico City; $\times$, 1970 World Rowing Championships, Ontario; $\circ$, 1970 Lucerne International Championships. After McMahon (1971).

be steady, and l is the characteristic length scale of the wetted surface. Obviously $P = AN$.

Thus, the velocity of the motion v is a function of the governing parameters N, A, G and ρ. The dimensions of the parameters in the class $RVLN$ (R is the dimension of density, V is the dimension of velocity, L is the dimension of length, and N is the dimension of the number of oarsmen; these four can be considered as independent dimensions) are

$$[v] = V, \quad [G] = \frac{L^3}{N}, \quad [A] = \frac{RV^3L^2}{N}, \quad [\rho] = R, \quad [N] = N. \quad (1.55)$$

So, obviously, $m = 0$ and $k = n = 4$. According to dimensional analysis we obtain

$$\Pi = \text{const}, \quad \Pi = \frac{v}{A^{1/3}\rho^{-1/3}G^{-2/9}N^{1/9}} \quad (1.56)$$

and the final result for the velocity of the boat is

$$v = \text{const} \frac{A^{1/3}}{\rho^{1/3}G^{2/9}}N^{1/9}. \quad (1.57)$$

The time for a fixed distance, say 2 000 m should be, according to (1.57) inversely proportional to $N^{-1/9}$. The validity of this result is well illustrated by Figure 1.9.

4. *Thermal convection in a horizontal fluid layer bounded by smooth, rigid, isothermal surfaces.* A temperature that exceeds the temperature T_0 maintained on the upper surface by an amount δT is maintained on the lower surface. As is well known, the phenomenon of convection is due to the fact that the density of a fluid decreases as it is heated; if this decrease is large enough, the less dense fluid floats from bottom to top. For small changes in temperature, the temperature dependence of the fluid density can be assumed to be linear: $\rho = \rho_0[1 + \alpha(T_0 - T)]$, where ρ_0 is the fluid density at temperature T_0, and α is the coefficient of volume expansion. The variation in the density of the fluid as it is heated is small, so that we only need take the density variation into account where it is combined with the action of the gravitational force. This approximation was suggested by the French scientist J. Boussinesq, and carries his name. The *Boussinesq approximation* is related to the assumption that all of the accelerations in convective flow are small compared with the gravitational acceleration. This is not so in strongly developed convection; here the Boussinesq approximation is no longer valid. If we adopt the Boussinesq approximation, the coefficient of volume expansion of the fluid, α, and the gravitational acceleration g obviously do not enter into consideration separately, but only as a product. The product αg is called the *buoyancy parameter*.

We shall now consider the governing parameters for the phenomenon of thermal convection in a layer. The properties of the phenomenon clearly must depend on the buoyancy parameter αg, on the thickness of the layer H, on the dynamical properties of the fluid: its viscosity μ and density ρ_0 at temperature T_0, on its specific heat capacity of c and thermal conductivity λ, and on the excess temperature of the lower layer δT. We shall neglect the variations with temperature in the viscosity, specific heat capacity, and thermal conductivity of the fluid, since we only intend to model the process of convection in a basic way.

In principle, the contribution of viscous energy dissipation to the thermal balance of the fluid should also be taken into account. To do this, one additional parameter must be included; the mechanical equivalent of heat J.

We now consider the dimensions of the governing parameters. The specific heat capacity c is, by definition, the quantity of heat necessary to increase the temperature of a unit mass of the fluid by one temperature unit. Thus, the dimension of heat capacity is

$$[c] = \frac{Q}{M\Theta},$$

where Q stands for the independent dimension of heat, and Θ stands for the independent dimension of temperature. The thermal conductivity of the fluid, λ, is, by the fundamental law of heat conduction (the Fourier law), the coefficient of proportionality in the expression for the heat flow in a quiescent fluid as a function of the temperature drop δT and the thickness of the layer H:

$$q = -\lambda \frac{\delta T}{H}.$$

Now, the heat flux is, by definition, the amount of heat that passes through a unit area of the plane layer boundary per unit time, so that $[q] = QL^{-2}T^{-1}$. From this result and the preceding equation, we find that

$$[\lambda] = \frac{Q}{LT\Theta}.$$

The dimension of the mechanical equivalent of heat is obviously equal to the dimension of mechanical energy divided by the independent dimension of thermal energy:

$$[J] = \frac{ML^2}{T^2Q}.$$

So, the dimensions of the governing parameters in the $LMT\Theta Q$ class are as follows:

$$[\alpha g] = \frac{L}{\Theta T^2}, \quad [H] = L, \quad [\mu] = \frac{M}{LT}, \quad [\rho] = \frac{M}{L^3},$$
$$[c] = \frac{Q}{M\Theta}, \quad [\delta T] = \Theta, \quad [\lambda] = \frac{Q}{LT\Theta}, \quad [J] = \frac{ML^2}{T^2Q}. \tag{1.58}$$

Clearly, $n = 8, m = 3$ and $k = 5$. Applying dimensional analysis, we obtain the following three similarity parameters:

$$\Pi_1 = \frac{\delta T}{(\alpha g)^{-1}H^{-3}\mu^2\rho^{-2}}, \quad \Pi_2 = \frac{\lambda}{\mu c}, \quad \Pi_3 = \frac{Jc}{\alpha g H}. \tag{1.59}$$

In what follows, we shall discuss convective motion in thin layers, where the parameter Π_3 is large ($\Pi_3 \gg 1$), so that the effect of this parameter on the similarity conditions may be neglected.[†] It is useful to estimate the value of the characteristic length $\Lambda = Jc/\alpha g$ in order to get an idea of how the extent to which this condition is restrictive. We have $J = 4.2 \times 10^7$ erg/cal, $c = 1$ cal/g°C and $\alpha = 2 \times 10^{-4}$ for water, and $g = 10^3$ cm/s^2, from which we find that $\Lambda \simeq 2 \times 10^8$ cm = 2000 km. Thus, when modelling convection in layers, even a layer one kilometre thick can be assumed to be thin. On the other hand, when modelling

---

[†] We shall see below that neglecting the effect of a certain parameter is always a strong assumption, no matter how large or small this parameter may be.

convection in the Earth's mantle, the parameter Π_3 is of order unity, and can not be neglected.

We should also note one significant fact that follows from equations (1.59) for the similarity parameters: if the contribution from the dissipation of the energy in the convective motion to the thermal balance is neglected, it turns out that the quantities λ and c enter into the discussion as a ratio rather than separately.

The parameter Π_1 is called the *Grasshof number*, and the following combinations of the parameters in (1.59) are frequently used in the literature:

$$\frac{\Pi_1}{\Pi_2} = \frac{\alpha g (\delta T) H^3}{\mu \rho^{-2} \lambda c^{-1}} = \text{Ra}, \quad \frac{1}{\Pi_2} = \frac{\mu c}{\lambda} = \text{Pr}. \tag{1.60}$$

The parameter Ra, the *Rayleigh number*, is named after the great English physicist who was the first to study the onset of convection in a horizontal layer theoretically. When the critical value of this parameter,

$$\text{Ra} = \text{Ra}_{\text{cr}} \simeq 657, \tag{1.61}$$

which does not depend on the second parameter[†], the *Prandtl number* Pr, is reached, the state with the fluid at rest in a horizontal layer becomes unstable, and the so-called regime of buoyancy-driven convection rolls sets in. In this regime, the layer breaks up into fluid rolls that rotate in opposite directions (Figure 1.10). Until the Rayleigh number reaches its critical value, the equilibrium state for a quiescent fluid layer is stable. The later changes in the convection regime in the horizontal layer are associated with the passage of the Rayleigh number through other critical values.

Figure 1.10. Uniformly heating from below a fluid in a vessel shaped like a rectangular parallelepiped with sides in the ratio of 10:4:1 produces flow with rotating rolls parallel to one of the sides. From van Dyke (1982).

The similarity parameters (1.59) indicate that if the modelling is carried out in the same fluid and same gravitational field as the prototype motion the following condition on the model's temperature difference

[†] The critical value given here is calculated under the assumption that the tangential stresses vanish at the boundaries of the layer.

$\delta T^{(M)}$ must be satisfied:

$$\delta T^{(M)} = \delta T^{(P)} \left(\frac{H^{(P)}}{H^{(M)}} \right)^3 ; \qquad (1.62)$$

this condition ensures that the model's convective motion is similar to that of the prototype. Furthermore, as may easily be shown, the dimensionless parameter for the heat flux q is of the form

$$\Pi = \frac{q}{(\alpha g)^{-1} H^{-4} \mu^3 \rho^{-2} c} . \qquad (1.63)$$

Thus, the rule for scaling the heat flux when modelling in a layer of the same fluid as the prototype motion, takes the following form:

$$q^{(P)} = q^{(M)} \left(\frac{H^{(M)}}{H^{(P)}} \right)^4 , \qquad (1.64)$$

so that the ratio of the heat fluxes in the prototype and in the model is inversely proportional to the fourth power of the modelling scale.

As was mentioned above, the influence of the similarity parameter Π_3 becomes appreciable for thick layers. Since this parameter is the ratio of the characteristic length scale of the fluid, $\Lambda = Jc/\alpha g$, to the layer thickness, it is, strictly speaking, impossible to model the phenomenon in a layer of the same fluid under identical conditions (compare this result with the second example).

The present example shows that one must be careful when determining the similarity parameters. For example, if we assume that the dimensions of mechanical energy and thermal energy are independent without having taken the governing parameter for the mechanical equivalent of heat into consideration, we will not notice the restrictions on the thickness of the model layer. Meanwhile, the phenomena for thick and thin layers are substantially different; they are not similar, and it is in general not possible to scale the heat fluxes using the simple relation in (1.64).

Furthermore, it is obvious that if the thermal and mechanical energy were measured in the same units, i.e., if we were to pass from the $LMT\Theta Q$ class to the $LMT\Theta$ class, the conclusions reached above would not be affected in any respect. Indeed, the difference between the total number of governing parameters and the number of governing parameters with independent dimensions would remain constant, even though the mechanical equivalent of heat had been removed from consideration. The fact that the contribution of viscous dissipation to the thermal balance is negligible would then be interpreted to mean that the phenomenon is not governed by the heat capacity c and thermal conductivity λ separately. The governing parameter would be their ratio, which appears in the so-called thermometric conductivity $\kappa = \lambda/\rho c$.

This would lead to a decrease in the number of governing parameters, and the disappearance of the similarity parameter Π_3.

The examples presented[†] show that dimensional considerations play a decisive role in establishing rules for modelling and criteria for similarity. The crucial step in modelling, as in any application of dimensional analysis to cases where an exact mathematical formulation of the problem is missing, lies in the proper choice of a system of governing parameters. Often the procedure is as follows. One takes as governing parameters all quantities that could possibly, in the investigator's opinion, have an influence on the phenomena, no matter how hypothetical. As governing parameters with independent dimensions one takes those governing parameters that are definitely known to be essential, and with respect to the remaining ones, one looks at the numerical values of the corresponding similarity parameters Π_i. If these values are very small or very large, the corresponding dimensional parameter b_i is considered inessential and is discarded.

In many cases one can actually proceed in this way. It is important to note, however, that in general this is not so, and one must be very careful about arguments such as the above. One should see in them not a proof of the possibility of disregarding one parameter or another, but a strong conjecture. This last assertion is essentially obvious: it is not necessarily true that a function $\Pi = \Phi(\Pi_1, \ldots, \Pi_i, \ldots, \Pi_m)$ tends to a definite and moreover finite limit for small or large values of the argument Π_i. Only the existence of such a limit (and in fact even a sufficiently rapid convergence to it) can justify neglecting a governing parameter when the corresponding similarity parameter is very large or very small. Subsequent discussion will show us that such crudeness of analysis can lead to serious mistakes.

[†] Very instructive also are similarity considerations for the atmospheres of planets (Golitsyn, 1973).

The construction of intermediate-asymptotic solutions using dimensional analysis. Self-similar solutions

2.1 Heat propagation from a concentrated instantaneous source

2.1.1 The equation of heat conduction. Initial and boundary conditions

We shall demonstrate the basic ideas in the application of dimensional analysis to problems in mathematical physics by discussing heat conduction in a long bar (Figure 2.1), the properties of the material (what these are will be mentioned below) and the cross-sectional area remaining constant along the bar. The sides of the bar are thermally isolated, so that inflow or outflow of heat through the sides of the bar may be neglected, and the temperature distribution across the bar is assumed to be uniform. Thus, the temperature θ depends only on the longitudinal coordinate of the cross section, x, and the time, t, and is independent of the transverse coordinates of the points in the cross section. For the reader's convenience we will briefly recall the derivation of the basic heat conduction equation.

Consider a section of the bar between adjacent cross sections with coordinates x and $x + dx$. The volume of the section is $S dx$, where S is the constant (by assumption) cross-sectional area of the bar. The change in the quantity of heat contained in the section corresponding to a change in temperature $d\theta$ is (up to small, higher-order terms, whose contribution is negligibly small for $dx \to 0$) equal to $\rho c d\theta S dx$, where ρ is the (constant) density of the bar material and c is the heat capacity per unit mass, which is also a constant property of the bar material. The change in the quantity of heat, if it occurs over a small time interval dt,

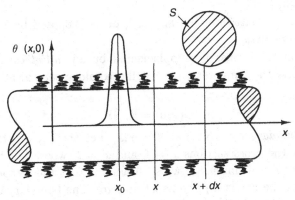

Figure 2.1. Derivation of the heat conduction equation.

where $d\theta = \partial_t \theta dt$, is given by (also up to small, higher-order quantities)

$$\rho c S \partial_t \theta dx dt \; .$$

Since heat is neither created nor annihilated within the bar, and it does not flow out through the sides, this change can only be produced by a difference between the heat flux into the section under consideration through cross section x and the heat flux out through cross section $x+dx$. But, within the time interval dt, a quantity of heat $Sqdt$ (where q is the heat flux, i.e., the quantity of heat flowing normally through unit cross-sectional area in unit time) flows into the section at x and, a quantity of heat $S(q + \partial_x q dx)dt$ flows out of the section at $x + dx$. Therefore, in time dt a quantity of heat

$$-S\partial_x q dx dt$$

accumulates in the section. Equating the two expressions for the change in the quantity of heat contained in the section and dividing by $Sdxdt$, we find that

$$\rho c \partial_t \theta + \partial_x q = 0 \; . \tag{2.1}$$

The Fourier law, according to which the heat flux q is proportional to the temperature gradient and directed opposite to it,

$$q = -\lambda \partial_x \theta \; , \tag{2.2}$$

is adopted as the basic law in the mathematical theory of heat conduction. In this equation, λ is the thermal conductivity (a property of the bar material), a quantity that is, by assumption, constant, as are the area S, the density ρ, and specific heat capacity c. Substituting (2.2) into (2.1), we obtain a *differential equation for the conduction of heat* :

$$\partial_t \theta = \kappa \partial_{xx}^2 \theta \; . \tag{2.3}$$

The constant $\kappa = \lambda/\rho c$ is, as proposed by Maxwell, called the *ther-*

mometric conductivity of the material, or as proposed by Kelvin, the *thermal diffusivity*.

Equation (2.3) must be supplemented by an *initial condition* that specifies how the temperature is distributed along the bar at the initial time,

$$\theta(x,0) = \theta_0(x),\qquad(2.4)$$

and by *boundary conditions* at its ends. Let the ends of the bar be located at the cross sections $x = 0$ and $x = l$, where l is the length of the bar. For definiteness, we shall assume that the temperature at the ends of the bar is kept constant at zero. The boundary conditions expressing this are

$$\theta(0,t) = \theta(l,t) = 0.\qquad(2.5)$$

2.1.2 Concentrated instantaneous heat source. The first self-similar intermediate stage

We shall now discuss an important special case, in which all the heat is concentrated at the initial time within a small region of length h around the cross section $x = x_0$ (Figure 2.1), and the initial temperature at the remaining sections is equal to the temperature at which the ends are kept, i.e., zero. Thus, it is assumed that the size of the region where the heat is originally released, h, is much smaller than the distance from the left-hand end of the bar, x_0, and the distance from the right-hand end of the bar, $l - x_0$, so that $h \ll x_0$ and $h \ll l - x_0$. Of course, it is possible to obtain a numerical solution to this problem for any initial temperature distribution: modern numerical methods allow one to do this rather easily on a small-size computer. The results of such a numerical investigation – the temperature distribution along the bar at various times – are shown for a typical case in Figure 2.2. The results of these numerical calculations reflect all the special features of the initial distribution.

We now proceed as follows.

We do not concern ourselves about the precise details of the initial temperature distribution in the region where the heat is released. They are only important during a small time interval at the beginning of the process, when the heated region, where the temperature is substantially different from zero, has a length that is still of order the size of the region where the heat was originally released. We shall only consider times when size of the heated region is much greater than h. For such times, it seems plausible that it may be assumed that all the heat was initially concentrated at the cross section $x = x_0$, i.e., that $h = 0$.

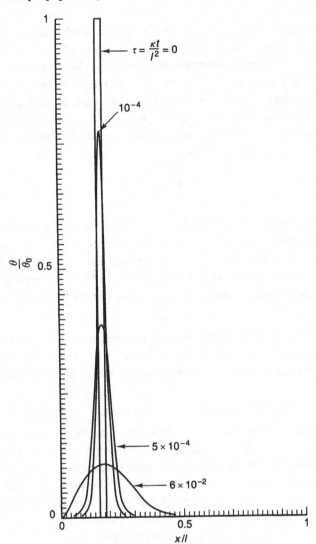

Figure 2.2. Temperature distributions obtained by numerical calculation of the conduction of heat from a narrow region, closer to the left-hand end of the bar which was initially heated to a temperature θ_0. It is apparent that neither end of the bar has any influence on the temperature distribution during the first intermediate stage, so that the bar may be assumed to be infinite.

Furthermore, it is clear that until the boundaries of the region where the temperature has changed substantially reach at least one of the ends of the bar, the bar may be assumed to be infinite. Indeed, the fact that the bar is finite and that the ends are maintained at zero temperature is of little significance during this time. This is readily apparent in the graphs of the numerical solution shown in Figure 2.2.

Thus, in this intermediate stage, the problem reduces to a highly idealized one – the propagation of heat in an infinite bar from an instantaneous heat source concentrated at one of its cross sections. Heat is neither created nor annihilated in the bar; therefore the total amount of heat in the infinite bar at any instant of time is equal to the initial amount of heat E, from which we have that

$$\rho c S \int_{-\infty}^{\infty} \theta dx = E. \qquad (2.6)$$

Thus, the temperature θ at time t at cross section x is determined by the following quantities: the time t, the parameters κ and $Q = E/\rho c S$, and the distance of the cross section from the source, $x - x_0$; the parameters E, ρ, c and S do not appear in the problem separately, and only their combination Q is a governing parameter. We thus have

$$\theta = f(t, \kappa, Q, x - x_0). \qquad (2.7)$$

Obviously, $n = 4$. We shall now give the dimensions of all the quantities involved in the problem in the $LT\Theta$ class of systems of units, where Θ stands for the dimension of temperature, which is independent of L and T:

$$[\theta] = \Theta, \quad [t] = T, \quad [\kappa] = L^2 T^{-1}, \quad [Q] = \Theta L, \quad [x - x_0] = L. \quad (2.8)$$

The dimensions of κ and Q are obtained from the obvious condition that the left- and right-hand sides of equations (2.3) and (2.6) have the same dimensions. The first three governing parameters, t, κ and Q have independent dimensions. The dimension of the governing parameter $x - x_0$ can be expressed in terms of the dimensions of the first three governing parameters using the following relation:

$$[x - x_0] = [t]^{1/2}[\kappa]^{1/2}[Q]^0 ;$$

the reader may easily verify this using (2.8). Thus, $k = 3$. Furthermore,

$$[\theta] = [Q][\kappa]^{-1/2}[t]^{-1/2} .$$

Thus, from dimensional analysis, we find that

$$\Pi = \Phi(\Pi_1), \quad \Pi = \frac{\theta}{Q(\kappa t)^{-1/2}}, \quad \Pi_1 = \frac{x - x_0}{(\kappa t)^{1/2}} . \qquad (2.9)$$

Hence, we find that

$$\theta = \frac{Q}{(\kappa t)^{1/2}}\Phi(\xi), \quad \xi = \frac{x - x_0}{(\kappa t)^{1/2}} . \qquad (2.10)$$

We shall now introduce the time-dependent temperature scale $\theta_0(t) =$

$Q(\kappa t)^{-1/2}$ and length scale $l_0(t) = (\kappa t)^{1/2}$; the temperature distribution (2.10) may then be written in the form

$$\theta = \theta_0(t)\Phi\left(\frac{x - x_0}{l_0(t)}\right),\qquad(2.11)$$

so that if this distribution is plotted in the transformed ('self-similar') coordinates $\theta/\theta_0(t) = \theta(\kappa t)^{1/2}/Q, (x - x_0)/l_0(t) = \xi$, it is an identical curve at all instants of time (Figure 2.3).

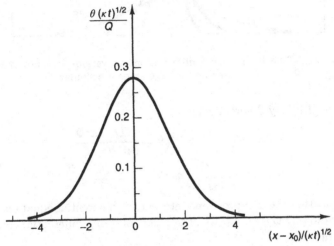

Figure 2.3. The temperature distributions during the first self-similar intermediate stage can be represented by a single curve in the reduced variables.

Thus, we have found that the solution to the idealized problem under consideration possesses the fundamental property of *self-similarity*: the temperature distributions along the bar at various instants of time can be obtained from one another by a similarity transformation (Figure 2.4). We emphasize once again that in the case under discussion, the self-similar nature of the phenomenon of heat propagation from an instantaneous concentrated source and the form of the self-similar variables were established using dimensional analysis alone.

Therefore, by virtue of equation (2.10), which was derived from an idealized formulation of the problem using dimensional analysis, obtaining the appropriate solution to a *partial differential equation* (2.3) with two independent variables, x and t, has been reduced to determining a function Φ of a *single* variable, $\xi = (x - x_0)/(\kappa t)^{1/2}$, composed of the independent variables $x - x_0$ and t. This makes it possible easily to obtain the solution analytically. Indeed, we obtain the following from

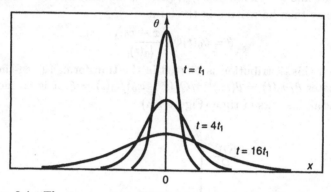

Figure 2.4. The temperature distributions at various instants of time during the first self-similar intermediate stage are similar.

equation (2.10) by differentiation:

$$\partial_x \theta = \frac{Q}{(\kappa t)} \frac{d\Phi}{d\xi}\,, \quad \partial_{xx}^2 \theta = \frac{Q}{(\kappa t)^{3/2}} \frac{d^2\Phi}{d\xi^2}\,,$$

$$\partial_t \theta = -\frac{1}{2t} \frac{Q}{(\kappa t)^{1/2}} \Phi(\xi) - \frac{1}{2t} \frac{Q}{(\kappa t)^{1/2}} \xi \frac{d\Phi}{d\xi}\,. \tag{2.12}$$

Substituting these expressions into partial differential equation (2.3), and dividing by $Q/(\kappa t)^{1/2}$, we obtain the following differential equation for the function Φ:

$$\frac{d^2\Phi}{d\xi^2} + \frac{\xi}{2} \frac{d\Phi}{d\xi} + \frac{\Phi}{2} = 0\,. \tag{2.13}$$

This is an *ordinary* differential equation, since it contains only one independent variable, ξ. For the idealized problem under discussion, the propagation of heat in an infinite bar, we have the following condition: the temperature θ at infinity is equal to zero for all times. By virtue of (2.10), the function Φ therefore also satisfies the condition

$$\Phi(\pm\infty) = 0\,. \tag{2.14}$$

Then, substituting solution (2.10) into condition (2.6), we find that

$$\int_{-\infty}^{\infty} \theta dx = \frac{Q}{\sqrt{\kappa t}} \int_{-\infty}^{\infty} \Phi\left(\frac{x - x_0}{\sqrt{\kappa t}}\right) dx$$

$$= Q \int_{-\infty}^{\infty} \Phi\left(\frac{x - x_0}{\sqrt{\kappa t}}\right) d\left(\frac{x - x_0}{\sqrt{\kappa t}}\right) = Q \int_{-\infty}^{\infty} \Phi(\xi)d\xi = Q\,, \tag{2.15}$$

from which we obtain a second condition on the function $\Phi(\xi)$:

$$\int_{-\infty}^{\infty} \Phi(\xi)d\xi = 1. \tag{2.16}$$

Equation (2.13) is an equation in total differentials, so it may easily be solved. Integrating (2.13) once, we find that

$$\frac{d\Phi}{d\xi} + \frac{\xi\Phi}{2} = \text{const.}$$

Setting $\xi = 0$, and using the obvious fact that the solution is symmetric, so that $d\Phi/d\xi = 0$ for $\xi = 0$, we find that const $= 0$. Integrating once again, we find that a solution of the following form satisfies condition (2.14):

$$\Phi = A\exp(-\xi^2/4), \tag{2.17}$$

where A is some constant to be determined. If we substitute (2.17) into the second condition imposed on the function $\Phi(\xi)$, equation (2.16), we find that

$$A\int_{-\infty}^{\infty} e^{-\xi^2/4}d\xi = 2A\int_{-\infty}^{\infty} e^{-z^2}dz = 2A\sqrt{\pi} = 1, \tag{2.18}$$

where we have used the well-known integral

$$\int_{-\infty}^{\infty} e^{-z^2}dz = \sqrt{\pi}.$$

Hence, we find that $A = 1/(2\pi^{1/2})$ and, finally,

$$\theta = \frac{Q}{2\sqrt{\pi\kappa t}}e^{-(x-x_0)^2/4\kappa t}. \tag{2.19}$$

Once again, we recall that this solution describes the temperature distribution in the region where it varies substantially only over some intermediate time interval. During this time interval the size of the heated region is much larger than that of the region in which the heat was initially released, but is still less than the distance to the ends of the bar.

We will have to reconsider later the above derivation of (2.19). While it is typical and seems entirely transparent, this derivation in fact contains underwater reefs that become apparent after a seemingly small and insignificant modification of the problem (see the following chapter).

2.1.3 Concentrated instantaneous heat source. The second self-similar intermediate stage

As is evident from (2.19), the temperature distribution obtained above falls off extremely rapidly with distance $x - x_0$ from the initially heated cross section, even though the temperature turns out to be formally non-zero at any distance from the source at an arbitrarily small time after the beginning of the process. Using tables of the exponential function, it may easily be shown that the temperature increase at a distance x_0 from the source only becomes appreciable after a time of order one hundredth to one tenth of x_0^2/κ. For example, the temperature at a distance of x_0 becomes of order one-hundredth of the maximum temperature at the source cross section $x = x_0$, this is given by

$$\theta_{max} = \frac{Q}{2\sqrt{\pi\kappa t}} \approx 0.28\frac{Q}{\sqrt{kt}}, \qquad (2.20)$$

at a time $\sim 0.1 x_0^2/\kappa$ from the beginning of the process. After this time, the influence of the end of the bar at $x = 0$ becomes appreciable (see Figure 2.2), and solution (2.19) is no longer valid. It is instructive to examine the case where the distance from the source to the right-hand end of the bar (at $x = l$) is at least one or one-and-a-half orders of magnitude greater than the distance to the left-hand end of the bar ($x = 0$), so that, for example, $l - x_0 = 30x_0$. In this case, the temperature increase at the end $x = l$ will be small, less than one-hundredth of the maximum temperature, until a time $\sim 100x_0^2/\kappa$. Thus, even though the bar may not be assumed to be infinite during the time interval $0.1x_0^2/\kappa < t < 100x_0^2/\kappa$, it can be assumed to be semi-infinite, for $0 \leq x < \infty$; the temperature is equal to zero at the left-hand end of the bar $x = 0$ and at $x = \infty$:

$$\theta(0, t) = 0, \quad \theta(\infty, t) = 0.$$

We multiply both sides of (2.3) by x and integrate from $x = 0$ to $x = \infty$:

$$\int_0^\infty \partial_t \theta x dx = \frac{d}{dt} \int_0^\infty \theta x dx = \kappa \int_0^\infty x \partial_{xx}^2 \theta dx. \qquad (2.21)$$

Integrating by parts and using the boundary conditions $\theta(0, t) = \theta(\infty, t) = 0$ at the ends of the semi-infinite bar, we find that

$$\int_0^\infty x \partial_{xx}^2 \theta dx = x \partial_x \theta \Big|_0^\infty - \int_0^\infty \partial_x \theta dx = \theta(0, t) - \theta(\infty, t) = 0.$$

From this relation and (2.21), it follows that the quantity

$$\int_0^\infty \theta x dx = M \qquad (2.22)$$

is independent of time. As may easily be determined from the initial conditions, $M = Qx_0$.

We shall now discuss times of order x_0^2/κ; now the size of the heated region is much greater than the distance from the source to the nearer end, x_0, but still much smaller than the distance from the far end, $l - x_0$. For x much greater than x_0, the influence of the fine features in the temperature distribution formed when the heated region reached the near end of the bar ($x = 0$) must disappear, so that the temperature depends only on the quantities $t - t_0$, κ, M and x:

$$\theta = f(t - t_0, \kappa, M, x),\qquad (2.23)$$

where t_0 is some effective initial time for the second stage in the process; we shall determine the value of this constant later. Equation (2.22) indicates that $[M] = \Theta L^2$, so that, applying dimensional analysis in much the same way as we did in obtaining the previous result, equation (2.10), we find that

$$\theta = \frac{M}{\kappa(t - t_0)}\Phi_1(\xi), \quad \xi = \frac{x}{\sqrt{\kappa(t - t_0)}}.\qquad (2.24)$$

(The reader may easily carry out the appropriate calculations for him-/herself).

Substituting (2.24) into (2.3), we find that the function $\Phi_1(\xi)$ satisfies the ordinary differential equation

$$\frac{d^2\Phi_1}{d\xi^2} + \frac{\xi}{2}\frac{d\Phi_1}{d\xi} + \Phi_1 = 0.\qquad (2.25)$$

Furthermore, the function Φ_1 obviously satisfies the following conditions at $\xi = 0$ and $\xi = \infty$:

$$\Phi_1(0) = 0, \quad \Phi_1(\infty) = 0.\qquad (2.26)$$

Condition (2.22) allows us to obtain, in much the same way that (2.16) was obtained, an integral relation that the function Φ_1 must satisfy:

$$\int_0^\infty \xi\Phi_1(\xi)d\xi = 1.\qquad (2.27)$$

Note that (as may easily be verified) the derivative $d\Phi/d\xi$ of function (2.17) satisfies both equation (2.25) and conditions (2.26), so that the solution to equation (2.25) that satisfies conditions (2.26) can be written in the following form:

$$\Phi_1 = B\xi e^{-\xi^2/4},\qquad (2.28)$$

where B is a constant. In order to determine this constant, we make use of (2.27), so that we have

$$B \int_0^\infty \xi^2 e^{-\xi^2/4} d\xi = 8B \int_0^\infty z^2 e^{-z^2} dz = 2B\sqrt{\pi} = 1, \qquad (2.29)$$

where we have once again used a well-known integral:

$$\int_0^\infty z^2 e^{-z^2/4} dz = \sqrt{\pi}/4 .$$

From this result and (2.28), we obtain the final relation for the solution describing the second intermediate stage:

$$\theta = \frac{Mx}{2\sqrt{\pi}[\kappa(t - t_0)]^{3/2}} e^{-x^2/4\kappa(t - t_0)} . \qquad (2.30)$$

The effective initial time for the beginning of the second stage, t_0, which was left undetermined, may be found in the following way.

For $t \gg t_0$, solution (2.30) may be expanded in a series in the small parameter t_0/t and written in the following form:

$$\theta = \frac{Q x_0 x e^{-x^2/4\kappa t}}{2\sqrt{\pi}(\kappa t)^{3/2}} \left[1 + \frac{3}{2}(t_0/t)\right] + O\left((t_0/t)^{7/2}\right) . \qquad (2.31)$$

We now note that in the time interval under consideration, the temperature distribution must correspond to that for an instantaneous heat source Q concentrated at the cross section $x = x_0$ in a semi-infinite bar. This solution may easily be obtained using the solution to the problem of an instantaneous source in an infinite bar discussed in the previous section. Namely, we construct the solution for an infinite bar with an instantaneous source of intensity Q at the cross section $x = x_0$ and one of intensity $-Q$ at $x = -x_0$:

$$\theta = \frac{Q}{2\sqrt{\pi\kappa t}} \left(e^{-(x-x_0)^2/4\kappa t} - e^{-(x+x_0)^2/4\kappa t}\right) . \qquad (2.32)$$

It is easily seen that the condition $\theta(0, t) = 0$ is satisfied for this solution. Expanding solution (2.32) in a series in the small parameter x_0/x to the same accuracy as (2.31), we find that

$$\theta = \frac{Q x_0 x e^{-x^2/4\kappa t}}{2\sqrt{\pi}(\kappa t)^{3/2}} - \frac{Q x_0^3 x e^{-x^2/4\kappa t}}{8\sqrt{\pi}(\kappa t)^{5/2}} . \qquad (2.33)$$

Comparing the second terms in (2.31) and (2.33), we find that $t_0 = -x_0^2/6\kappa$.

The solution obtained above (2.30) is sometimes called the *thermal dipole solution*, and it obviously also possesses the property of self-similarity. It is not difficult to find the cross section $x = x_*(t)$ where the temperature from solution (2.30) reaches its maximum value. Setting $\partial_x \theta$ equal to zero, we find that $\xi = \sqrt{2}$ at this cross section, so that

$x = x_*(t) = [2\kappa(t - t_0)]^{1/2}$. Substituting this expression into (2.30), we obtain the value of the maximum temperature:

$$\theta_* = \theta_{\max} = \frac{Me^{-1/2}}{\sqrt{2\pi}\kappa(t - t_0)} \simeq 0.25\frac{M}{\kappa(t - t_0)}. \qquad (2.34)$$

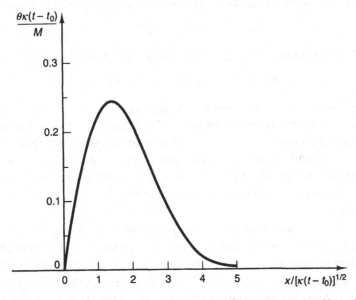

Figure 2.5. The temperature distributions for the second self-similar intermediate stage can be represented by a single curve in the reduced variables.

The solution we have constructed is shown in Figure 2.5. It is appropriate for describing the phenomenon at times greater than x_0^2/κ, but less than times of order $0.1(l - x_0)^2/\kappa$, when the heat has begun to affect the far end. The results obtained are instructive: for times much greater than $0.1h^2/\kappa$ but less than $0.1x_0^2/\kappa$, the distribution approaches the self-similar distribution given by (2.19):

$$\theta \simeq \frac{Q}{2\sqrt{\pi\kappa t}}e^{-(x-x_0)^2/4\kappa t},$$
$$0.1h^2/\kappa \ll t < 0.1x_0^2/\kappa. \qquad (2.35)$$

During the time interval from x_0^2/κ to $\sim 0.1(l-x_0)^2/\kappa \simeq 100x_0^2/\kappa$, the temperature distribution approaches the self-similar distribution (2.30):

$$\theta \simeq \frac{Mx}{2\sqrt{\pi}[\kappa(t - t_0)]^{3/2}}e^{-x^2/4\kappa(t-t_0)}, \qquad (2.36)$$
$$x_0^2/\kappa < t < 100x_0^2/\kappa$$

In the intermediate interval $0.1x_0^2/\kappa < t < x_0^2/\kappa$ the solution is not self-similar. It corresponds to an intermediate stage between two different self-similar regimes.

Thus, the exact self-similar solutions to special, highly idealized and extremely schematicized problems are approximate representations of the solution to a more general problem, and may be applied to high accuracy on certain fairly broad (two orders of magnitude) time intervals.

2.2 Phenomena at the initial stage of a nuclear explosion

2.2.1 Very intensive thermal waves

As we have seen in the example in the previous section, dimensional analysis allows one to obtain, by means of a completely standard method, exact special solutions to problems of mathematical physics that reduce to initial, boundary, or mixed problems for partial differential equations or systems of such equations. These special solutions are expressed in terms of solutions of boundary-value problems for ordinary differential equations.

Another indication of the general method of applying dimensional analysis to obtain exact special solutions is provided by the example of the thermal and gas-dynamic phenomena arising at the initial stages of an atomic explosion in a gas[†]. We shall discuss the corresponding solutions here for two reasons. Firstly, the application of dimensional analysis to essentially nonlinear problems is well demonstrated by these solutions. Secondly, and this is more important for us, we shall indicate explicitly here as we did in the previous section some assumptions that are not ordinarily mentioned and which are, as a matter of fact very strong hypotheses made in formulating the corresponding problems. These hypotheses turn out to be valid for the problems considered in the present chapter. However, as we shall see later, apparently small complications of the problems that at first glance leave the considerations of dimensional analysis unaltered make these hypotheses inapplicable, and we encounter a paradox whose resolutions will lead us to self-similar solutions of a new type.

Thus, at the very first stage of an atomic explosion, immediately fol-

---

[†] A more detailed consideration of the physics of gas-dynamic phenomena in very intense explosions can be found in the monograph of Zeldovich and Raizer (1966, 1967). For similar phenomena arising from the action of a focussed laser impulse on matter, see the monograph of Raizer (1977).

lowing the release of energy, the hot gas is still at rest. Strong thermal waves propagate through the motionless gas. The radiative transfer of energy takes place with a speed many times exceeding the speed of sound; hence at this stage the hydrodynamic transfer of matter can be neglected, and the thermal conductivity of the gas is determined basically by radiation. The coefficient of thermal conductivity λ can be considered as a power function of the temperature θ:

$$\lambda = \lambda_0 \theta^n .\tag{2.37}$$

The value of n is roughly 5. The dependence of the specific heat c on temperature is substantially weaker, and to a first approximation can be neglected. We write the equation for the conservation of energy in the form

$$c\partial_t\theta + \operatorname{div}\mathbf{q} = 0,$$

where $\mathbf{q} = -\lambda \operatorname{grad} \theta$ is the heat flux and t is the time. We have

$$\operatorname{div}\mathbf{q} = -\operatorname{div}\lambda \operatorname{grad}\theta = -\lambda_0 \operatorname{div}\theta^n \operatorname{grad}\theta$$

$$= -\frac{\lambda_0}{n+1} \operatorname{divgrad}\theta^{n+1} = -\frac{\lambda_0}{n+1}\Delta(\theta^{n+1}).$$

In the case of interest, spherically symmetric wave propagation, we have, by virtue of the symmetry of the problem, $\Delta(\theta^{n+1}) = r^{-2}\partial_r r^2 \partial_r \theta^{n+1}$ (r being the distance from the centre), and the equation of heat propagation finally assumes the form

$$\partial_t\theta = \kappa r^{-2}\partial_r(r^2\partial_r\theta^{n+1}).\tag{2.38}$$

Here $\kappa = \lambda_0/(n+1)c$ is a constant.

We consider a solution to this equation under the following initial conditions and condition at infinity:

$$\theta(r,0) = 0 \quad (r \neq 0); \qquad 4\pi c \int_0^\infty \theta(r,0)r^2 dr = E,\tag{2.39}$$

$$\theta(\infty,t) = 0 \quad (t > 0).$$

These conditions correspond to the instantaneous release, at the initial moment and at the point that is the centre of the explosion, of a definite finite amount of heat E, with the initial temperature equal to zero everywhere except at the centre of the explosion[†].

For the solution θ the governing parameters will obviously be the independent variables r and t and the constant parameters κ and $Q = E/c$ (because the parameters E and c enter the problem statement not

---

[†] The asymptotic meaning of the solution under such initial conditions will be considered in detail below.

separately but only as this ratio) that appear in the equation and the initial conditions:

$$\theta = f(t, \kappa, Q, r).\qquad(2.40)$$

The dimensions of the governing parameters are as follows:

$$[t] = T, \quad [\kappa] = L^2 T^{-1} \Theta^{-n}, \quad [Q] = \Theta L^3, \quad [r] = L,\qquad(2.41)$$

where Θ is a symbol for the temperature dimension.

We now apply dimensional analysis. It is evident that in this case $n = 4$, $m = 1$ and $k = 3$. Choosing as governing parameters with independent dimensions t, κ and Q, we obtain by dimensional analysis

$$\Pi = \Phi(\Pi_1),\qquad(2.42)$$

where

$$\Pi = \frac{\theta}{[Q^2(\kappa t)^{-3}]^{1/(3n+2)}}, \quad \Pi_1 = \frac{r}{[Q^n \kappa t]^{1/(3n+2)}} = \xi.$$

Hence we find

$$\theta = [Q^2(\kappa t)^{-3}]^{1/(3n+2)}\Phi(\xi).\qquad(2.43)$$

Calculating the required derivatives of θ with respect to t and r with the help of (2.43), and substituting into (2.38), we obtain for the function $\Phi(\xi)$ the ordinary differential equation

$$\frac{d^2 \Phi^{n+1}}{d\xi^2} + \frac{2}{\xi}\frac{d\Phi^{n+1}}{d\xi} + \frac{1}{3n+2}\xi\frac{d\Phi}{d\xi} + \frac{3}{3n+2}\Phi = 0.\qquad(2.44)$$

Equation (2.43) shows that for any t,

$$4\pi c \int_0^\infty \theta(r,t)r^2 dr = 4\pi E \int_0^\infty \xi^2 \Phi(\xi)d\xi = \text{const.}$$

From this and (2.39) we get the conditions

$$\int_0^\infty \Phi(\xi)\xi^2 d\xi = 1/4\pi, \quad \Phi(\infty) = 0.\qquad(2.45)$$

To this we also add the requirement of continuity of the function Φ and *of the derivative* $d\Phi^{n+1}/d\xi \sim \Phi^n d\Phi/d\xi$. This follows from the continuity at any instant of time $t > 0$ of the temperature, which is proportional to Φ, and of the heat flux $\mathbf{q} = -\lambda \text{grad}\theta = -[\lambda_0/(n+1)]\text{grad}\theta^{n+1}$, which is proportional to $d\Phi^{n+1}/d\xi$. The last requirement is nontrivial – it shows that for $\Phi \neq 0$ the derivative $d\Phi/d\xi$ must be continuous; at the same time, at points where Φ vanishes, the derivative $d\Phi/d\xi$ can suffer a finite or even infinite discontinuity, provided only that $d\Phi^{n+1}/d\xi$ be continuous.

Integration of (2.44) gives a solution satisfying the second condition of (2.45) in the form

$$\Phi = K(\xi_f^2 - \xi^2)^{1/n} , \ (\xi \leq \xi_f) \ ; \ \Phi \equiv 0 \ (\xi \geq \xi_f), \quad (2.46)$$

where $K = [n/2(n+1)(3n+2)]^{1/n}$. To determine the remaining constant ξ_f we use the first condition of (2.45). We have

$$K \int_0^{\xi_f} (\xi_f^2 - \xi^2)^{1/n} \xi^2 d\xi \equiv K \xi_f^{\frac{3n+2}{n}} \int_0^1 (1 - \zeta^2)^{1/n} \zeta^2 d\zeta = 1/4\pi, \quad (2.47)$$

whence, and using the expression of the integral in terms of beta functions (Abramowitz and Stegun, 1970), we find

$$\xi_f = [2\pi K B(3/2, (n+1)/n)]^{-n/(3n+2)} , \quad (2.48)$$

Here B is the symbol for Euler's beta function.

Thus the temperature distribution finally assumes the form

$$\theta = K \left[\frac{E^2}{c^2 \kappa^3 t^3} \right]^{1/(3n+2)} \left(\xi_f^2 - \frac{r^2}{(E^n c^{-n} \kappa t)^{2/(3n+2)}} \right)^{1/n} \quad (2.49)$$

for $r \leq r_f(t) = \xi_f(n)[(E/c)^n \kappa t]^{1/(3n+2)}$, and $\theta \equiv 0$ for $r \geq r_f(t)$.

From (2.49) follows the simple relation

$$\frac{\theta}{\theta(0,t)} = \left(1 - \frac{r^2}{r_f^2} \right)^{1/n} , \quad (r \leq r_f),$$

$$\theta \equiv 0 , \quad (r \geq r_f). \quad (2.50)$$

This function is shown in Figure 2.6 for $n = 5$. It is of essential importance that for $n > 0$, in contrast with the linear case, one has a finite speed of heat propagation; the perturbation zone is bounded, $r_f(t) < \infty$ for any finite t. Passing to the limit $n \to 0$ we recover the known solution of the linear equation of heat conduction for an instantaneous point source. In this case $r_f(t) = \infty$ for any $t > 0$. The solution discussed above was obtained by Zeldovich and Kompaneets (1950) and by Barenblatt (1952). The latter paper considered the mathematically equivalent problem of gas or ground-water filtration. Later the solution (2.50) was obtained by Pattle (1959). In the paper by Barenblatt and Zeldovich (1957a) a solution to the very intense thermal-wave problem, similar to the dipole-type solution considered in section 1.3, was obtained.

The solution (2.50) is in fact not a classical solution to the differential equation (2.38). Indeed, equation (2.38) contains space derivatives of the second order; meanwhile even the first space derivative of the solution (2.50) has a discontinuity. Therefore an important mathematical question appeared when the solution (2.50) was obtained: in what sense is (2.50) a solution to (2.38) and is it unique? These questions were answered in the fundamental paper by Oleynik, Kalashnikov and Chzhou

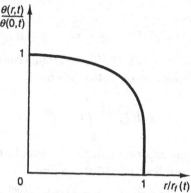

Figure 2.6. Distribution of temperature behind a very intense thermal wave in the reduced variables $\theta(r,t)/\theta(0,t)$, $r/r_f(t)$.

Yui-lin (1958). They introduced the natural class of weak (generalized) solutions of the equation (2.38), and proved the existence and uniqueness of such solutions. The peculiar property of the solution (2.50) is as was mentioned the finite speed of heat propagation. In rigorous mathematical language this property means that if at some time instant t_0 a weak solution $u(x,t)$ has 'compact support', i.e., is represented by a finite function, then the solution will have a compact support at any $t \geq t_0$. This property was rigorously proved in the paper by Oleynik, Kalashnikov and Chzhou Yui-lin (1958). (Note also an earlier paper, Barenblatt and Vishik (1956).) Later, the investigation of the solutions to equation (2.38), known in mathematical literature as the porous medium equation (PME), and its generalizations became the subject of many investigations by mathematicians. The fundamental review by Kalashnikov (1987) is highly recommended in this respect.

2.2.2 Very intense blast waves

The solution obtained in subsection 2.2.1 describes the phenomenon of a very intense explosion only at the initial thermal stage. As time passes, the speed of radiative transfer of energy decreases and quickly becomes small compared with the speed of sound. There arises in the heated gas an intense shock wave, which outstrips the thermal wave and initiates the transition to the subsequent gas-dynamic stage. At this stage it is necessary to consider the motion of the gas, which can be considered adiabatic. We recall the well-known equations (see Kochin, Kibel' and Roze, 1964; Batchelor, 1967; Germain, 1986b; Landau and Liftschitz, 1987) for adiabatic motion of a gas in the case of spherical symmetry in which we are interested. The first equation, Newton's law written for a unit volume of gas, is

$$\frac{dv}{dt} \equiv \partial_t v + v\partial_r v = -\frac{1}{\rho}\partial_r p\,.$$

Here v is the radial component of velocity, p the pressure, ρ the density of the gas, r the radial coordinate measured from the centre of the explosion, and t the time. The only force acting is the pressure drop in the radial direction, and the mass of a unit volume is equal to the density of the gas. Further, the conservation law for the mass of the gas is satisfied:

$$\partial_t \rho + \mathrm{div}\rho\mathbf{v} = 0\,.$$

In the case of spherical symmetry, when the radial velocity v is the only non-zero velocity component, $\mathrm{div}\rho\mathbf{v} \equiv r^{-2}\partial_r(r^2\rho v) = (2/r)\rho v + \partial_r(\rho v)$. Finally, by virtue of the adiabaticity of the motion, one has the equation of conservation of entropy in a fluid particle:

$$\frac{ds}{dt} \equiv \partial_t s + v\partial_r s = 0\,.$$

Here s is the entropy of a unit mass, which in the case considered of a thermodynamically and calorifically ideal gas is equal to $s = c_v \ln(p/\rho^\gamma)$; c_v is the specific heat of the gas at constant volume, and γ is the ratio of the specific heats at constant pressure and constant volume. Thus the basic equations of motion for the gas can be written as

$$\partial_t v + v\partial_r v + \partial_r p/\rho = 0\,,$$
$$\partial_t p + \partial_r(\rho v) + 2\rho v/r = 0\,, \qquad (2.51)$$
$$\partial_t(p/\rho^\gamma) + v\partial_r(p/\rho^\gamma) = 0\,.$$

We consider here an exact solution to the idealized problem of the gas motion arising from the instantaneous release at the centre of the explosion of a finite amount of energy E. The gas is assumed to be initially at rest, its pressure equal to zero, and the initial density of the gas equal to ρ_0 everywhere except at the centre of the explosion[†]. A classical, i.e., smooth, solution of this problem does not exist, and we shall seek a piecewise-smooth solution: the perturbed domain, inside which the solution varies continuously and is described by (2.51), is bounded by a shock wave, which is a sphere of radius $r_f(t)$ on which the properties of the motion – the pressure, density and velocity change discontinuously. Outside this sphere the state of rest of the gas is preserved, and the initial pressure of the gas is also equal to zero by assumption. Thus the conditions of conservation (continuity of flux) of mass, momentum and

---

[†] The asymptotic meaning of this solution will also be considered below.

energy at the front of the shock wave are written in the following form (where the index f denotes the value of a quantity immediately behind the shock wave, i.e., for $r = r_f - 0$):

$$\rho_f(v_f - D) = -\rho_0 D,$$

$$\rho_f(v_f - D)^2 + p_f = \rho_0 D^2,$$

$$\rho_f(v_f - D)\left[\frac{\gamma}{\gamma - 1}\frac{p_f}{\rho_f} + \frac{(v_f - D)^2}{2}\right] = -\rho_0\frac{D^3}{2}.$$

Here $D = dr_f/dt$ is the speed of propagation of the shock wave through the ambient gas. (We recall that the energy flux is equal to the product of the mass flux and the sum of the kinetic energy per unit mass and the enthalpy per unit mass.) The last two relations are conveniently written in the form

$$\rho_f(v_f - D)v_f + p_f = 0, \quad \rho_f(v_f - D)\left[\frac{p_f}{(\gamma - 1)\rho_f} + \frac{v_f^2}{2}\right] + p_f v_f = 0.$$

Solving the continuity equations for the flux of mass, momentum and energy, we find convenient relations for the density, pressure and velocity behind the shock wave expressed in terms of the initial density and shock wave speed:

$$p_f = \frac{2}{\gamma + 1}\rho_0 D^2, \quad \rho_f = \frac{\gamma + 1}{\gamma - 1}\rho_0, \quad v_f = \frac{2}{\gamma + 1}D. \qquad (2.52)$$

Further, the energy of a unit volume of gas is equal to $\rho(v^2/2 + c_v T) = \rho[v^2/2 + p/(\gamma - 1)\rho]$ (T being the absolute temperature). Hence the initial conditions at $t = 0$ for the problem of a point explosion can be written in the form

$$\rho(r, 0) = \rho_0, \quad p(r, 0) \equiv 0, \quad v(r, 0) \equiv 0 \quad (r \neq 0),$$

$$4\pi \int_0^\infty \rho\left[\frac{v^2}{2} + \left(\frac{1}{\gamma - 1}\right)\frac{p}{\rho}\right]r^2 dr = E. \qquad (2.53)$$

Here E is the energy released at the centre at the initial moment. Finally, we have an obvious condition: the absence of influx of matter and energy at the central point after the instantaneous explosion for $t > 0$:

$$v(0, t) \equiv 0. \qquad (2.54)$$

Analysis of equations (2.51) and the conditions (2.52), (2.53) and (2.54) shows that the properties p, ρ and v of the gas motion depend on the governing parameters

$$t, \; E, \; \rho_0, \; r, \; \gamma, \qquad (2.55)$$

whose dimensions in the class of systems of units of measurement MLT are respectively

$$T, \; ML^2 T^{-2}, \; ML^{-3}, \; L, \; 1. \qquad (2.56)$$

The radius of the shock front depends on the same parameters (2.55) with the exception of r. Thus $n = 5$, $m = 2$ and $k = 3$, and taking the first three as the governing parameters with independent dimensions we obtain relations for the similarity parameters of the problem under consideration:

$$\Pi_1 = r(Et^2/\rho_0)^{-1/5} = \xi, \quad \Pi_2 = \gamma.$$ (2.57)

From this and dimensional analysis it follows that[†]

$$p = \rho_0 \frac{r^2}{t^2} P(\xi, \gamma), \quad \rho = \rho_0 R(\xi, \gamma), \quad v = \frac{r}{t} V(\xi, \gamma),$$

$$r_f = \xi_f(\gamma)(Et^2/\rho_0)^{1/5}, \quad D = \frac{2}{5}\xi_f(Et^{-3}/\rho_0)^{1/5}.$$ (2.58)

Substituting (2.58) into (2.51), we obtain for the functions, P, V and R the following system of ordinary differential equations

$$\left(V - \frac{2}{5}\right) R \frac{dV}{d\ln\xi} + \frac{dP}{d\ln\xi} - RV + RV^2 + P = 0,$$

$$\frac{dV}{d\ln\xi} + \left(V - \frac{2}{5}\right)\frac{d\ln R}{d\ln\xi} + 3V = 0,$$ (2.59)

$$\frac{d}{d\ln\xi}\left(\ln\frac{P}{R^\gamma}\right) - \frac{2(1-V)}{V - 2/5} = 0.$$

Substitution of the representation (2.58) of the solution into the boundary conditions (2.52) on the shock wave gives the initial conditions for the system (2.59) of ordinary differential equations:

$$P(\xi_f - 0) = \frac{8}{25(\gamma+1)}, \quad V(\xi_f - 0) = \frac{4}{5(\gamma+1)},$$

$$R(\xi_f - 0) = \frac{\gamma+1}{\gamma-1}.$$ (2.60)

Further, relations (2.58) imply that the bulk energy of gas in the perturbed region is constant in time, i.e., is an integral of the motion:

$$4\pi \int_0^\infty \rho \left[\frac{v^2}{2} + \frac{p}{(\gamma-1)\rho}\right] r^2 dr = 4\pi \int_0^{r_f} \rho \left[\frac{v^2}{2} + \frac{p}{(\gamma-1)\rho}\right] r^2 dr$$

$$= 4\pi\rho_0 \frac{E}{\rho_0} \int_0^{\xi_f} R(\xi)\left[\frac{V^2(\xi)}{2} + \frac{P(\xi)}{(\gamma-1)R(\xi)}\right]\xi^4 d\xi = \text{const.} \quad (2.61)$$

[†]   Here, following tradition, we have deviated somewhat from the formal rule for applying dimensional analysis. For example, for the pressure we should write

$$p = E^{2/5}t^{-6/5}\rho_0^{3/5}\Phi(\Pi_1, \Pi_2).$$

The notation of (2.58) is obtained if we write $\Phi = \Pi_1^2 P$, and analogously for the velocity.

(outside the shock wave the integrand is equal to zero). By virtue of the last initial condition in (2.53), the constant on the right-hand side of (2.61) is equal to E, whence

$$\int_0^{\xi_f} R(\xi) \left[\frac{V^2(\xi)}{2} + \frac{P(\xi)}{(\gamma - 1)R(\xi)} \right] \xi^4 d\xi = \frac{1}{4\pi}. \qquad (2.62)$$

The solution of (2.59) under the conditions (2.60) is experienced, as in the problem of very intense thermal waves, in the following explicit form:

$$\left(\frac{\xi_f}{\xi} \right)^5 = C_1 V^2 \left(1 - \frac{3\gamma - 1}{2} V \right)^{\nu_1} \left(\frac{5}{2} \gamma V - 1 \right)^{\nu_2},$$

$$R = C_2 \left(\frac{5}{2} \gamma V - 1 \right)^{\nu_3} \left(1 - \frac{3\gamma - 1}{2} V \right)^{\nu_4} \left(1 - \frac{5}{2} V \right)^{\nu_5}, \qquad (2.63)$$

$$P = C_3 R \left(1 - \frac{5}{2} V \right) V^2 \left(\frac{5}{2} \gamma V - 1 \right)^{-1}.$$

Here

$$C_1 = \left[\frac{5}{4} (\gamma + 1) \right]^2 \left[\frac{5(\gamma + 1)}{7 - \gamma} \right]^{\nu_1} \left(\frac{\gamma + 1}{\gamma - 1} \right)^{\nu_2},$$

$$C_2 = \left(\frac{\gamma + 1}{\gamma - 1} \right)^{\nu_3 + \nu_5 + 1} \left[\frac{5(\gamma + 1)}{7 - \gamma} \right]^{\nu_4}, \quad C_3 = \frac{\gamma - 1}{2},$$

$$\nu_1 = \frac{13\gamma^2 - 7\gamma + 12}{(3\gamma - 1)(2\gamma + 1)}, \quad \nu_2 = - \frac{5(\gamma - 1)}{2\gamma + 1}, \quad \nu_3 = \frac{3}{2\gamma + 1}, \qquad (2.64)$$

$$\nu_4 = - \frac{13\gamma^2 - 7\gamma + 12}{(2 - \gamma)(3\gamma - 1)(2\gamma + 1)}, \quad \nu_5 = - \frac{1}{2 - \gamma}.$$

It is easy to verify that the solution (2.63) satisfies the condition (2.54). The dependence of the remaining constant ξ_f on γ is determined by substituting (2.63) into (2.62), in principle completely analogously to the way this was done in the previous problem. Calculation shows, for example, that for $\gamma = 1.4$, $\xi_f = 1.033$, i.e., it is close to unity.

The solution obtained also has the property of self-similarity: instantaneous 'photographs' of it are always identical: only the length scale as well as the scales of pressure and velocity change. In particular, consider the pressure p as a function of the distance r from the centre of the explosion; this is shown schematically for various instants of time by the curves in Figure 2.7. These curves are similar to one another. If we introduce a time-dependent scale for the distance from the centre of the explosion (for example, the radius of the wave front r_f) and a time-dependent scale for the pressure (for example, the pressure at the

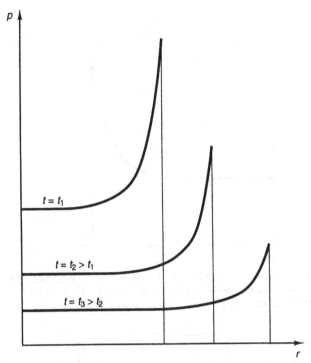

Figure 2.7. Air pressure as a function of radius at various instants of time for the motion of air following an atomic explosion. The pressure distributions at various instants of time are similar to one another.

wave front p_f), all the curves in Figure 2.7 will lie on a single curve in the reduced variables $r/r_f(t)$, $p/p_f(t)$ (Figure 2.8). Analogous curves for density and velocity distributions in the reduced variables are given in Figure 2.9.

The solution discussed in this section was obtained by Taylor (1941, 1950a,b, 1963) and von Neumann (1941, 1963) practically simultaneously: Taylor presented his manuscript on Friday, 27 June 1941 and von Neumann apparently wanted to check his calculations over the weekend and presented his paper on Monday 30 June 1941. Moreover, Taylor solved the system of ordinary differential equations (2.59) numerically; von Neumann noticed an integral of the system (2.59): the energy conservation law for an arbitrary sphere bounded by the variable radius $r = \text{const} \, r_f$, where const < 1. This integral allowed him to obtain the solution in the explicit form (2.63). Later this solution was obtained by Sedov (1946); it is therefore sometimes called the Taylor–von Neumann–Sedov solution.

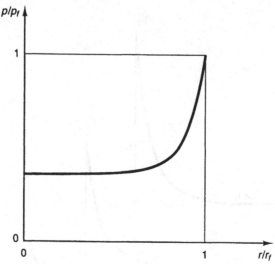

Figure 2.8. The same function as in Figure 2.7, represented as a single curve in the reduced (self-similar) variables $r/r_f(t)$, $p/p_f(t)$ (r_f is the radius of the front, and p_f is the pressure at the front).

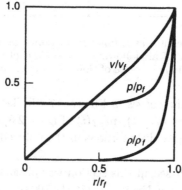

Figure 2.9. Distributions of the gas density, pressure and the velocity behind a very intense blast wave in the self-similar variables ρ/ρ_f, v/v_f, r/r_f.

2.3 Self-similarity. Intermediate asymptotics

2.3.1 Self-similarity

The solutions to the problems of an instantaneous heat source, a thermal dipole, very intense thermal waves, and very intense blast waves, considered earlier in this chapter, have the very important feature of self-similarity. *A time dependent phenomenon is called self-similar if the spatial distributions of its properties at different times can be ob-*

tained from one another by a similarity transformation. Thus, if we choose time-dependent scales $r_0(t)$ for the spatial variable and $u_0(t)$ for any property u of the phenomenon (u can be a vector quantity), then the distribution of u at various instants can be expressed in the form

$$u = u_0(t)f\left(\frac{r}{r_0(t)}\right). \tag{2.65}$$

Hence it follows that if we describe this distribution in self-similar co-ordinates $u/u_0(t)$, $r/r_0(t)$, then the distributions for any value of time in the range considered is represented by a single curve. Thus, in the instantaneous heat source problem,

$$r_0(t) = (\kappa t)^{1/2}, \quad u_0(t) = \theta_0(t) = Q/(\kappa t)^{1/2}, \tag{2.66}$$

for a thermal dipole,

$$r_0(t) = [\kappa(t - t_0)]^{1/2}, \quad u_0(t) = \theta_0(t) = M/\kappa(t - t_0); \tag{2.67}$$

for very intense thermal waves

$$r_0(t) = \left[\left(\frac{E}{c}\right)^n \kappa t\right]^{1/(3n+2)}, \quad u_0(t) = \left[\left(\frac{E}{c}\right)^2 (\kappa t)^{-3}\right]^{1/(3n+2)} \tag{2.68}$$

(here $u_0(t)$ is the temperature scale); for very intense blast waves

$$r_0(t) = (Et^2/\rho_0)^{1/5},$$
$$u_0(t) = \rho_0 \quad \text{(for density)},$$
$$u_0(t) = \rho_0^{3/5} E^{2/5} t^{-6/5} \quad \text{(for pressure)}, \tag{2.69}$$
$$u_0(t) = (Et^{-3}/\rho_0)^{1/5} \quad \text{(for velocity)}.$$

As already mentioned in the Introduction, self-similar solutions are encountered in many branches of mathematical physics. Obtaining a self-similar solution has always been regarded as a success by researchers. The point is that, in many cases, self-similarity allows one to reduce a problem in mathematical physics involving partial differential equations (which are frequently nonlinear, so that this is especially important) to one involving ordinary differential equations. According to the hierarchy of difficulties that existed in the pre-computer era, this made certain studies easier to carry out. Moreover, self-similar solutions have been widely used as standards for evaluating all kinds of approximation methods, irrespective of the immediate urgency of the problems. The appearance of computers changed the attitude towards self-similar solutions, but did not reduce the need for them: self-similarity continues to attract even more attention than before, but now as a deep physical property of a process, which indicates that it stabilizes itself in a certain way. The statement that a phenomenon has stabilized or is entering a steady-state regime is, clearly, highly informative. The statement that a phenomenon is entering a self-similar regime is every bit as informative.

Self-similar solutions are always solutions to limiting, 'idealized', problems where the parameters in the problem with the same dimensions as the independent variables are equal to either zero or infinity. Thus, in the idealized way of formulating the problem described above, the bar undergoing heat conduction was assumed to be either infinite or semi-infinite. The region in which the heat was initially released was assumed to be infinitely small, and the release of heat was assumed to be instantaneous. If this had not been the case, self-similarity would not have existed. Therefore for a long time self-similar solutions were treated by most researchers as though they were merely isolated 'exact' solutions to special problems: elegant, sometimes useful, but extremely limited in significance. It was only gradually realized that these solutions were actually of much broader significance. In fact, as we saw in the discussion of the theory of heat conduction, self-similar solutions turn out not only to describe the behaviour of physical systems under some special conditions, but also to describe the 'intermediate-asymptotic' behaviour of solutions to broader classes of problem in the regions where these solutions have ceased to depend on the details of the initial conditions or boundary conditions but where the system is still far from its final equilibrium state. This situation is common, and greatly increases the significance of self-similar solutions.

2.3.2 Intermediate asymptotics

It is again essential for us to emphasize that self-similar solutions are of basic value not only and not mainly as exact solutions of isolated, albeit urgent, specific problems but above all as intermediate-asymptotic representations of the solutions of much wider classes of problem. We have seen this above for the instantaneous heat source and thermal dipole problems. Now consider this point for the very intense thermal wave problem. Of course the release of energy in a nuclear explosion does not actually take place at a point but in a finite domain of some radius r_*, it is not spherically symmetric, and the initial temperature T_0 is not equal to zero. Hence, the governing parameters in (2.40) should also include the parameters r_* and T_0 and the polar angles φ and ψ. It follows immediately that in addition to the parameter Π_1 the function Φ in (2.42) will be determined by the four other dimensionless parameters

$$\Pi_2 = r_*/[(E/c)^n \kappa t]^{\frac{1}{(3n+2)}}, \qquad \Pi_3 = T_0/[(E/c)^2(\kappa t)^{-3}]^{\frac{1}{(3n+2)}}$$
$$\Pi_4 = \varphi, \qquad\qquad\qquad \Pi_5 = \psi \qquad\qquad\qquad . \quad (2.70)$$

It is intuitively clear, however, and well confirmed by numerical computations, that the asymmetry of the region of initial release is important only in the very first moments, when the thermal wave has spread only to a distance of the order of the size of the initial region of heat release. At these distances the various details of the initial heat discharge influence the solution; these are different from case to case, are never recorded and are of no particular interest. We shall abandon their consideration, i.e., we shall be interested in the propagation of strong thermal waves only at the stage when the wave has travelled a distance $r_f(t)$, large compared with the size of the initial region of heat discharge. This means that $r_f(t) \gg r_*$, and from this, and the fact that $r_f(t)$ is of order $[(E/c)^n \kappa t]^{1/(3n+2)}$ it follows that we must here have $t \gg r_*^{3n+2}/\kappa(E/c)^n$. But for such t the parameter Π_2 is much smaller than unity. One ordinarily assumes that if some similarity parameter has a value much smaller or much larger than unity then the dependence on that parameter, and consequently also on the corresponding dimensional parameter, can be neglected. In this special case this turns out to be correct, so that for $r \gg r_*$ and $t \gg r_*^{3n+2}/\kappa(E/c)^n$ the dependence of the solution on the parameters Π_2, Π_4, Π_5 is unimportant.

Further, since the explosion is very intense, the temperature in the region traversed by the thermal wave is at first very high, much greater than the initial temperature T_0. But the temperature near the centre of the wave is of order $[(E/c)^2(\kappa t)^{-3}]^{1/3n+2}$, whence it follows that for $t \ll (Ec)^{2/3}/\kappa T_0^{(3n+2)/3} = T_2$ the parameter $\Pi_3 \ll 1$, and the initial temperature is unimportant. Keeping in mind that for such t, $r_f \ll (E/cT_0)^{1/3}$, we find that for a sufficiently intense and sufficiently concentrated explosion (large E and small r_*) the characteristic upper and lower time scales of the problem,

$$T_1 = r_*^{3n+2}/\kappa(E/c)^n, \quad T_2 = (E/c)^{2/3}/\kappa T_0^{\frac{3n+2}{3}}. \tag{2.71}$$

and the spatial scales of the problem,[†]

$$L_1 = r_*, \quad L_2 = (E/cT_0)^{1/3}, \tag{2.72}$$

are strongly separated from each other, i.e., are such that $T_1 \lll T_2$ and $L_1 \lll L_2$.[‡] The self-similar solution we have obtained describes the phenomenon of a strong and concentrated explosion well at times and distances from the centre large enough to make the influence of the

---

[†] As a matter of fact, the scales T_2 and L_2 are bounded also by the beginning of the gas motion; at the outset the gas is at rest.

[‡] The symbol $a \lll b$ means that there exists a range of values of a quantity x such that $x \gg a$, but $x \ll b$.

asymmetry of the initial conditions and the size of the domain of original heat release disappear, and at the same time small enough so as to make the original temperature negligible:

$$\frac{r_*^{3n+2}}{\kappa(E/c)^n} \ll t \ll \frac{(E/c)^{2/3}}{\kappa T_0^{(3n+2)/3}}, \tag{2.73}$$

$$r_* \ll r \ll (E/cT_0)^{1/3}.$$

We therefore say that the self-similar solution is an intermediate asymptotics of the phenomenon described. By *intermediate asymptotics* in general one means the following. *Suppose in the problem there are two constant governing quantities $X_i^{(1)}$ and $X_i^{(2)}$ having the dimensions of a certain independent variable x_i. An intermediate asymptotics is an asymptotic representation of the property as $x_i/X_i^{(1)} \to 0$ while $x_i/X_i^{(2)} \to \infty$.*

The situation is quite analogous in the problem of describing the gas-dynamic stage of a very intense explosion. In this case we must take into consideration that the energy release occurs not at a point but in a sphere of radius r_0 (r_0 corresponding to the time when the intense shock wave outstrips the thermal wave), and that outside this sphere the ambient gas of density ρ_0 is under a pressure that is not zero but has some finite value p_0. The solution discussed above represents an intermediate asymptotics describing the gas-dynamic stage of the explosion for

$$T_1 = \left(\frac{\rho_0 r_0^5}{E}\right)^{1/2} \ll t \ll \left(\frac{\rho_0 E^{2/3}}{p_0^{5/3}}\right)^{1/2} = T_2,$$

$$L_1 = r_0 \ll r \ll \left(\frac{E}{p_0}\right)^{1/3} = L_2, \tag{2.74}$$

i.e., for times, and at distances from the centre of the explosion, sufficiently large that the influence of the size of the region of initial energy discharge vanishes and at the same time sufficiently small that the influence of the counter-pressure p_0 is not yet felt. We shall give some figures. Under the conditions of the first American atomic explosion at Alamogordo, $\rho_0 \sim 10^{-3}$ g/cm^3, $E \sim 10^{21}$ ergs, $p_0 \sim 10^6$ dynes/cm^2, and $r_0 \sim 10^3$ cm $= 10$ m, whence for the temporal and spatial bounds on the domain of applicability of the self-similar intermediate asymptotics we find $T_1 \sim 10^{-4}$ s, $T_2 \sim 1$ s, $L_1 \sim 10^3$ cm, and $L_2 \sim 10^5$ cm. (One should note that as a matter of fact the upper bound on the applicability of the self-similar intermediate asymptotics is actually lower, because of the influence of viscous erosion of the front).

The situation is just the same in the general case. Self-similar solutions are always solutions of idealized problems in which constant parameters that have the dimensions of the independent variables appearing in the problem assume zero or infinite values; consequently, self-similar solutions correspond to singular initial or boundary conditions, as we see in the examples considered in this chapter. Hence self-similar solutions are always intermediate-asymptotic solutions of non-idealized problems.[†]

The idea has been widespread that obtaining self-similar solutions is always connected with dimensional analysis, so that by applying dimensional analysis to the formulation of an idealized problem that has some self-similar solutions, one can always obtain the form of the solution, i.e., relations for the self-similar variables; then after obtaining the exact solution it is easy to find the class of non-idealized problems for which the self-similar solution considered is an intermediate asymptotics. This is actually the situation for some self-similar solutions. The examples considered in the present chapter have demonstrated this, and have indicated a general approach that is applicable in similar cases. It is an essential point, however, that the cases in which the construction of self-similar solutions is exhausted by dimensional analysis constitute, as is sometimes said, only the visible tip of the iceberg. As a rule, the situation is different: there exist extensive classes of problems for which, although a self-similar intermediate asymptotics exists, it cannot be obtained from the original formulation of the problem by applying dimensional considerations. The form of the self-similar variables in these cases is obtained from the solution of nonlinear eigenvalue problems, and sometimes even from some additional considerations. We emphasize again that it is not a question here of exceptions but rather of the rule; the set of self-similar solutions that cannot be obtained from similarity considerations is considerably richer than the set of self-similar solutions whose form is completely determined by dimensional considerations. Subsequent examination will clarify the situation here: modifying the problems considered in this chapter apparently slightly, and moreover in such a way that at first glance all similarity considerations used, and hence also everything deduced from them, must remain valid, we will arrive at a contradiction. Resolving the contradiction will lead us to a new class of self-similar solutions.

One final note concerns the very meaning of intermediate asymptotics and, in particular, self-similar solutions. Sometimes we can listen to or read the opinion that now, in the computer era all this matter is of lesser importance because the solutions of all these problems can be obtained numerically for all conceivable initial and boundary conditions. Please,

---

[†] More precisely, stable self-similar solutions (see chapter 8).

it is often said, formulate the problem whose solution you need and you will obtain without special effort a set of tables of all the numerical values you need with the desired (and, of course, paid for) accuracy. To understand this point properly we have to recall that the crucial step in any field of research is to establish what is the minimum amount of information that we actually require about the phenomenon being studied. All else should be put aside in the study. This is precisely the path taken by the natural sciences since Newton. Features that are not repeated from case to case and cannot be experimentally recorded are always deliberately thrown out. Thus, the researcher is actually not so much interested in a table of values of the function f describing the way in which the property a being studied depends on the governing parameters $a_1, \ldots, a_k, b_1, \ldots, b_m$,

$$a = f(a_1, \ldots, a_k, b_1, \ldots, b_m),$$

as in the principal physical laws that determine the major features of this relationship. These principal laws, the basic features of phenomena can often be seen and understood when analysing the intermediate asymptotics.

The general idea of intermediate asymptotics is well expressed by the Russian poet Alexander Blok in his poem 'Возмездие' ('The retribution'):

Сотри случайные черты,

И ты увидишь – мир прекрасен.

(Obliterate the accidental features, and you will see: the world is splendid.) We shall use also an example from the field of graphic arts to illustrate this idea: at a sufficiently large (but not too large) distance, everyone will recognise Leonardo's *Mona Lisa* in Figure 2.10. If one examines this picture close up, it becomes clear that it is composed of 560 monochromatic squares distributed in a particular order (Harmon, 1973). Specialists in the printing trades have a numbering system for colours such that all of the colours (even the fine tints) in Figure 2.10 have unique numbers[†].

Thus, a table approximately like Table 2, with 560 entries, would be an 'exact' representation of Figure 2.10: an analogue of the set of tables with the results of numerical calculations. Here, exactly, is your *Mona Lisa*! Obviously, this kind of 'exactness' does not increase our understanding of the phenomenon (in the present case, our artistic perception

---

[†]   The author is grateful to the Swedish architect Mrs. E. Bark for supplying him with the Swedish standard for colour determination.

Figure 2.10. Leonardo's *Mona Lisa* is an example of intermediate asymptotics! Indeed, at some intermediate distance from this figure, everyone will recognize the *Mona Lisa*. Up close, however, the image disappears – it turns out to consist of 560 monochromatic squares distributed in a particular way. On the other hand, at large distances from this figure the image naturally disappears again.

Table 2.1

Square number	Row number	Column number	Number for the colour of the square
1	1	1	2040-G 20Y
2	1	2	4050-G 20Y
⋮	⋮	⋮	⋮
560	28	20	2040-G 20Y

of the picture) – it kills it! And it is the same in any scientific study. Thus, the primary thing in which the investigator is interested is the development of the phenomenon for intermediate times and distances away from the boundaries such that the effects of random initial features or fine details in the spatial structure of the boundaries have disappeared but the system is still far from its final equilibrium state. This is precisely where the underlying laws governing the phenomenon (which are what we are in fact primarily interested in) appear most clearly. Therefore intermediate asymptotics are of primary interest in every scientific study.

The concept of intermediate asymptotics was formally introduced into mathematical physics by Ya.B. Zeldovich and the author (see Barenblatt and Zeldovich 1971, 1972, Barenblatt 1959b, Zeldovich and Raizer 1966, 1967) although it was used implicitly long before.

Self-similarities of the
second kind: first examples

3.1 Flow of an ideal fluid past a wedge

3.1.1 The problem statement and a direct
application of dimensional analysis

In the heat conduction and gas dynamic problems discussed in the
preceding chapter, we were able to establish that the intermediate-
asymptotic solutions were self-similar and to determine the self-similar
variables using dimensional analysis alone. However, this is very frequently not the case. It often turns out that a problem has a self-similar
intermediate-asymptotic solution, but that dimensional analysis alone
is insufficient to obtain it. We shall presently see this for a simple example – the flow of an ideal fluid (one having no internal friction) past
a wedge. Even this simple example will help us to find out why dimensional analysis on its own is sufficient for constructing self-similar
solutions in some cases but not in others. Further examples will demonstrate other, more complicated situations. Thus, in this section, we shall
consider the steady symmetric flow of an ideal fluid of constant density
past a wedge-shaped body with uniform velocity U at infinity perpendicular to the leading edge of the wedge (Figure 3.1). We select a system
of rectangular coordinates as shown in Figure 3.1(a), with the x-axis in
the plane of symmetry of the wedge, parallel to the flow velocity, the
z-axis along the leading edge of the wedge, and the y-axis perpendicular
to these two axes. It may be assumed that the motion is identical in
all planes perpendicular to the leading edge of the wedge. Thus, the
z-component of the velocity, u_z, can be set equal to zero, while the lon-

gitudinal and transverse components of the velocity, u_x and u_y, depend
only on the coordinates x and y. The region in which the fluid moves
is very large; we shall assume that it is infinite. It is also an important
assumption; however, for the present case it is not critical.

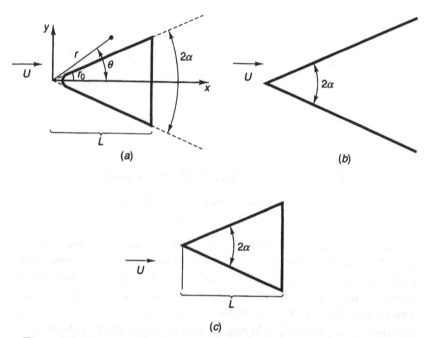

Figure 3.1. Flow of an ideal incompressible fluid past a wedge: (*a*) original
formulation of the problem; (*b*) initial schematization – an infinite sharp
wedge; and (*c*) a sharp wedge of finite size.

We shall now briefly recall for the reader's convenience the derivation
of a well-known hydrodynamical relation, the equation of continuity.
This equation reflects the fact that the fluid is neither created nor annihilated in the flow. Consider the fluid balance in a rectangular volume
element with sides dx and dy along the x and y axes, and unit thickness along the z-axis (Figure 3.2). In a unit time, a quantity of fluid
$u_x dy$ flows into the volume element through the left-hand boundary,
and a quantity of fluid $(u_x + \partial_x u_x dx)dy$ flows out through the right-
hand boundary; an amount of fluid $u_y dx$ flows in through the lower
boundary, and an amount $(u_y + \partial_y u_y dy)dx$ flows out through the upper
boundary. (We have neglected small, higher-order quantities that arise

owing to the fact that the velocity is not constant along the boundaries: their relative contribution goes to zero as the size of the element is decreased.) Thus, the difference between the volumes of inflowing fluid and outflowing fluid is $(\partial_x u_x + \partial_y u_y)dxdy$. However, fluid is neither created nor annihilated within the element, and the fluid density is constant. Therefore, the total difference must be equal to zero; from this condition, we obtain the equation of continuity:

$$\partial_x u_x + \partial_y u_y = 0. \tag{3.1}$$

Furthermore, it is known from hydrodynamics that a flow of this type, uniform at infinity, past a body involving a fluid with no internal friction is a potential flow. This means that both components of the flow velocity are given by the appropriate derivatives of a single function – the velocity potential ϕ:

$$u_x = \partial_x \phi, \qquad u_y = \partial_y \phi. \tag{3.2}$$

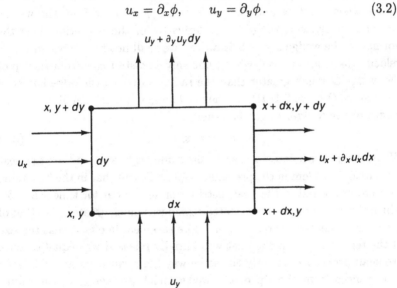

Figure 3.2. Derivation of the equation of continuity.

Substituting equations (3.2) into the equation of continuity (3.1), we obtain Laplace's equation for the potential ϕ:

$$\partial^2_{xx}\phi + \partial^2_{yy}\phi = 0. \tag{3.3}$$

We now transform to the polar coordinates r and θ (Figure 3.1(a)), which are convenient for the present problem, using the formulae $x = r\cos\theta$ and $y = r\sin\theta$. The relations for the radial and tangential components of the velocity, u_r and u_θ, in terms of the velocity potential are of the form

$$u_r = \partial_r \phi, \qquad u_\theta = (1/r)\partial_\theta \phi, \tag{3.4}$$

and Laplace's equation (3.3) takes the form

$$\partial_{rr}^2 \phi + (1/r)\partial_r \phi + (1/r^2)\partial_{\theta\theta}^2 \phi = 0. \tag{3.5}$$

We shall now derive the boundary conditions for the problem at hand. They should reflect the fact that the velocity component perpendicular to the wedge surface vanishes at the surface and the fact that the flow is unperturbed far from the wedge. Recall that in the flow of an ideal fluid past a body at rest, only the velocity component perpendicular to the surface of the body vanishes at its surface. The velocity component tangential to the surface is generally speaking non-vanishing: the flow of an ideal fluid does not satisfy the non-slip condition. The boundary conditions for the present problem are therefore of the form $\partial_n \phi = 0$ at the wedge surface and $\partial_x \phi = U$ and $\partial_y \phi = 0$ far from the wedge, i.e., at infinity, where $\partial_n \phi$ is the derivative of the potential along the normal to the wedge surface. However, we shall not be interested in the velocity distribution everywhere, but only at distances r from the tip of the wedge O much greater than the radius to which the edge has been blunted, r_0 (Figure 3.1(a)), on the one hand, and much smaller than the length of the wedge, L, on the other:

$$r_0 \ll r \ll L. \tag{3.6}$$

We now try to proceed in exactly the same way as we did for the heat conduction problem in the previous chapter. Recall that in the beginning we were interested in the heat distribution in the bar for time scales for which the current size of the heated region was much greater than that of the initially heated region. We therefore assumed, in effect, that the size of the region in which the heat was initially released was equal to zero. We shall proceed in exactly the same way here: since we are interested in distances from the tip of the wedge much greater than the radius to which it has been blunted, r_0, we shall neglect the fact that the tip of the wedge has been blunted, and assume that the wedge is sharp. Furthermore, in the preceding chapter we were interested in the heat distribution in the bar for time scales such that the size of the heated region was still smaller than the distances to the ends of the bar, and the bar could be thus assumed to be infinite in this intermediate stage. We shall proceed in exactly the same way here: since we are interested in distances from the tip of the wedge much less than the length of the wedge L, we shall assume that the wedge is infinite in the x-direction. This idealized version of the body past which the fluid is flowing (an infinite sharp wedge) is indicated by the dashed line in Figure 3.1(a). The angle θ is constant along the boundaries of this wedge, so that the normal component of the velocity, $u_\theta = (1/r)\partial_\theta\phi$, must go to zero there.

The boundary condition for the potential along the boundaries therefore takes the form

$$\partial_\theta\phi = 0 \quad \text{for } \theta = \alpha \quad \text{and} \quad \theta = 2\pi - \alpha. \tag{3.7}$$

Following the same reasoning, the velocity of the flow should remain unperturbed far from the wedge; thus, the relations

$$\phi = Ux, \quad \partial_x\phi = U, \quad \partial_y\phi = 0 \tag{3.8}$$

must be satisfied at $x, y = \infty$ outside the wedge. In the special case $\alpha = 0$ the wedge becomes an infinitely thin plate parallel to the flow direction. Such plate does not perturb the ideal flow, so for this case the potential remains equal to Ux.

The potential ϕ for the flow past an infinite sharp wedge with non-zero opening angle (Figure 3.1(b)) may depend on the following governing parameters (i.e., quantities that appear in Laplace's equation (3.5) and in the boundary conditions (3.7) and (3.8)): the velocity of the incident flow U, the radius r, the polar angle θ, and the opening angle of the wedge 2α; thus

$$\phi = f(U, r, \theta, \alpha). \tag{3.9}$$

The dimensions of the governing parameters are, obviously,

$$[U] = LT^{-1}, \quad [r] = L, \quad [\theta] = 1, \quad [\alpha] = 1. \tag{3.10}$$

It follows from relations (3.2) that ϕ has the dimension L^2T^{-1} – a product of the dimensions of velocity and length:

$$[\phi] = L^2T^{-1}.$$

Thus, $n = 4$, $m = 2$ and $k = 2$, and, by the general rule, dimensional analysis yields

$$\phi = Ur\Phi(\theta, \alpha). \tag{3.11}$$

Differentiating relation (3.11) for the potential, we obtain

$$\partial^2_{rr}\phi = 0, \quad \partial_r\phi = U\Phi(\theta, \alpha), \quad \partial^2_{\theta\theta}\phi = Ur(d^2\Phi/d\theta^2).$$

Substituting these equations into Laplace's equation (3.5), we find that the function Φ must satisfy the simple differential equation

$$\frac{d^2\Phi}{d\theta^2} + \Phi = 0. \tag{3.12}$$

The solution to this equation is well known: $\Phi = A\cos\theta + B\sin\theta$, where A and B are arbitrary constants. Thus, using equation (3.11), we find that $\phi = AUr\cos\theta + BUr\sin\theta = AUx + BUy$. Differentiating,

we obtain $u_x = AU$ and $u_y = BU$, i.e., the velocity distribution in
the uniform flow. Condition (3.8) implies that the flow must remain
unperturbed at infinity, so that $A = 1$ and $B = 0$. Thus, $\phi = Ux$. We
have arrived at a paradoxical conclusion: the presence of the wedge of
a finite opening angle does not change the uniform flow! But for the
non-zero opening angle α the unperturbed uniform flow (with potential
$\phi = Ur\cos\theta = Ux$) obviously does not satisfy condition (3.7) which
states that the flow does not cross the surface of the wedge. Thus, no
solution exists for the infinite-wedge problem with finite opening angle.
We repeat once again that if such a solution existed, it would have to
be determined by the parameters U, r, θ and α alone. However, the
only such solution is the potential for a uniform flow, which does not
satisfy the boundary conditions along the faces of the wedge for non-zero
opening angle of the wedge.

3.1.2 Resolution of the paradox

What kind of paradox has arisen here? It would seem that we have
proceeded in exactly the same way as in the preceding chapter. We were
interested in the velocity field for 'intermediate' distances from the tip
of the wedge – i.e., much larger than the radius to which the wedge was
blunted, but much smaller than the length of the wedge. Thus, just as we
reduced the problem in the previous chapter to discussing the conduction
of heat from a concentrated, instantaneous source in an infinite bar, we
reduced this problem to one of determining the flow past an infinite
sharp wedge. However, even though we were completely successful in
the heat-conduction problem, we have reached a contradiction in the
problem of flow past a wedge. In order to resolve this paradox, let us
examine the solution to the somewhat more complicated problem of the
flow past a sharp wedge with finite length L (Figure 3.1(c)).

The solution to this problem is well-known in hydrodynamics. It is of
the form

$$\phi = \text{Real part of } U\zeta, \tag{3.13}$$

where $\zeta = F(z)$ is an analytic function of the complex variable $z = x + iy$
that carries out a conformal mapping of the exterior of the triangle (the
cross section of the wedge in the flow) onto the exterior of the segment
$0 \le x \le a$ of the x-axis such that $d\zeta/dz \to 1$ as z goes to infinity. We
emphasize that the size of the segment, a, is not specified beforehand,
but is determined by the parameters of the triangle: its altitude L and
the angle at its apex, 2α. From the theory of functions of a complex

variable it is known (the Christoffel–Schwartz formula), that

$$z = \int\limits_0^\zeta t^{-\alpha/\pi}(t-a)^{-1/2}(t-b)^{1/2+\alpha/\pi}dt,\qquad(3.14)$$

where

$$a = F(L) = \text{const.},\quad b = F(L + iL\tan\alpha) = \text{const.}$$

We now substitute (3.14) into (3.13), keeping in mind the fact that we are interested in values of the radius r such that $r \ll L$. Setting $a = L\delta$, $b = L\mu$ and $t = L\tau$ (where $|\tau| \ll 1$) and expanding the integrand in (3.14), we obtain

$$z = re^{i\theta} = L\left(\frac{\zeta}{L}\right)^{1-\frac{\alpha}{\pi}} e^{i\alpha}\mu^{(\frac{1}{2}+\frac{\alpha}{\pi})}\delta^{-\frac{1}{2}}\left(\frac{\pi}{\pi-\alpha}\right),$$

$$\zeta = L\left(\frac{z}{L}\right)^{\frac{\pi}{\pi-\alpha}} e^{-\left(\frac{\pi\alpha i}{\pi-\alpha}\right)}\mu^{-(\frac{1}{2}+\frac{\alpha}{\pi})\frac{\pi}{\pi-\alpha}}\delta^{\frac{\pi}{2(\pi-\alpha)}}\left(\frac{\pi}{\pi-\alpha}\right)^{-\left(\frac{\pi}{\pi-\alpha}\right)}$$

$$= Lr^{\frac{\pi}{\pi-\alpha}}L^{-\left(\frac{\pi}{\pi-\alpha}\right)}\beta e^{i\left(\frac{\pi\theta}{\pi-\alpha}-\frac{\alpha\pi}{\pi-\alpha}\right)},$$

so that

$$\phi = Ur\cos[(\lambda+1)\theta+\gamma]\beta(r/L)^\lambda,\qquad(3.15)$$

where

$$\lambda = \frac{\alpha}{\pi-\alpha},\quad \gamma = -\frac{\pi\alpha}{\pi-\alpha},$$

and β is a dimensionless constant.

It may easily be verified that the expression in (3.15) satisfies Laplace's equation (3.5) and boundary condition (3.7) at the faces of the wedge, although it does not satisfy the condition (3.8) at infinity for $\alpha > 0$. Passing to the limit $L \to \infty$ in the solution (3.13)–(3.14) with $U = \text{const}$ merely yields $\phi = 0$, i.e., fluid at rest. As equation (3.15) indicates, to obtain the correct limiting solution (i.e., one that exhibits the desired asymptotic behaviour), the velocity of the incident flow, U, must also tend to infinity as the wedge grows longer (i.e., as $L \to \infty$) in such a way that the product

$$B = UL^{-\lambda}\qquad(3.16)$$

remains constant.

It now becomes clear what has happened. The problem of flow past a wedge with finite length (for which a solution exists) has another governing parameter in addition to those for the problem of flow past an infinite wedge (for which, as it turned out, no solution exists): the length of the wedge L. Thus, in place of (3.9) and (3.11), respectively, we have

$$\phi = f(U, r, \theta, \alpha, L) = Ur\Phi\left(\theta, \alpha, \frac{L}{r}\right). \qquad (3.17)$$

Since we were interested in distances from the sharp tip of the wedge $r \ll L$ for which the dimensionless parameter L/r is large, we simply set the latter equal to infinity, and attempted to determine the limiting flow, having identified this limiting flow with the solution to the problem of flow past an infinite wedge by a uniform stream with velocity U. In doing so, we tacitly assumed that such a solution, the limit of solutions (3.17) for $L/r \to \infty$, existed, was finite, and non-zero. This last assumption was in fact the one that turned out to be false.

3.1.3 Self-similar solutions of the first and second kind

We shall now discuss expression (3.15) for the flow potential in more detail. We saw in the previous chapter that the self-similar solution for an instantaneous concentrated heat source in an infinite bar can be used to represent the non-self-similar solution to the equation of heat conduction for the propagation of heat in a finite bar on some intermediate time interval. Analogously, relation (3.15) is an approximate representation of the velocity potential for the flow past a finite wedge for intermediate distances from the leading edge of the wedge.

As may easily be seen, potential distribution (3.15) is self-similar: the distributions at various distances from the leading edge of the wedge can be obtained from one another by a similarity transformation. Furthermore, as noted above, (3.15) is a solution to Laplace's equation that satisfies boundary conditions (3.7) on the faces of the wedge. However, if we write solution (3.15) in self-similar form,

$$\phi = \beta(UL^{-\lambda})r^{\lambda+1}\Phi_1(\theta, \alpha), \qquad (3.18)$$

we see that this solution for $\alpha > 0$, $\lambda > 0$ does not have the form for the flow past an infinite wedge that we would have expected from dimensional considerations (see (3.11)).

Note that once we have specified the form of the self-similar solution (3.18) the exponent λ can be determined by solving an *eigenvalue problem*. Namely, by substituting (3.18) into Laplace's equation (3.5), we can obtain an ordinary differential equation for the function Φ_1, following much the same procedure used to obtain (3.12):

$$\frac{d^2\Phi_1}{d\theta^2} + (\lambda+1)^2\Phi_1 = 0, \qquad (3.19)$$

from which we obtain $\Phi_1 = \beta\cos[(\lambda+1)\theta + \gamma]$, where β and γ are arbitrary constants. But the normal velocity $u_\theta = (1/r)\partial_\theta\phi$ must vanish along the line of symmetry and the faces of the wedge. We therefore require that $d\Phi_1/d\theta$ vanishes along the rays $\theta = \alpha, \theta = \pi$ and $\theta = 2\pi - \alpha$, and only along these rays within the region of motion. We find that

$$\lambda = \frac{\alpha}{\pi - \alpha}, \quad \gamma = -\frac{\pi\alpha}{\pi - \alpha} + m\pi, \qquad (3.20)$$

where m is an arbitrary integer that is positive, negative, or zero. Substituting these relations into equation (3.18) for the self-similar solution, we recover solution (3.15), which we had obtained earlier as an asymptotic representation of the solution to the more complicated problem of the flow past a wedge of finite length.

Thus, the self-similar solution obtained above, which describes the flow near the leading edge of the wedge, departs in a fundamental way from the self-similar solution to the problem discussed in the previous chapter, involving an instantaneous heat source. There are several differences, and they are instructive. First, the exponent λ cannot now be determined using dimensional analysis; it is necessary to solve an eigenvalue problem in order to determine it. Strictly speaking, (3.17) is all that dimensional analysis is capable of providing; the fact that the dimensional parameter L/r appears in the expression for the solution as a factor raised to some power does not follow from dimensional analysis. Furthermore, unlike the size h of the region of initial heat release and the distance to the ends of the bar x_0 and $l - x_0$ which did not appear in the solution for the instantaneous heat source, the dimensional parameter L, the length of the wedge, explicitly appears in solution (3.18). Finally, the constant β remains undetermined in the final solution: there is no integral conservation law from which it could be determined in the problem of flow past a wedge, as there was in the heat conduction problem.

Self-similar solutions of this type began to appear in various physics and mechanics problems long ago, following the work of Guderley (1942). Zeldovich (1956) suggested that they be identified as a special class. In contrast to self-similar solutions like those that we encountered in solving the instantaneous heat source problem and the very intense thermal and blast wave problems, in which dimensional analysis turned out to be sufficient to construct the complete solution, solutions like that for flow past a wedge are called *self-similar solutions of the second kind*. The name *self-similar solutions of the first kind* is reserved for self-similar solutions that can be constructed using dimensional analysis alone. We would like to shed some light on the nature of these differences, and thus understand why everything can be obtained simply, using dimensional analysis alone, in some cases, while in other cases dimensional analysis is not sufficient to determine the exponents and it is necessary to solve an eigenvalue problem. This will be done in chapter 5 after some additional examples of similar solutions of the second kind have been considered.

3.2 Filtration in an elasto-plastic porous medium: the modified instantaneous heat source problem

3.2.1 The problem statement

The modification of the instantaneous heat source problem that we shall consider consists in changing the equation for the temperature θ in those regions where the body is cooling: instead of the classical equation of heat conduction (2.3), θ is now taken to satisfy an equation with a discontinuous coefficient of thermal diffusivity:

$$\partial_t \theta = \kappa \partial_{xx}^2 \theta \quad (\partial_t \theta \geq 0),$$
$$\partial_t \theta = \kappa_1 \partial_{xx}^2 \theta \quad (\partial_t \theta \leq 0),$$

(3.21)

where κ_1 is a constant that is in general different from κ, so that the coefficient of thermal diffusivity depends upon whether the body is heating or cooling at a given point. It is essential that the steplike behaviour of the coefficient of thermal diffusivity is connected with the difference in the specific heat for heating and cooling, and that the thermal conductivity does not depend on the direction of the change in temperature. Therefore the variable thermal diffusivity is put outside the space derivative, and the condition of continuity of heat flux requires the continuity of the derivative $\partial_x \theta$.

Thus we are interested in a solution of (3.21) that is continuous with continuous derivatives with respect to both independent variables. As was proved by Kamenomostskaya (Kamin) (1957), a solution to the initial-value problem for (3.21) with an arbitrary sufficiently smooth function $\theta(x,0)$, that decreases monotonically, and sufficiently rapidly, with increasing $|x|$ exists, is unique, and has a continuous derivative with respect to t and two continuous derivatives with respect to x.

Equation (3.21) occurs in the theory of the filtration of an elastic fluid in an elasto-plastic porous medium. A short derivation of it is given below; the reader who is not interested in the actual physical meaning of the modified problem can skip this section without damage to his understanding of what follows. The equation (3.21) also describes diffusion in two-component media where the second component consists of particles admitting osmotic diffusion, so that the first substance can diffuse into the particles but not out of them (see Ginsburg, Entov and Theodorovich, 1992).

The equation for the conservation of fluid mass in the filtration (slow motion) of fluids in porous media has the form

$$\partial_t(m\rho) + \text{div}\rho\mathbf{v} = 0 \,.$$

Here m is the porosity of the medium, that is, the relative volume occupied in the medium by the pores; ρ is the density of the fluid; $\mathbf{v}$ is the velocity of filtration, equal in magnitude to the mass flux of the fluid passing through a section of unit area of the porous medium normal to the flow, divided by the fluid density; and t is the time. This equation can be derived in the same way as the equation of heat balance in chapter 2. The velocity of filtration is proportional to the pressure gradient; this constitutes Darcy's law, which is the basis for the theory of filtration (and is analogous in its formulation to Fourier's law in the theory of heat conduction):

$$\mathbf{v} = -\frac{k}{\mu}\text{grad}p \,.$$

Here k is the so-called coefficient of permeability, determining the resistance of the porous medium to the fluid leaking through it, and μ is the coefficient of viscosity of the fluid. The fluid is assumed to be weakly compressible, so that its density grows linearly with increasing pressure:

$$\rho/\rho_0 = 1 + \beta_f(p - p_0) \,,$$

where β_f is the coefficient of compressibility of the fluid, and p_0 and ρ_0 are the reference pressure and density of the fluid. The porous medium is also considered to be weakly compressible. Its porosity m can, as experiments show, be considered to a first approximation to depend only on σ, the first invariant of the stress tensor[t] (equal to one-third of the sum of the principal stresses) acting on the skeleton of the porous medium: $m = m(\sigma)$. If the porous medium is linearly elastic, then

$$m/m_0 = 1 - \beta_r(\sigma - \sigma_0) \,,$$

where β_r is the coefficient of compressibility of the porous medium, σ_0 is the reference value of σ (for increasing stress the medium compresses, $\beta_r > 0$), and m_0 is the corresponding value of the porosity. Under the conditions in a deep-lying porous stratum the total stress state of the fluid–porous–medium system is fixed, since the fluid and the porous skeleton together restrain the burden of higher-lying strata. Hence $\sigma + p = \sigma_0 + p_0$, whence $\sigma - \sigma_0 = -(p - p_0)$. Substituting these relations into the equation for the conservation of fluid and discarding higher–order terms in $\beta(p - p_0)$, we find (for details see Shchelkachev,

[t] Here it is convenient to assume a compressive stress as positive.

1959; Collins, 1961; Barenblatt, Entov and Ryzhik, 1990) that for the filtration of an elastic fluid in an elastic porous medium under the conditions in a deep-lying porous stratum the pressure of the fluid $p(r,t)$ satisfies the classical linear equation of heat conduction,

$$\partial_t p = \kappa \Delta p. \tag{3.22}$$

Here κ is the so-called 'coefficient of piezoconductivity', analogous to the coefficient of thermal diffusivity and equal to $k/\mu(m_0\beta_f + \beta_r)$.

Now, as often happens in practice, let the porous medium be irreversibly deformable. Then (for details see Barenblatt and Krylov, 1955; Barenblatt, Entov and Ryzhik, 1990)

$$\partial_t m = -m_0\beta_r\partial_t\sigma = m_0\beta_r\partial_t p$$

for increasing σ (decreasing fluid pressure, because the total stress state of the fluid–porous–medium system remains unchanged, $\sigma + p = \sigma_0 + p_0$, $\partial_t\sigma = -\partial_t p$) and

$$\partial_t m = -m_0\beta_{r1}\partial_t\sigma = m_0\beta_{r1}\partial_t p$$

for decreasing σ (increasing fluid pressure), where β_{r1} is not equal to β_r. Hence, the equation for the excess fluid pressure, i.e., the difference between the initial and instantaneous pressures $\theta(r,t) = p_0 - p(r,t)$, assumes the form

$$\partial_t\theta = \kappa(\partial_t\theta)\Delta\theta, \tag{3.23}$$

where $\kappa(\partial_t\theta)$ is a step function: $\kappa(\partial_t\theta) = \kappa$ for $\partial_t\theta > 0$ and $\kappa(\partial_t\theta) = \kappa_1$ for $\partial_t\theta < 0$. The coefficients κ and κ_1 are determined by the properties of the fluid and the deformation properties of the medium. They are different for loading of the stratum by the burden of the higher-lying strata (a drop in fluid pressure), and for unloading of the stratum (a subsequent increase in fluid pressure)[†]:

$$\kappa = \frac{k}{\mu(m_0\beta_f + \beta_r)}, \quad \kappa_1 = \frac{k}{\mu(m_0\beta_f + \beta_{r1})}.$$

Thus the analogue of thermal conductivity, the quantity k/μ, is identical for loading and unloading, whereas the analogue of specific heat, the quantity $m_0\beta_f + \beta_r$, is different for loading and unloading.

In particular, for the one-dimensional problem of rectilinear parallel fluid motion (filtration to a drainage gallery or from it), (3.23) assumes the form of the basic equation (3.21).

[†] It is assumed that at each point in the porous medium the processes of loading and unloading take place only once. One can also consider more complicated processes, but we shall not do so here.

We shall now try to find the solution to the problem of the instantaneous removal of a finite mass of a fluid from a small region of an elasto-plastic porous stratum. It would seem that obtaining this solution reduces, in view of the linear dependence of the fluid density on pressure, to constructing the solution to (3.21) for a problem of instantaneous point-source type. We shall try to construct such a solution with the help of dimensional analysis; but later discussion will show that the matter is actually more complicated.

Thus a solution of (3.21) is sought, satisfying an initial condition and a condition at infinity similar to those in section 2.2

$$\theta(x,0) \equiv 0 \quad (x \neq 0), \quad \int_{-\infty}^{\infty} \theta(x,0)dx = Q, \quad \theta(\pm\infty,t) \equiv 0. \quad (3.24)$$

3.2.2 Direct application of dimensional analysis to the modified instantaneous heat source problem

As already mentioned above, for the case $\kappa_1 = \kappa$ (the classical equation for heat conduction or for filtration in an elastic porous medium) such a solution exists, is self-similar, and takes the form (2.19). It would seem that for $\kappa_1 \neq \kappa$ dimensional considerations would proceed exactly the same as for the case $\kappa_1 = \kappa$, because the list of governing parameters in the modified problem has been increased, compared with the classical instantaneous heat source problem, only by the dimensionless constant parameter $\epsilon = (\kappa_1/\kappa) - 1$. Hence it would seem at first glance that the desired solution must be expressed in the form

$$\theta = \frac{Q}{\sqrt{\kappa t}}\Phi(\xi,\epsilon), \quad \xi = \frac{x}{\sqrt{\kappa t}}, \quad (3.25)$$

where the function Φ is continuous, with a continuous derivative with respect to ξ, and even: $\Phi(-\xi,\epsilon) = \Phi(\xi,\epsilon)$. Further, the loading domain $(\partial_t\theta \geq 0)$ must correspond by virtue of the self-similarity of the problem to

$$|x| \geq x_0(t) \equiv \xi_0\sqrt{\kappa t},$$

where ξ_0 is a constant depending on ϵ; and for the unloading domain $(\partial_t\theta \leq 0)$,

$$0 \leq |x| \leq x_0(t).$$

However, for $\kappa_1 \neq \kappa$ there does not exist a solution of (3.21), in the form (3.25), which is continuous and has a continuous derivative with respect to x (continuity of the fluid flux), which satisfies the natural conditions of symmetry, and which vanishes at infinity. In order to see this, we substitute (3.25) into (3.21) and obtain for Φ the following ordinary differential equation with discontinuous coefficient at the highest derivative:

$$(1+\epsilon)\frac{d^2\Phi}{d\xi^2} + \frac{1}{2}\frac{d}{d\xi}\xi\Phi = 0, \quad (0 \leq \xi \leq \xi_0), \quad \epsilon = \frac{\kappa_1}{\kappa} - 1,$$

$$\frac{d^2\Phi}{d\xi^2} + \frac{1}{2}\frac{d}{d\xi}\xi\Phi = 0, \quad (\xi_0 \leq \xi < \infty), \tag{3.26}$$

where the point $\xi = \xi_0$ corresponds to the vanishing of the quantity $d(\xi\Phi)/d\xi$ which, as is easily seen from (3.25), is proportional to the derivative $\partial_t\theta$. Integrating, we have

$$(1+\epsilon)\frac{d\Phi}{d\xi} + \frac{\xi\Phi}{2} = c_1, \quad (0 \leq \xi \leq \xi_0),$$

$$\frac{d\Phi}{d\xi} + \frac{\xi\Phi}{2} = c_2, \quad (\xi_0 \leq \xi < \infty). \tag{3.27}$$

By virtue of symmetry and the absence of influx at $x = 0$ for times $t > 0$, $d\Phi/d\xi = 0$ for $\xi = 0$ and, as $\xi \to \infty$ the function $\xi\Phi$ tends to zero (the total amount of removed fluid is finite at each instant and Φ must be integrable). Hence $c_1 = c_2 = 0$. Integrating the preceding equations we obtain

$$\Phi = c_3 \exp\left(-\frac{\xi^2}{4(1+\epsilon)}\right), \quad (0 \leq \xi \leq \xi_0),$$

$$\Phi = c_4 \exp\left(-\frac{\xi^2}{4}\right), \quad (\xi_0 \leq \xi < \infty), \tag{3.28}$$

where c_3 and c_4 are new constants. The condition of continuity of the function $\theta(x, t)$ and its derivative with respect to x reduces to the requirement of continuity of Φ and $d\Phi/d\xi$ at $\xi = \xi_0$, and from this and the previous equation we get a linear system of homogeneous algebraic equations for determining c_3 and c_4:

$$c_3 \exp\left(-\frac{\xi_0^2}{4(1+\epsilon)}\right) = c_4 \exp\left(-\frac{\xi_0^2}{4}\right),$$

$$c_3\frac{\xi_0}{(1+\epsilon)}\exp\left(-\frac{\xi_0^2}{4(1+\epsilon)}\right) = c_4\xi_0\exp_0\left(-\frac{\xi_0^2}{4}\right). \tag{3.29}$$

For $\epsilon \neq 0$, i.e., $\kappa_1 \neq \kappa$, this system evidently has no nontrivial solution for any finite ξ_0, since its determinant is different from zero. Thus, it is proved that there exists no nontrivial solution in the form of (3.25) of the problem posed. Meanwhile, the trivial solution obviously does not satisfy the initial condition (3.24).

3.2.3 Numerical experiment. Self-similar intermediate asymptotics

In order to resolve the paradox that has arisen, let us appeal to the

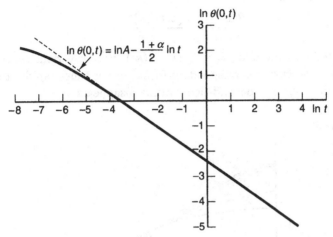

Figure 3.3. The decay of the maximum of the function $\theta(x,t)$ rapidly follows a scaling law.

results of a numerical experiment.[†] The equation (3.21) was integrated at various initial conditions for various values of the parameter $\epsilon = (\kappa_1/\kappa) - 1$. The initial conditions (3.24) are represented by a generalized function, so they could not be introduced to the computer program directly. Therefore they were simulated by various initial distributions $\theta(x,0)$ 'concentrated' in a small but finite region near $x = 0$, for example

$$\theta(x,0) = 10 \qquad (-0.1 \leq x \leq 0.1),$$

$$\theta(x,0) \equiv 0 \qquad (0.1 \leq |x| < \infty).$$

The results of the numerical experiment appeared to be instructive. First of all it was obtained that the decay of the quantity $\theta(0,t)$, the maximum of the function $\theta(x,t)$, rapidly approaches a scaling, power law type of behaviour (Figure 3.3). More exactly, some time after the beginning of computation the following relation holds:

$$\theta(0,t) = A(\kappa t)^{-(1+\alpha)/2}$$

where A and α are certain constants. Numerical calculations showed that, under varying initial conditions, only the constant A varied, the constant α being a function of the parameter $\epsilon = (\kappa_1/\kappa) - 1$ only.

The quantity $\theta_0(t) = A(\kappa t)^{-(1+\alpha)/2}$ can be considered as a natural

[†] The numerical experiment was performed by V.M. Uroev, while at Moscow University.

time-dependent scale for the variable $\theta(x,t)$. Let us take as a linear scale the quantity $l_0(t) = (\kappa t)^{1/2}$ and let us consider (Figure 3.4) the dependence of the quantity

$$\frac{\theta(x,t)}{\theta_0(t)} = \frac{\theta(x,t)(\kappa t)^{(1+\alpha)/2}}{A}$$

on the reduced variable $\xi = x/l_0(t) = x(\kappa t)^{-1/2}$, for various times. It is seen that the curves corresponding to increasing times rapidly tend to coincidence, so their dependence on time disappears.

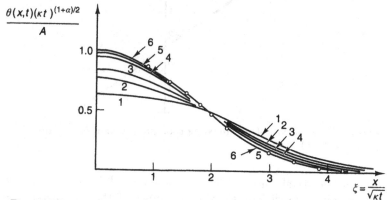

Figure 3.4. Transition to self-similar intermediate asymptotics of the solution to the non-self-similar problem (3.21) with $\epsilon = 1$ and initial data $u(x,0) = 10$ $(0 \le x \le 0.1)$, $u(x,0) \equiv 0$ $(x > 0.1)$. Curves 1–6 correspond respectively to $t = 0.001, 0.002, 0.003, 0.015, 0.040$, and finally 0.225 and all greater values. The open circles are the values of the function determined by solving the nonlinear eigenvalue problem.

Thus, this numerical experiment showed that the solution to the Cauchy problem for equation (3.21) rapidly converges to the self-similar intermediate asymptotics

$$\theta(x,t) = \frac{A}{(\kappa t)^{(1+\alpha)/2}} \Phi\left(\frac{x}{\sqrt{\kappa t}}, \epsilon\right) \qquad (3.30)$$

where the exponent α depends on the parameter $\epsilon = \kappa_1/\kappa - 1$ i.e. on the ratio of coefficients in the equation (3.21) only, whereas the other parameter A depends also on the initial conditions.

3.2.4 Self-similar limiting solution

Let us now try to understand why the asymptotics of the solution to the Cauchy problem (3.30), although also self-similar, differs from the form (3.25) predicted by dimensional analysis.

As we mentioned before, the initial condition (3.24) that led us to a solution in the form (3.25) has a limiting character, is described by a generalized function, and cannot be put directly into a numerical computation.

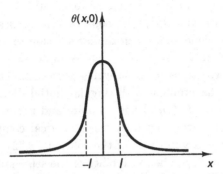

Figure 3.5. To resolve the paradox of the non-existence of the required solution of (3.21), a non-self-similar problem with non-concentrated initial data is considered, whose solution certainly exists.

In fact in a numerical experiment a solution was calculated that satisfied an initial condition (Figure 3.5)

$$\theta(x,0) = \frac{Q}{l} u_0 \left(\frac{x}{l} \right) , \qquad (3.31)$$

where l is some length scale characterizing the size of the region from which fluid was removed at the initial instant and where

$$Q = \int\limits_{-\infty}^{\infty} \theta(x,0)dx, \qquad \int\limits_{-\infty}^{\infty} u_0(\zeta)d\zeta = 1 ,$$

$u_0(\zeta)$ being an even dimensionless function that is finite (becoming identically equal to zero for sufficiently large absolute values of its argument). For such an initial condition one can be certain that a solution to the Cauchy problem exists, is unique, and has continuous derivatives with respect to x up to the second order and a continuous derivative with respect to t; this follows from the general theorems proved by Kamenomostskaya (Kamin) (1957). However, a new dimensional governing parameter l has appeared in this problem, and the solution is no longer self-similar. In fact, the standard procedure, based on dimensional analysis and demonstrated above in several cases, gives here

$$\theta = \frac{Q}{\sqrt{\kappa t}} F(\xi, \eta, \epsilon) , \qquad \eta = \frac{l}{\sqrt{\kappa t}} , \qquad (3.32)$$

The self-similar special solution of instantaneous source type for the case $\kappa_1 = \kappa$ ($\epsilon = 0$) considered above corresponds to the singular initial condition obtained from (3.31) for $l = 0$. But a solution of instantaneous source type is broader than a simple exact special solution to a single problem. In fact (3.32) is valid also for $\epsilon = 0$. We see that $\eta \to 0$ as $t \to \infty$. Choosing x appropriately, we can carry out this passage to the limit so that $\xi = x/(\kappa t)^{1/2}$ remains constant; so, asymptotically we obtain the well-known self-similar solution indicated above. Thus, as has already been remarked, a self-similar solution to the problem with singular initial data is, for $\kappa = \kappa_1$, an asymptotics for large times of a wider class of solutions of initial-value problems. The non-existence of a solution to the problem with singular initial data means that for $\kappa \neq \kappa_1$ the function $F(\xi, \eta, \epsilon)$ has no finite and non-zero limit as $\eta \to 0$. Nevertheless, as was shown by the numerical computations, there exists a self-similar asymptotics of the solution (3.32), although in the form (3.30), not (3.25). The availability of the self-similar intermediate asymptotics (3.30) at $t \to \infty$ suggests that, at $\eta \to 0$ for the function $F(\xi, \eta, \epsilon)$ an asymptotic representation[†] is valid:

$$F(\xi, \eta, \epsilon) = \eta^\alpha f(\xi, \epsilon) + o(\eta^\alpha). \qquad (3.33)$$

Therefore as $t \to \infty$ the asymptotic form of the solution of the problem considered cannot be expressed in the form (3.25), but has the form

$$\theta = \frac{Ql^\alpha}{(\kappa t)^{(1+\alpha)/2}} f(\xi, \epsilon), \qquad (3.34)$$

where the function $f(\xi, \epsilon)$ is finite and different from zero.

We now observe that we can also make η tend to zero for finite ξ by passage to the limit as $l \to 0$ with fixed x and t. As is well known, in the classical case with $\kappa_1 = \kappa$, $\epsilon = 0$, such a passage to the limit again yields a solution of instantaneous–source type. Equation (3.34) shows that if this passage to the limit is carried out leaving Q fixed, then for $\alpha \neq 0$ the limit of the solution will be equal to either zero or infinity, depending on whether α is positive or negative. For $\alpha \neq 0$, in order to obtain, in the limit $l \to 0$ with x and t fixed, the same limiting expression to the solution of the problem that is obtained for finite l and $t \to \infty$ it is necessary to proceed to the limit $l \to 0$ with Q simultaneously tending either to infinity or zero, depending on the sign of α, and moreover with the product Ql^α remaining finite. The self-similar solution obtained by such a passage to the limit does not have the form (3.25) but can be expressed in the form

$$\theta = \frac{A}{(\kappa t)^{(1+\alpha)/2}} \Phi(\xi, \epsilon), \quad A = \beta \lim_{l \to 0} Ql^\alpha; \quad x_0(t) = \xi_0 \sqrt{\kappa t}. \qquad (3.35)$$

[†] The symbol $O(x)$ denotes, as usual, a quantity of order x; the symbol $o(x)$ denotes a quantity small compared with x.

Here β is a dimensionless constant that depends on the normalization of the function $\Phi(\xi, \epsilon)$, and the parameter α is the 'trace' of the parameters Q and l after the limiting process. The parameter α can be determined by carrying out, for example by means of numerical calculation, the limiting passage from a solution of the non-self-similar problem to the self-similar asymptotics. In the direct construction of a self-similar solution by substitution of (3.35) into the basic equation and initial conditions, the parameter α is unknown and subject to determination. Thus, the determination of the parameter α appears explicitly in the statement of the problem, constituting a part of the determination of the self-similar solution.

We now note that for the solution (3.35) the 'moment'

$$\int_{-\infty}^{\infty} |x|^{\alpha} \theta(x, t) dx \qquad (3.36)$$

is finite, different from zero, and invariant in time if the integral

$$\int_{-\infty}^{\infty} |\xi|^{\alpha} \Phi(\xi, \epsilon) d\xi \qquad (3.37)$$

is finite and non-zero. The solution (3.35) itself corresponds to singular initial data; however, this singularity is no longer the classical delta function, as in the case $\epsilon = 0$.

3.2.5 The nonlinear eigenvalue problem

The function $\Phi(\xi, \epsilon)$ is conveniently normalized by the relation

$$\Phi(0, \epsilon) = 1. \qquad (3.38)$$

Substituting (3.35) into (3.21), we obtain for the function $\Phi(\xi, \epsilon)$ an ordinary differential equation with a discontinuous coefficient at the highest derivative

$$(1 + \epsilon)\frac{d^2\Phi}{d\xi^2} + \frac{\xi}{2}\frac{d\Phi}{d\xi} + \left(\frac{1+\alpha}{2}\right)\Phi = 0, \quad (0 \le \xi \le \xi_0),$$

$$\frac{d^2\Phi}{d\xi^2} + \frac{\xi}{2}\frac{d\Phi}{d\xi} + \left(\frac{1+\alpha}{2}\right)\Phi = 0, \quad (\xi_0 \le \xi < \infty). \qquad (3.39)$$

Here ξ_0 is the point at which $d^2\Phi/d\xi^2$ vanishes, or, what is by virtue of (3.39) the same, at which the relation

$$\xi\frac{d\Phi}{d\xi} + (1 + \alpha)\Phi = 0 \qquad (3.40)$$

is satisfied, where the quantity on the left–hand side is proportional to $\partial_t u$. The function $\Phi(\xi, \epsilon)$ is even by virtue of the natural symmetry of the solution. Because of the absence of influx at $x = 0$ for times $t > 0$, it satisfies the boundary condition

$$\frac{d\Phi(0, \epsilon)}{d\xi} = 0 \,. \tag{3.41}$$

Moreover, $\Phi(\xi, \epsilon)$ along with its first derivative with respect to ξ, must be continuous everywhere, and in particular for $\xi = \xi_0$. (We recall that this follows from the fact that the fluid pressure and flux are continuous.) A solution to equation (3.39) can be expressed simply in terms of well-known special functions: the so-called confluent hypergeometric functions or the parabolic cylinder functions related to them (Abramowitz and Stegun, 1970). For $0 \leq \xi \leq \xi_0$, a solution of (3.39) satisfying (3.41) has the form

$$\Phi = C \left[\exp\left(-\frac{\xi^2}{8(1+\epsilon)} \right) \right] \left[D_\alpha \left(\frac{\xi}{\sqrt{2(1+\epsilon)}} \right) + D_\alpha \left(-\frac{\xi}{\sqrt{2(1+\epsilon)}} \right) \right] \,,$$
$$\tag{3.42}$$

where C is a constant and D_α is the symbol for the parabolic cylinder function. From (3.38) we obtain

$$C = \frac{1}{2D_\alpha(0)} = \frac{\Gamma((1-\alpha)/2)}{2^{1+\alpha/2}\sqrt{\pi}} \,,$$

here Γ is the symbol for the Γ-function.

For $\xi \geq \xi_0$ a solution of (3.39) for which the integral of (3.37) converges can be expressed in the form

$$\Phi = F \exp\left(-\frac{\xi^2}{8} \right) D_\alpha \left(\frac{\xi}{\sqrt{2}} \right) \,, \tag{3.43}$$

where F is a constant; a second linearly independent solution decays as $\xi^{-\alpha-1}$ at infinity, and the integral (3.37) diverges[†]. Requiring that the condition

$$\xi \frac{d\Phi}{d\xi} + (1+\alpha)\Phi = 0 \qquad (\xi = \xi_0 \pm 0)$$

be satisfied, and using the recursion relations for the derivatives of the parabolic cylinder functions and the expressions for parabolic cylinder functions in terms of confluent hypergeometric functions (Abramowitz and Stegun, 1970), we obtain

$$D_{\alpha+2}\left(\frac{\xi_0}{\sqrt{2}} \right) = 0, \quad M\left(-\frac{\alpha}{2} - 1, \frac{1}{2}, \frac{\xi_0^2}{4(1+\epsilon)} \right) = 0, \tag{3.44}$$

where $M(a, b, z)$ is the symbol for the confluent hypergeometric function.

[†] Like the solution of the initial-value problem for the classical equation of heat conduction, the solution of the present problem must decrease at infinity faster than any power of x so that the integral (3.37) will converge.

These equations must determine the dependence on ϵ of the parameter α and the quantity ξ_0. Further, the condition of continuity of the function Φ for $\xi = \xi_0$ determines the constant F:

$$
\begin{aligned}
F &= C \left[D_\alpha \left(\frac{\xi_0}{\sqrt{2(1+\epsilon)}} \right) + D_\alpha \left(-\frac{\xi_0}{\sqrt{2(1+\epsilon)}} \right) \right] \\
&\quad \times \left[D_\alpha \left(\frac{\xi_0}{\sqrt{2}} \right) \right]^{-1} \exp \left[\frac{\xi_0^2}{8} \left(1 - \frac{1}{(1+\epsilon)} \right) \right] \\
&= \frac{\Gamma((1-\alpha)/2)}{2^{1+\alpha/2}\sqrt{\pi}} \left[D_\alpha \left(\frac{\xi_0}{\sqrt{2(1+\epsilon)}} \right) + D_\alpha \left(-\frac{\xi_0}{\sqrt{2(1+\epsilon)}} \right) \right] \\
&\quad \times \left[D_\alpha \left(\frac{\xi_0}{\sqrt{2}} \right) \right]^{-1} \exp \left[\frac{\xi_0^2}{8} \left(1 - \frac{1}{(1+\epsilon)} \right) \right] .
\end{aligned}
\tag{3.45}
$$

By virtue of (3.40) the requirement of continuity of the derivative $d\Phi/d\xi$ is satisfied automatically.

Thus, assuming the existence of a self-similar intermediate asymptotics to the solution of the original non-self-similar initial-value problem in the form (3.35), we have arrived at the classical situation of a nonlinear eigenvalue problem (nonlinear because the coordinate ξ_0 of the point of discontinuity of the coefficient at the highest derivative in (3.39) is unknown in advance and must be found in the course of solving the problem). For arbitrary α the basic equation (3.39) does not have a solution of the required smoothness. However, if the system (3.44) is solvable, then for an α satisfying (3.44) the solution (3.35) satisfies all the requirements.

To complete our investigation of the solution, it remains to elucidate the solvability of the system of transcendental equations (3.44) that determine α and ξ_0. Solving the first equation with respect to $\xi_0/2^{1/2}$, we obtain the monotonically increasing function of α represented in Figure 3.6(a) by curve I. Solving the second equation with respect to

$$\xi_0/[2(1+\epsilon)]^{1/2}$$

we get the monotonically decreasing function of α represented in Figure 3.6(a) by curve II. For any given ϵ the corresponding dependence of $\xi_0/2^{1/2}$ on α is obtained by simple expansion or contraction of curve II along the vertical axis. For $\epsilon = 0$, i.e. $\kappa_1 = \kappa$, curves I and II intersect at $\alpha = 0$, in accord with the known results for the classical case, giving the ordinate of the point of inflection of the function $\Phi(\xi, 0) = \exp(-\xi^2/4)$, as $\xi_0 = 2^{1/2}$. For $\epsilon \neq 0$, the point of intersection of curves I and II is unique, and the corresponding variation of α with ϵ is shown in Figure 3.6(b). The function $\alpha(\epsilon)$ obtained from the solution of the nonlinear eigenvalue problem practically coincides with the function obtained numerically by solving the non-self-similar Cauchy problem. It is evident that the quantity α is positive for $\epsilon > 0$ and negative for $\epsilon < 0$.

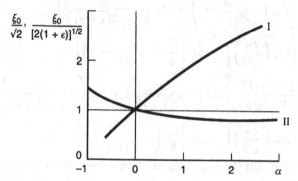

Figure 3.6(a). Investigation of the solvability of the system of transcendental equations (3.44). The dependence of $\xi_0/2^{1/2}$ on α, determined from the first equation (curve I) and the dependence of $\xi_0/[2(1 + \epsilon)]^{1/2}$ on α, determined from the second equation (curve II). Curve I is monotonically increasing, curve II is monotonically decreasing; the intersection point of the curves exists and is unique.

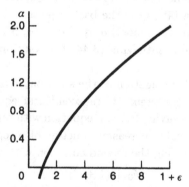

Figure 3.6(b). Dependence of the eigenvalue α on $\epsilon = \kappa_1/\kappa - 1$. For $\epsilon < 0$, α is negative; for $\epsilon > 0$, α is positive. For $\epsilon = 0$ (the classical linear equation of heat conduction), $\alpha = 0$.

In figure 3.4 the open circles are the values of the function $\Phi(\xi, \epsilon)$ obtained by solving the nonlinear eigenvalue problem. These open points fall nicely on the curve of $\theta(x, t)(\kappa t)^{(1+\alpha)/2}/A$ that corresponds to $t \to \infty$. This gives a numerical confirmation that the self-similar solution (3.35) is indeed an asymptotic representation of the solution to the non-self-similar Cauchy problem.

Thus, the construction of a limiting self-similar solution – an asymptotic representation of the solution to the Cauchy problem (3.31) for equation (3.21) – reduces to the solution of a nonlinear eigenvalue problem. The solution to this last problem determines the self-similar asymptotics only up to the constant $A = \beta Q l^{\alpha}$ or, what is the same, up to the dimensionless constant β. In the classical case $\epsilon = 0$, when $\alpha = 0$ this constant is found from the integral conservation law

$$\int_{-\infty}^{\infty} \theta(x,t)dx = \int_{-\infty}^{\infty} \theta(x,0)dx = Q, \qquad (3.46)$$

which is valid also for non-self-similar motions. This conservation law does not hold for $\kappa_1 \neq \kappa$ ($\epsilon \neq 0$); it is then replaced by the nonintegrable relation

$$\frac{d}{dt}\int_{-\infty}^{\infty} \theta(x,t)dx = 2(\kappa_1 - \kappa)(\partial_x \theta)_{x=x_0(t)} \neq 0, \qquad (3.47)$$

which is easy to obtain if one integrates (3.21) with respect to x from $x = -\infty$ to $x = \infty$ and takes into account the fact that κ suffers a discontinuity at $x = \pm x_0(t)$. Hence one cannot define the constant A from the initial conditions using the integral conservation law; A is a more complicated functional of the initial pressure distribution, i.e., the function $\theta(x,0)$. We note that if in place of $\theta(x,0)$ one takes as initial distribution the function $\theta(x,t_1)$ corresponding to any moment of time $t = t_1 > 0$, then the constant A is unchanged, so in this sense A is an 'integral' of (3.21).

An important step was performed by Kamin, Peletier and Vasquez (1991). They rigorously proved analytically the existence of the self-similar solution (3.35) and showed that for every initial condition of the considered class there exists a value A such that this solution is an asymptotics of the solution to the non-self-similar Cauchy problem. They proved also that $\alpha(\epsilon)$ is a monotonically increasing function. Numerical computations are unable to reveal weak terms in the asymptotics, such as logarithmic ones, therefore a rigorous result was of special value.

The self-similar asymptotics that was obtained is no longer a solution to the instantaneous point source problem. In fact, the amount of fluid Q that must be removed at the initial instant from a region with characteristic length l must change as this length decreases if one wants to obtain one and the same limiting representation of the solution for large t; Q increases for $\epsilon > 0$ and decreases for $\epsilon < 0$, in such a way that the product Ql^{α} is constant.

We have to make a comment concerning the similarity rules. The solution obtained gives for the coordinate of the point of discontinuity of the thermal diffusivity the scaling law

$$x_0(t) = \xi_0 \sqrt{\kappa t},$$ (3.48)

and for the variation in the value of θ at the maximum point the scaling law

$$\theta_{\max} = \frac{A}{(\kappa t)^{(1+\alpha)/2}}.$$ (3.49)

The first of these relations is obtained easily from 'naive' considerations of similarity, i.e., by applying dimensional analysis proceeding from the concept of an instantaneous point source. For the second relation, this is impossible to do in principle, despite the fact that the similarity rule (3.49) has a power-type, scaling form and is completely determined by the dimensions of the quantity A. The point is that the dimensions of the quantity A are unknown in advance, and to determine them it is necessary to solve the nonlinear eigenvalue problem formulated above.

The problem presented above was formulated and solved in the paper by Barenblatt and Sivashinsky (1969).

Self-similarities of the second kind: further examples

4.1 Modified very intense explosion problem

4.1.1 Statement of the modified problem

The self-similar solutions and intermediate asymptotics of the type considered in the previous chapter were the simplest self-similarities of the second kind. To illustrate more complicated possibilities we shall consider in this chapter two instructive problems in gas dynamics. First, we shall make what seems an insignificant modification in the very intense explosion problem considered in chapter 2. We assume that at the front of the shock wave there occurs for some reason a loss of energy (for example, due to radiation), or an influx of energy (for example, due to chemical reaction). In this case the flux of energy at the shock front is not preserved, and the equation of energy balance at the shock front assumes the modified form

$$\rho_f(v_f - D)\left[\frac{\gamma}{\gamma - 1}\frac{p_f}{\rho_f} + \frac{(v_f - D)^2}{2}\right] - \rho_f(v_f - D)q = -\frac{\rho_0 D^3}{2},$$

where q is the intensity of loss ($q < 0$) or deposition ($q > 0$) of energy in unit time in a unit mass of gas passing through the front. Here, as before, p is the pressure, ρ the density, v the speed of the gas, and D the speed of propagation of the shock wave, and subscript f denotes quantities just behind the wave front, i.e., for $r = r_f - 0$ where r_f as before is the radius of the shock wave. We also assume as before that ahead of the wave the gas is at rest at zero pressure and has density ρ_0.

Performing the same transformations as were used in chapter 2 in considering the classical very intense explosion problem, we write the equation of energy balance across the front in the form

$$\rho_f(v_f - D)\left[\frac{p_f}{(\gamma - 1)\rho_f} + \frac{v_f^2}{2}\right] + p_f v_f - \rho_f(v_f - D)q = 0. \qquad (4.1)$$

In the model problem considered below it is assumed that the intensity of energy loss or deposition per unit mass is proportional to the internal energy or the temperature at the shock front:

$$q = kT_f = Cp_f/\rho_f,$$

where k and C are constants. This is necessary in order that the asymptotics obtained be self-similar. We emphasize that our concern here is more with a qualitative mathematical model of a very intense explosion wave with variable energy flux at the front than with a completely adequate analysis of a physical phenomenon.

It is convenient to introduce the new notation

$$C = \frac{\gamma_1 - \gamma}{(\gamma_1 - 1)(\gamma - 1)}.$$

For $\gamma_1 = 1$ we have $C = -\infty$, which means that all the thermal energy of the gas particles is absorbed at the front. For γ_1 increasing from unity to γ the constant C grows from $-\infty$ to zero, and the fraction of lost energy decreases. The case $\gamma_1 = \gamma$ corresponds to the absence of energy loss or deposition at the front – the classical very intense explosion. For $\gamma_1 > \gamma$ there is a deposition of energy at the wave front.

Using the relation assumed for q and the new notation we can rewrite (4.1) in the form

$$\rho_f(v_f - D)\left[\frac{p_f}{(\gamma_1 - 1)\rho_f} + \frac{v_f^2}{2}\right] + p_f v_f = 0, \qquad (4.2)$$

i.e., in the same form as for the classical very intense explosion problem, but with a different adiabatic exponent: in place of exponent γ in (4.2) we have γ_1, the effective adiabatic exponent at the front, which takes into account the loss or deposition of energy. The conditions of continuity of mass and of momentum flux at the wave front, just as for the classical problem, assume the form

$$\rho_f(v_f - D) = -\rho_0 D, \qquad \rho_f(v_f - D)v_f + p_f = 0, \qquad (4.3)$$

and these conditions do not contain the adiabatic exponent γ. As in the classical problem of a very intensive explosion (chapter 2), the conditions at the front (4.2), (4.3) can be reduced to the form

$$p_f = \frac{2}{\gamma_1 + 1}\rho_0 D^2, \quad v_f = \frac{2}{\gamma_1 + 1}D, \quad \rho_f = \frac{\gamma_1 + 1}{\gamma_1 - 1}\rho_0. \qquad (4.4)$$

The equations of gas motion in the region of continuous motion remain unchanged:

$$\partial_t v + v \partial_r v + \frac{1}{\rho} \partial_r p = 0 \,,$$

$$\partial_t \rho + \partial_r(\rho v) + \frac{2}{r} \rho v = 0 \,, \qquad (4.5)$$

$$\partial_t \left(\frac{p}{\rho^\gamma} \right) + v \partial_r \left(\frac{p}{\rho^\gamma} \right) = 0 \,.$$

The condition of no influx of matter or energy at the centre of the explosion for $t > 0$ also preserves its form:

$$v(0,t) = 0 \,. \qquad (4.6)$$

Thus we have obtained seemingly almost the same problem as before, the only one difference being that the adiabatic exponent in the conditions at the shock front is different from that entering the energy equation for the motion of the gas in the region of continuous motion.

4.1.2 Direct application of dimensional analysis

We now attempt, just as before, to construct a self-similar solution to an ideal problem, corresponding to the instantaneous release of a finite amount of energy E at a point – the centre of the explosion. It would seem that nothing in our arguments has to change. In fact, the only new governing parameter in the problem we are considering compared with the classical very intense point explosion problem is the constant dimensionless parameter γ_1, so that the dimensional considerations remain as before, and at first glance it would seem that the desired solution must, for the same reason as in chapter 2, be representable in the form

$$p = \rho_0 \frac{r^2}{t^2} P(\xi, \gamma, \gamma_1), \quad \rho = \rho_0 R(\xi, \gamma, \gamma_1), \quad v = \frac{r}{t} V(\xi, \gamma, \gamma_1) \,,$$

$$\xi = r \left(\frac{Et^2}{\rho_0} \right)^{-1/5}, \quad r_f(t) = \xi_0(\gamma, \gamma_1) \left(\frac{Et^2}{\rho_0} \right)^{1/5} \,. \qquad (4.7)$$

The functions P, V, R must satisfy the same system of ordinary differential equations as in the classical problem, (2.59), because the gas–dynamic equations in the region of continuous motion have not changed and the form of self-similar solution being sought remains essentially the same.

Furthermore, due to (4.7), the speed of the shock wave $D = dr_f/dt = 2r_f/5t$. Substituting into (4.4) the remaining relations in (4.7), we obtain

$$P(\xi_0, \gamma, \gamma_1) = \frac{8}{25(\gamma_1 + 1)}, \quad R(\xi_0, \gamma, \gamma_1) = \frac{(\gamma_1 + 1)}{(\gamma_1 - 1)},$$

$$V(\xi_0, \gamma, \gamma_1) = \frac{4}{5(\gamma_1 + 1)} \tag{4.8}$$

The only difference between these relations for the boundary values of the functions P, V, R at the front $\xi = \xi_0$ and those for the classical very intense explosion problem is that instead of the adiabatic exponent γ we have γ_1.

However, a solution to the modified ideal point explosion problem in the form (4.7), for $\gamma_1 \neq \gamma$, in a reasonable class of functions does not exist. To demonstrate this, we simply note that if the solution had the form (4.7), the bulk energy of the gas in the perturbed region would be constant. This can be proved in the same way as before (see relation (2.61)). However, for $\gamma_1 \neq \gamma$ the bulk energy of gas in the perturbed region $\mathcal{E}$ would vary owing to the loss or influx of energy at the wave front:

$$\frac{d\mathcal{E}}{dt} = -4\pi r_f^2 \rho_f (v_f - D)q = -\frac{4\pi r_f^2 \rho_f (v_f - D)(\gamma_1 - \gamma)p_f}{(\gamma_1 - 1)(\gamma - 1)\rho_f}. \tag{4.9}$$

So, only the trivial solution, for which the energy is equal to zero can satisfy both requirements. The trivial solution, however, obviously does not satisfy the initial condition. The contradiction obtained demonstrates the non-existence of a solution to the problem having the form (4.7) for $\gamma_1 \neq \gamma$, i.e., the non-existence of a solution to ideal point source problem.

4.1.3 Numerical experiment. Self-similar intermediate asymptotics

In order to resolve the contradiction discussed above , we will again, as in the analogous situation described in the previous chapter, depart from the exact statement of the ideal point source problem. We recall that the solution corresponding to a point explosion makes sense only if it is an asymptotics for a solution corresponding to the initial release of energy in a small but finite domain. Hence we turn to consideration of the problem in which the energy at time $t = 0$ is released not at a point, but within a sphere of a certain finite radius r_0; in other respects the problems coincide.

Thus in a numerical experiment the following problem was solved. There is an unbounded space filled with gas. At the initial instant, outside a sphere of radius r_0 the density of the gas is constant and

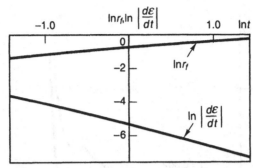

Figure 4.1. The dependences of $\ln r_f(t)$ and $\ln |d\mathcal{E}/dt|$ on $\ln t$ obtained by numerical solution of the non-self-similar problem rapidly approach straight lines corresponding to self-similar intermediate asymptotics.

equal to ρ_0 and the pressure is equal to zero. Inside the sphere the distributions of the flow properties, i.e., the pressure p, velocity v and density ρ of the gas, correspond to the solution of the classical problem of a very intense explosion for the same energy E and the same values of the other parameters at some instant $t = t_0 > 0$ such that $r_f(t_0) = r_0$. Thus we assume that for $-t_0 \leq t < 0$ there occurs the same gas motion as in a very intense explosion without radiation or release of energy and that at $t = 0$, when $r_f = r_0$, the radiation or energy release at the front is switched on. For the subsequent evolution of the motion the properties of the flow in the region of continuous motion are described by the system (4.5) of equations for the adiabatic motion of a gas. At the front of the shock wave the conditions have the form (4.4), the same conditions as for the classical problem of a very intense explosion; but we emphasize again that the effective adiabatic exponent γ_1 in the conditions at the shock front differs from the adiabatic exponent γ that enters the equation of entropy conservation in the region of continuous motion. Furthermore, the condition (4.6) of no influx of matter or energy at the centre of the explosion for $t > 0$ is to be satisfied.

The results of the numerical calculation are presented in Figures 4.1–4.3 and in Table 4.1.

The calculations were performed for two values of γ, one of them close to 1, $\gamma = 1.1$, and the other close to 2, $\gamma = 1.9$. In each case calculations were done for various values of γ_1 in the interval $1 \leq \gamma_1 < 2\gamma + 1$. Along with the quantities p, ρ, v and r_f the quantity $\mathcal{E}$, the total energy of the gas in the perturbed domain, was calculated. The basic result is that

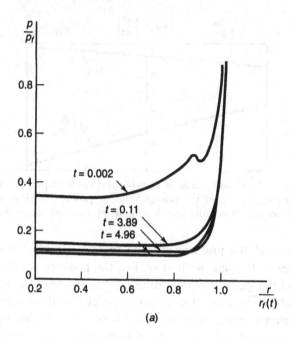

(a)

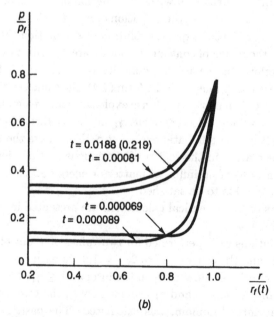

(b)

Figure 4.2. The dependences of p/p_f on r/r_f for increasing times t, obtained by numerical solution of the non-self-similar problem, approach the self-similar intermediate asymptotics. (a) $\gamma = 1.9$, $\gamma_1 = 1.1$; (b) $\gamma = 1.1$, $\gamma_1 = 1.9$.

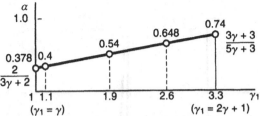

Figure 4.3. Dependence of the time exponent α in the relation $r_f \sim t^\alpha$ obtained by numerical solution of the non-self-similar problem (solid lines), and as an eigenvalue of the nonlinear eigenvalue problem (4.18), (4.22), (4.23) (open dots), on the effective adiabatic exponent γ_1 at the wave front, for $\gamma = 1.1$. For $\gamma_1 < \gamma$, $\alpha < 2/5$; for $\gamma_1 = \gamma$ (an adiabatic very intense explosion), $\alpha = 2/5$; for $\gamma_1 = 2\gamma + 1$ (detonation with variable speed of propagation of the detonation wave; the Chapman–Jouguet condition is satisfied) $\alpha = (3\gamma + 3)/(5\gamma + 3)$.

Table 4.1

		Time exponent, α	
γ	γ_1	non-self-similar problem	eigenvalue problem
1.1	1.9	0.54	0.54
1.9	1.1	0.29	0.28

the solution rapidly approaches a self-similar asymptotics of the form

$$p = \rho_0 \frac{r^2}{t^2} P\left(\frac{r}{r_f}, \gamma, \gamma_1\right), \quad \rho = \rho_0 R\left(\frac{r}{r_f}, \gamma, \gamma_1\right),$$

$$v = \frac{r}{t} V\left(\frac{r}{r_f}, \gamma, \gamma_1\right), \quad r_f = \left(\frac{At^2}{\rho_0}\right)^{\alpha/2} \tag{2.10}$$

where A and α are constants. Here the constant α turns out to depend only on γ and γ_1 (the graph of the function $\alpha(\gamma_1)$ for $\gamma = 1.1$, obtained from numerical calculations is shown in Figure 4.3) and not on the initial conditions (i.e., α is independent of t_0, E and ρ_0). The constant A, however, does turn out to depend also on the initial conditions. The approach to the self-similar regime is seen in the dependence of $\ln r_f(t)$ on $\ln t$, which rapidly approaches the straight line $\ln r_f(t) \sim \alpha \ln t$, and in the dependence of $\ln |d\mathcal{E}/dt|$ on $\ln t$, which rapidly approaches the straight line $\ln |d\mathcal{E}/dt| \sim (5\alpha - 3) \ln t$ (for $\gamma = 1.9, \gamma_1 = 1.1$ see Figure 4.1). It is also seen in the dependences of the quantity p/p_f on r/r_f for different instants of time (Figures 4.2(a) and 4.2(b)).

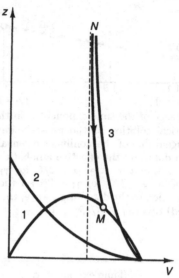

Figure 4.4. A phase portrait: the integral curves of the first-order equation (4.18) for $1 < \gamma_1 < 2\gamma + 1$. The number 1 denotes the curve $z = -\gamma V(V - \alpha)$ and 2 and 3 denote the curves $z = (V - \alpha)^2$ and $z = V(V - 1)(V - \alpha)(3V - \kappa)^{-1}$.

4.1.4 Self-similar limiting solution

Let us now clarify how this seemingly unexpected self-similar intermediate asymptotics (4.10) has appeared in the numerical experiment.

In addition to the governing parameters for the classical very intense point explosion problem (see conditions (2.53)), we must include, for the problem solved in the numerical experiment, the initial radius of the shock wave, r_0, so that there appear not one but two dimensionless independent variables,

$$\xi = r \left(\frac{Et^2}{\rho_0} \right)^{-1/5} , \quad \eta = r_0 \left(\frac{Et^2}{\rho_0} \right)^{-1/5} ; \qquad (4.11)$$

and according to dimensional analysis the velocity, density and pressure of the gas are then expressed in the form

$$v = \frac{r}{t} V(\xi, \eta, \gamma, \gamma_1) , \quad \rho = \rho_0 R(\xi, \eta, \gamma, \gamma_1) , \quad p = \rho_0 \frac{r^2}{t^2} P(\xi, \eta, \gamma, \gamma_1) . \tag{4.12}$$

For $\gamma = \gamma_1$ the solution to the very intense point explosion problem is on the one hand a solution of an ideal singular limiting problem corresponding to $r_0 = 0$ at $t = -t_0$ and on the other hand the limit of the solution

(4.12) as $t \to \infty$. For $\gamma_1 \neq \gamma$, as we have just explained, there exists no non-trivial solution of the problem corresponding to $r_0 = 0$. However, as we have emphasized repeatedly, we are interested not in a solution of the ideal limiting problem, but in an asymptotic representation for large t of the solution to the non-self-similar problem with $r_0 \neq 0$. For increasing t and fixed r, both ξ and η tend to zero. The appearance of the self-similar intermediate asymptotics (4.10) can be explained by the fact that as ξ, $\eta \to 0$ the leading terms in the asymptotic representations of the functions P, V and R have the form

$$P = P\left(\frac{\xi}{\eta^\beta}\right), \quad V = V\left(\frac{\xi}{\eta^\beta}\right), \quad R = R\left(\frac{\xi}{\eta^\beta}\right), \qquad (4.13)$$

where β is a positive number depending on γ and γ_1. Therefore the limiting solution is self-similar, because

$$\frac{\xi}{\eta^\beta} = \frac{r}{Bt^\alpha}$$

where

$$B = r_0 \left(\frac{E}{\rho_0}\right)^{(1-\beta)/5}, \quad \alpha = \frac{2(1-\beta)}{5}.$$

The class of self-similar solutions of the gas-dynamic equations to which the limiting solution (4.13) of this problem belongs was indicated by Bechert (1941) and Guderley (1942), and later considered by Sedov (1945) and others. It will be called the Bechert–Guderley class.

For later analysis it is convenient to renormalize the self-similar independent variable and take it in the form

$$\zeta = \text{const}\, \frac{\xi}{\eta^\beta} = r\left(\frac{At^2}{\rho_0}\right)^{-(1-\beta)/5}, \quad A = \sigma E r_0^{5\beta/(1-\beta)}, \qquad (4.14)$$

where the constant parameter σ is chosen so that at the front of the shock wave the value of the new self-similar variable ζ will be equal to unity. Then the asymptotic law of propagation of the shock wave can be written in the form

$$r_f = \left(\frac{A}{\rho_0}\right)^{(1-\beta)/5} t^{2(1-\beta)/5}. \qquad (4.15)$$

The variables ξ and η can be made to tend to zero in another way also. For fixed r and t, E tends to infinity or zero, and r_0 to zero; however, in order to obtain the same asymptotics as for our non-self-similar solution at large times, the product $Er_0^{5\beta/(1-\beta)}$ must remain constant:

$$Er_0^{5\beta/(1-\beta)} = \text{const}. \qquad (4.16)$$

Thus it is obtained that for $\gamma_1 \neq \gamma$ the self-similar limiting solution corresponds not to a point explosion, i.e., not to the release of a finite amount of energy at the initial instant at the centre of the explosion, but to the release within a finite region of radius r_0 of an amount of energy E that, according to (4.16), tends to zero or infinity, depending on the sign of β, as $r_0 \to 0$.

For given γ and γ_1, we can determine the parameter β or, what is the same thing, α, in either of two ways. In the first, we follow, for example numerically as we did previously, the evolution of the non-self-similar solution of the original problem up to its transition to a self-similar asymptotics. In the second, we use the fact that the self-similar asymptotics is itself a solution of the gas-dynamic equations that satisfies certain conditions, and attempt to construct that solution and simultaneously to determine the exponent α.

We will follow now the second way. Thus we seek the desired limiting solution in the form

$$p = \rho_0 \frac{r^2}{t^2} P(\zeta, \gamma, \gamma_1), \quad \rho = \rho_0 R(\zeta, \gamma, \gamma_1),$$

$$v = \frac{r}{t} V(\zeta, \gamma, \gamma_1), \quad r_f = \left(\frac{A}{\rho_0}\right)^{\alpha/2} t^\alpha, \quad \zeta = \frac{r}{r_f}. \tag{4.17}$$

We substitute this solution into (4.5) and then obtain, using a common technique (Guderley, 1942; see also Sedov, 1959), the first-order equation

$$\frac{dz}{dV} = -\frac{z}{V-\alpha}\left\{ \frac{[2(V-1)+3(\gamma-1)V][(V-\alpha)^2 - z]}{[(3V-\kappa)z - V(V-1)(V-\alpha)]} + \gamma - 1 \right\}, \tag{4.18}$$

where

$$\kappa = \frac{2(1-\alpha)}{\gamma}, \quad z = \frac{\gamma P}{R},$$

and two other first-order equations,

$$\frac{d\ln\zeta}{dV} = \frac{z - (V-\alpha)^2}{V(V-1)(V-\alpha) + (\kappa - 3V)z}, \tag{4.19}$$

and

$$(V-\alpha)\frac{d\ln R}{d\ln\zeta} = -3V - \frac{V(V-1)(V-\alpha) + (\kappa - 3V)z}{z - (V-\alpha)^2}. \tag{4.20}$$

Thus if the necessary solution of the first-order equation (4.18) has been constructed, the solution of (4.19) and (4.20) is then obtained in quadratures.

It is essential here that the desired solution of the first-order equation (4.18) must pass through two points: the image of the shock-wave front and the image of the centre of symmetry. Substituting (4.17) into the conditions (4.4) at the shock-front wave, we find

$$P(1,\gamma,\gamma_1) = \frac{2\alpha^2}{\gamma_1 + 1}, \quad V(1,\gamma,\gamma_1) = \frac{2\alpha}{\gamma_1 + 1},$$

$$R(1,\gamma,\gamma_1) = \frac{\gamma_1 + 1}{\gamma_1 - 1}. \tag{4.21}$$

Hence the image of the front in the Vz plane will be the point

$$V = \frac{2\alpha}{\gamma_1 + 1}, \quad z = \frac{2\alpha^2\gamma(\gamma_1 - 1)}{(\gamma_1 + 1)^2}. \tag{4.22}$$

The image of the centre of symmetry $\zeta = 0$ in the Vz plane will be a singular point of equation (4.18) of saddle type,

$$V = \frac{2(1 - \alpha)}{3\gamma}, \quad z = \infty. \tag{4.23}$$

Here we use the condition of no influx of matter or energy at the centre of the explosion for $t > 0$. Furthermore the self-similar variable ζ must increase monotonically from zero to unity in the course of moving from the image of the centre of symmetry to the image of the front. In general it is impossible to satisfy these conditions for arbitrary α; we cannot pass a solution of a first-order equation through two arbitrary points. We shall see, however, that there exists one exceptional value of α for which this is possible. Thus we have again arrived at a *nonlinear eigenvalue problem*, namely to construct a solution of the first-order equation (4.18) passing through the two points (4.22) and (4.23) and to determine the value of the parameter α for which such a solution exists – that is, the eigenvalue of the problem.

4.1.5 Qualitative investigation of the nonlinear eigenvalue problem

We consider the phase portrait – the picture of the integral curves in the first quadrant – which is the part of the Vz plane of interest to us. In the case $\gamma < 2$, $1 \le \gamma_1 \le 2\gamma + 1$, the phase portrait is shown in Figure 4.4, where the curves numbered 1, 2 and 3 correspond respectively to the equations

$$z = -\gamma V(V - \alpha) \tag{4.24}$$

(which is the locus of points of the shock front),

$$z = (V - \alpha)^2, \quad z = V(V - 1)(V - \alpha)(3V - \kappa)^{-1}, \tag{4.25}$$

The points of intersection of the two curves (4.25) are singular points of equation (4.18). For $\gamma_1 < 2\gamma + 1$ all these singular points are situated below the curve (4.24). One can show, using a standard technique in the qualitative theory of differential equations (which will be demonstrated in chapter 7 in a different example), that for such γ_1 we can find one value of α, and moreover for each pair γ, γ_1 only one, such that point M, which is the image of the shock front, and point N, which is the image of the centre of symmetry, lie on the same integral curve. This curve is the separatrix of two families of integral curves, – the single curve passing through the image of the centre of symmetry, the latter being a singular point of saddle type for (4.18). The graph of the function $\alpha(\gamma_1)$ is shown for $\gamma = 1.1$ in Figure 4.3; it is a monotonically increasing curve passing through the points

$$\alpha = \frac{2}{3\gamma + 2}, \ \gamma_1 = 1; \ \alpha = \frac{2}{5}, \ \gamma_1 = \gamma; \ \alpha = \alpha_* = \frac{3\gamma + 3}{5\gamma + 3}, \ \gamma_1 = 2\gamma + 1.$$
$$(4.26)$$

For $\gamma_1 > 2\gamma + 1$ the singular point (of nodal type) is located above the curve (4.24), as shown in Figure 4.5, and one consequently has a range of possible values of α. It will be shown that for such γ_1 the solution of the original non-self-similar problem also becomes non-unique. Hence, to select a unique solution $\gamma_1 > 2\gamma + 1$ requires an additional condition for the non-self-similar problem too.

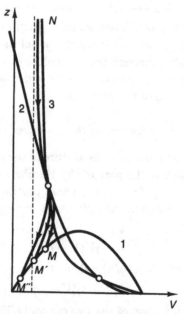

Figure 4.5. A phase portrait: the integral curves of (4.18) for $\gamma_1 > 2\gamma+1$. A singular point of nodal type is located above curve 1.

The points (4.26) are of special interest. The first of them corresponds to motion with complete loss of thermal energy at the front of the wave; the compression at the front (the ratio of the density just behind the front to the original density of the gas) obtained in this case is infinite, and the relative speed of the gas and the front is equal to zero. The second point corresponds to the classical very intense explosion. The third point is very curious; it corresponds to an influx of energy at the front and satisfaction of the so-called Chapman–Jouguet condition: the gas speed relative to the front is equal to the local speed of sound. The speed of sound at the front is in fact equal to

$$\sqrt{\gamma p_f / \rho_f} = D\sqrt{2\gamma(\gamma_1 - 1)}/(\gamma_1 + 1),$$

where D is, as before, the speed of the front. Now the speed of the gas relative to the front is given by $|v_f - D| = (\gamma_1 - 1)D/(\gamma_1 + 1)$. For $\gamma_1 = 2\gamma + 1$ these two speeds coincide, and, since $\alpha = \alpha_*$, the image of the front in the Vz plane coincides with a singular point of nodal type, lying for this γ_1 on the curve (4.24), so that the integral curve sought in this case joins two singular points of (4.18), a saddle point that is the image of the centre of symmetry, and a nodal point that is the image of the front.

Thus, for energy loss at the front ($\gamma_1 < \gamma$) and small energy release at the front ($\gamma < \gamma_1 < 2\gamma + 1$) the solution under investigation differs only slightly in its characteristics from the solution corresponding to the classical very intense explosion ($\gamma_1 = \gamma$): a singular point of nodal type is situated under the locus of points of the front, the motion in the perturbed domain is everywhere subsonic, etc. For $\gamma_1 = 2\gamma + 1$, $\alpha = \alpha_*$, the solution constructed represents a motion of detonation type, but with variable speed of propagation of the front

$$r_f = At^{(3\gamma+3)/(5\gamma+3)};$$

the speed of sound is achieved at the front of the wave, a node being the image of the front. Here the influx of energy at the front and the temperature at the front turn out to depend on time. It is remarkable that for $\gamma_1 = 2\gamma + 1$ the solution obtained is not uniquely determined, so the classical solution to the problem of a spherical detonation wave (Zeldovich, 1942; Zeldovich and Kompaneets, 1960), corresponding to $\alpha = 1$, i.e., to constant speed of propagation of the wave, constant influx of energy at the front, and constant temperature at the front, also satisfies all the conditions of the problem posed. The family of solutions with intermediate α, i.e., $(3\gamma + 3)/(5\gamma + 3) < \alpha < 1$ also exists. Indeed, for $\gamma_1 = 2\gamma + 1$ and such α the singular point of nodal type does not coincide any more with the image of the front and is found above the curve (4.24). The image of the front is on the intersection of the curve (4.24) and the first curve of (4.25). There exists an integral curve going from the image of the front to the singular point of nodal type and after it to the singular point of saddle type.

It is interesting to note that among the solutions corresponding to $\alpha_* \le \alpha \le 1$ only the solution corresponding to $\alpha = \alpha_*$ gives a finite acceleration at the front. For solutions with other values of α this acceleration is infinite.

Thus, for $1 < \gamma_1 < 2\gamma+1$ a self-similar solution is constructed that can be an intermediate asymptotics of the solution to the original non-self-similar problem. Only the constant A remains undetermined or, what is the same, the dimensionless constant σ. In the case $\gamma_1 = \gamma$ the quantity $A = \sigma E$ and the constant σ are found from the law of conservation of total energy,

$$\mathcal{E} = 4\pi \int\limits_0^\infty \left[\rho \frac{v^2}{2} + \frac{p}{(\gamma-1)} \right] r^2 dr = \text{const},$$

which is also valid for the non-self-similar stage of the motion. For $\gamma_1 \ne \gamma$ there is no such conservation law; the equation for the conservation of energy assumes, as was already noted, the non-integrable form

$$\frac{d\mathcal{E}}{dt} = -4\pi r_f^2 D \frac{(\gamma-\gamma_1)}{(\gamma-1)(\gamma_1-1)} \rho_0 \frac{p_f}{\rho_f} \ne 0. \tag{4.27}$$

The only method of determining the constant σ is to follow the evolution of the solution of the non-self-similar problem to a self-similar intermediate asymptotics as was done before.

For comparison a numerical solution was also evaluated for the nonlinear eigenvalue problem formulated above for a system of ordinary equations. Namely, the system of ordinary equations (4.18)–(4.20) was solved numerically for the initial conditions (4.21), the exponent α being calculated by a trial-and-error method so as to satisfy the condition of no influx of matter or energy at the centre for $t > 0$. Calculation was stopped when the quantity $P\zeta^2 = Rz\zeta^2/\gamma$ near $\zeta = 0$ had become constant to within an accuracy of 1%. The results of a comparison of the values of α obtained by numerical calculation the solution to the non-self-similar problem for the system of partial differential equations and by direct computation of the nonlinear eigenvalue problem are given in Table 4.1.

A comparison of the eigenfunctions from the nonlinear eigenvalue problem with the limiting distributions obtained after the stabilization of the solution to the non-self-similar problem also indicates good agreement.

Thus, numerical integration for the initial conditions we have taken before confirms that the asymptotics of the solution to the original non-self-similar problem is indeed the self-similar solution constructed in this and previous subsections. Like the self-similar solutions considered in chapter 3 this self-similar asymptotic solution is distinguished by two properties. First, the exponent α of time in the relation for the self-similar variable cannot be found from similarity considerations, but requires for its determination the solution of a nonlinear eigenvalue problem, i.e., it is found from the condition for the existence of a global self-similar solution. Furthermore, the solution is determined here to within a constant appearing in the self-similar variable, which can be found only by matching the self-similar intermediate asymptotics to a non-self-similar solution of the original problem; here there is no integral conservation law whose use directly determines the value of this constant from the initial data of the original non-self-similar problem.

The self-similar solution considered here was obtained by Barenblatt and Sivashinskii (1970) (see also Barenblatt and Zeldovich, 1971, 1972; Barenblatt, Guirguis, Kamel, Kuhl, Oppenheim and Zeldovich, 1980). Numerical calculations were carried out by Andrushchenko, Barenblatt and Chudov (1975). Ia.G. Sapunkov and A.K. Oppenheim and coauthors obtained by another approach a self-similar solution of the problem considered in this section for the special case, $\gamma_1 = 2\gamma + 1$, satisfying the Chapman–Jouguet condition (Sapunkov, 1967; Oppenheim, Kuhl, and Kamel, 1972; Oppenheim, Kuhl, Lundström and Kamel, 1971; Oppenheim, Lundström, Kuhl and Kamel, 1972).

4.2 The von Weizsäcker–Zeldovich problem: an impulsive loading

4.2.1 Statement of the impulsive loading problem

A problem of the same type as above that has for our purposes very instructive peculiarities is that of an impulsive loading, studied in the papers of C.F. von Weizsäcker, Ya.B. Zeldovich and their associates (von Weizsäcker, 1954; Hain and Hörner, 1954; Häfele, 1955; Meyer, 1955; Zeldovich, 1956; Adamsky, 1956; Zhukov and Kazhdan, 1956). To illustrate those peculiarities we briefly present the problem here; a more detailed account is given in the monograph by Zeldovich and Raizer (1967).

We suppose that space is divided initially into two halves by an impenetrable plane wall at $x = 0$ (x being the coordinate normal to the wall). The half-space $x \geq 0$ is occupied by a quiescent ideal gas of density ρ_0 at zero pressure; in the half-space $x \leq 0$ there is a vacuum. At the initial instant $t = 0$ a pressure $p = p_0$ is created on the right-hand side of the wall (by, for example, an explosion) and varies according to a certain law $p = p_0 f(t/\tau)$ up to some time $t = \tau$, after which the wall disappears instantaneously. The problem consists in investigating the motion arising for $t > \tau$, which obviously develops as follows. A plane shock wave $x = x_f(t)$ propagates to the right in the quiescent gas. In some region behind the shock wave the compressed gas continues to advance to the right. At a certain plane $x = x_0(t)$ the instantaneous speed of the gas particles becomes equal to zero, and all gas particles situated to the left of this plane move to the left; there occurs an expansion into the vacuum of the gas compressed by the shock wave.

The solution to this problem reduces to the solution to the same system of gas-dynamic equations as in the previous problem, but now for the rectilinear case:

$$\partial_t v + v\partial_x v + \frac{1}{\rho}\partial_x p = 0\,,$$

$$\partial_t \rho + \partial_x(\rho v) = 0\,, \qquad (4.28)$$

$$\partial_t\left(\frac{p}{\rho^\gamma}\right) + v\partial_x\left(\frac{p}{\rho^\gamma}\right) = 0\,.$$

The boundary conditions at the shock wave $x = x_f(t)$ are the same as in the very intense explosion problem:

$$\rho_f(v_f - D) = -\rho_0 D\,,$$

$$\rho_f(v_f - D)v_f + p_f = 0\,, \qquad (4.29)$$

$$\rho_f(v_f - D)\left[\frac{v_f^2}{2} + \frac{p_f}{(\gamma - 1)\rho_f}\right] + p_f v_f = 0\,,$$

where $D = dx_f/dt$. The initial conditions at time $t = \tau$ correspond for $x < 0$ to a vacuum and for $x > 0$ to the state of motion that has developed at time $t = \tau$ in the half-space filled with quiescent gas of density ρ_0 at zero pressure owing to the maintenance on the boundary during the time interval τ of a pressure varying according to the law $p(0,t) = p_0 f(t/\tau)$.

It is evident that the density, pressure, and speed of the gas depend on the dimensional quantities:

$$t, \ p_0, \ \rho_0, \ \tau, \ x,$$
(4.30)

and that the coordinate of the shock front depends on all these quantities except the last. Applying the standard procedure of dimensional analysis, we obtain

$$x_f(t) = \sqrt{p_0/\rho_0} \ \tau \xi_f(\Pi_1),$$
(4.31)

$$\Pi_\rho = \Phi_\rho(\Pi_1, \Pi_2), \quad \Pi_p = \Phi_p(\Pi_1, \Pi_2), \quad \Pi_v = \Phi_v(\Pi_1, \Pi_2).$$
(4.32)

Here

$$\Pi_1 = \frac{t}{\tau}, \quad \Pi_2 = \frac{x}{\sqrt{p_0/\rho_0} \ \tau},$$
$$\Pi_\rho = \frac{\rho}{\rho_0}, \quad \Pi_p = \frac{p}{p_0}, \quad \Pi_v = \frac{v}{\sqrt{p_0/\rho_0}}.$$
(4.33)

As is evident, the solution to the problem posed turns out to be non-self-similar. This results from the fact that the problem contains a characteristic time τ and a characteristic length scale $(p_0/\rho_0)^{1/2}\tau$.

4.2.2 Numerical experiment. Self-similar asymptotics

Numerical calculations reveal, however, that the solution to the formulated problem has an instructive property. Namely, the dependence of the coordinate of the wavefront on time rapidly (i.e., after some time interval of order τ after the start of expansion into the vacuum) approaches a scaling, power-law asymptotics, so that

$$\xi_f(\Pi_1) = \xi_0(\gamma)\Pi_1^\alpha,$$
(4.34)

where $\xi_0(\gamma)$ is some function of γ and the exponent α also depends on γ. Furthermore, the density at the front rapidly approaches a constant value, and the pressure and speed of the gas at the front rapidly approach the scaling laws

$$\frac{p_f}{p_0} \sim \Pi_1^{-2(1-\alpha)}, \quad \frac{v_f}{\sqrt{p_0/\rho_0}} \sim \Pi_1^{-(1-\alpha)}.$$
(4.35)

Finally, it turns out that if one constructs the distributions of density, pressure, and speed in reduced coordinates, taking x_f as the scale of length and p_f, ρ_f, and v_f as scales for the flow properties, then those distributions just as rapidly become independent of time (see Figure 4.6). In other words, it turns out that the solution of the problem

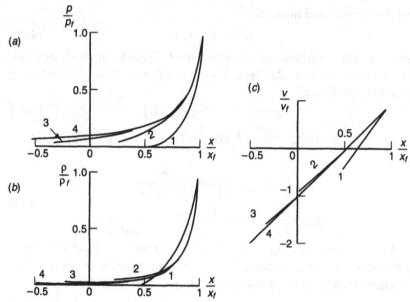

Figure 4.6. Dependence of (a) p/p_f, (b) ρ/ρ_f, and (c) v/v_f on x/x_f, obtained by numerical solution of the non-self-similar problem of an impulsive load for $\gamma = 1.4$, which rapidly approach the dependence corresponding to a self-similar intermediate asymptotics (curve 4). From Zhukov and Kazhdan (1956). (1) $t = 1.6$, (2) $t = 5.0$, (3) $t = 15.0$, (4) Self-similar intermediate asymptotics

rapidly approaches the self-similar asymptotics

$$\xi_f = \xi_0(\gamma)\Pi_1^\alpha, \quad \Phi_\rho = \Phi_{1\rho}\left(\frac{\Pi_2}{\Pi_1^\alpha}\right),$$

$$\Phi_p = \Pi_1^{-2(1-\alpha)}\Phi_{1p}\left(\frac{\Pi_2}{\Pi_1^\alpha}\right), \Phi_v = \Pi_1^{-(1-\alpha)}\Phi_{1v}\left(\frac{\Pi_2}{\Pi_1^\alpha}\right). \tag{4.36}$$

We note that the approach of the solution to the self-similar asymptotics does not occur uniformly in the whole region of motion $-\infty < x \leq x_f(t)$ but only close to the front $x_f(t)$, in a region that increases with time since the start of the expansion. Further, numerical calculation (Zhukov and Kazhdan, 1956) showed that the solution approaches a self-similar asymptotics of the form (4.36) with one and the same exponent α independently of whether the pressure at the wall in the time interval τ is constant or changes according to various laws.

4.2.3 Self-similar limiting solution

It is natural to try to construct the limiting self-similar solution directly.

Again we seek it in the Bechert–Guderley class of solutions:

$$p = \rho_0 \frac{x^2}{t^2} P(\xi), \quad \rho = \rho_0 R(\xi), \quad v = \frac{x}{t} V(\xi),$$

$$x_f = A t^\alpha, \quad \xi = x/A t^\alpha. \tag{4.37}$$

Here A and α are constants. For the functions P, V and R we obtain a system of ordinary differential equations, which reduce to one first-order equation

$$\frac{dz}{dV} = \frac{z}{\Delta} \{ [2(V-1) + (\gamma-1)V](V-\alpha)^2$$

$$- (\gamma-1)V(V-1)(V-\alpha) - [2(V-1) + \kappa(\gamma-1)]z \}, \tag{4.38}$$

$$\Delta = (V-\alpha)[V(V-1)(V-\alpha) + (\kappa-V)z],$$

with

$$\kappa = \frac{2(1-\alpha)}{\gamma}, \quad z = \frac{\gamma P}{R}, \tag{4.39}$$

and to two other first-order equations,

$$\frac{d\ln\xi}{dV} = \frac{z - (V-\alpha)^2}{V(V-1)(V-\alpha) + (\kappa-V)z}, \tag{4.40}$$

and

$$\frac{d\ln R}{d\xi} = -\frac{V(V-1)(V-\alpha) + (\kappa-V)z}{(V-\alpha)[z - (V-\alpha)^2]} - \frac{V}{V-\alpha}. \tag{4.41}$$

Thus if the desired solution of (4.38) is known, the solution of (4.40), (4.41) can be found by quadratures.

The desired solution of (4.38) must pass through two singular points: the image of the shock front

$$V = \frac{2\alpha}{\gamma+1}, \quad z = \frac{2\alpha^2\gamma(\gamma-1)}{(\gamma+1)^2} \tag{4.42}$$

and the image of the free boundary, a singular point of saddle type

$$V = \kappa, \quad z = \infty. \tag{4.43}$$

Here the variable ξ must increase monotonically upon moving from the singular point (4.43) to the image of the shock front (4.42). Thus the mathematical problem turns out in this case to be close to the modified very intense explosion problem considered at the beginning of this chapter[‡]. We have again arrived at the necessity of obtaining an integral curve of a first-order equation, of the same type as before, passing through two points, one of which is a saddle-type singular point. This

---

[‡] We recall that the problem of an impulsive loading was first solved significantly before the modified very intense explosion problem.

is impossible in general but, as in the previous problem, one can show that for each γ there exists a value of α, an eigenvalue of the problem, for which the integral curve of (4.38) passing through the image of the shock front also goes through the saddle, the image of the free boundary. Values of α corresponding to various values of γ over its complete range, $1 \leq \gamma \leq \infty$, are given in Table 4.2. It is seen that for all γ in the interval $1 < \gamma < \infty$ we have the inequality

$$\tfrac{1}{2} < \alpha < \tfrac{2}{3} .$$

Table 4.2

γ	1.0	1.1	7/5	5/3	2.8	∞
α	1/2	0.569	3/5	0.611	0.627	0.64

As in the previous problem, it turns out here that values of the exponent α determined by direct construction of a limiting self-similar solution of the impulsive loading problem agree well with values obtained by numerical calculation of an asymptotic solution of the non-self-similar problem.

Evidently the limiting self-similar solution (4.37) is determined by direct construction only to within a constant A; comparison of (4.37) and (4.34) gives

$$A = \xi_0(\gamma)\sqrt{p_0/\rho_0}\ \tau^{1-\alpha} . \tag{4.44}$$

Thus if we want to obtain the same asymptotics while reducing the duration τ of the impulse acting on the gas, we must correspondingly increase the pressure according to the law

$$p_0 = \text{const}\,\tau^{-2(1-\alpha)} . \tag{4.45}$$

4.2.4 Laws of conservation of energy and momentum in the impulsive loading problem

The mass of gas involved at each moment in the flow through unit area of the boundary is finite. Hence conservation laws of momentum and energy apply; these are valid also at the non-self-similar stage of the motion. Therefore, the idea naturally occurs of using these laws to determine the exponent α and the constant A of the limiting self-similar solution, as was done in chapter 2 for the self-similar solutions of the first kind considered there.

The gas is initially quiescent and at zero pressure, so its momentum and energy are zero. The total momentum J of the gas involved in the motion is equal at any instant to the impulse of the pressure load:

$$J = \beta p_0 \tau, \quad \beta = \int_0^1 f(\lambda)d\lambda. \tag{4.46}$$

Hence we obtain the momentum conservation law in the form

$$\beta p_0 \tau = \int_{-\infty}^{x_f} \rho v dx. \tag{4.47}$$

As time increases the solution tends to a self-similar one. Hence, it seems that passing to the limit under the integral sign, we can substitute into (4.47) the expressions for the density and speed from the self-similar solution (4.37) and obtain the relation

$$\beta p_0 \tau = \rho_0 A^2 t^{2\alpha-1} \int_{-\infty}^1 R(\xi)V(\xi)d\xi. \tag{4.48}$$

Since the integral on the right-hand side is obviously independent of time it is necessary, in order for the left-hand side to be independent of time also, to have the relation $\alpha = 1/2$, after which it would appear that one could find the constant A from (4.48).

However, we also have the energy conservation law. According to it the work per unit area performed by the loading of gas is equal to

$$\int_0^\tau p(0,t)v(0,t)dt = \delta p_0^{3/2} \rho_0^{-1/2} \tau, \tag{4.49}$$

where δ is a numerical constant. But the energy of the gas about to enter the motion is zero because its speed and pressure are equal to zero. Hence the energy of the gas actually involved in the motion is at any instant equal to the work performed by the impulsive loading:

$$\delta p_0^{3/2} \rho_0^{-1/2} \tau = \int_{-\infty}^{x_f} \rho \left[\frac{v^2}{2} + \frac{p}{(\gamma-1)\rho} \right] dx. \tag{4.50}$$

Again it seems that passing to the limit under the integral sign and substituting the expressions for speed, density and pressure from the limiting self-similar solution, we obtain

$$\delta p_0^{3/2} \rho_0^{-1/2} \tau = \rho_0 A^3 t^{3\alpha-2} \int_{-\infty}^1 R \left[\frac{V^2}{2} + \frac{P}{(\gamma-1)R} \right] \xi^2 d\xi. \tag{4.51}$$

At first glance it follows from this that $\alpha = 2/3$ and that (4.51) allows one to determine the constant A also. Thus a paradox arises, consisting in the fact that the exponents α in the self-similar variable determined from the laws of conservation of momentum ($\alpha = 1/2$) and of energy ($\alpha = 2/3$) do not agree with each other or with the exponent α ($1/2 < \alpha < 2/3$) determined by direct construction of a limiting self-similar solution, or by its numerical calculation.

The resolution of this paradox is trivial, and at the same time instructive. The fact is that the integral in the momentum equation (4.48) is equal to zero, and the integral in the energy equation (4.51) is equal to infinity, so that from these relations it is impossible to determine either the exponent or the constant A. The transition to the limit under the integral sign in the conservation laws (4.47) and (4.50) was itself inadmissible, because the convergence of the integrands to the limit is non-uniform over the domain of integration. Usually we leave such fine points as uniform convergence under the integral sign to pure mathematicians assuming that everything will be in order. Here it is not so, and that is the essence of the problem!

In fact the limiting self-similar motion is obtained by transition to the limit over the entire domain $-\infty < x \leq x_f$ with the duration τ of the impulse tending to zero and the pressure on the boundary tending to infinity according to the law $p_0 = \text{const}\, \tau^{-2(1-\alpha)}$. Here the total momentum $\beta p_0 \tau$ tends to zero as $\text{const}\, \tau^{2\alpha-1}$ and the energy $\delta p_0^{3/2} \rho_0^{-1/2} \tau$ to infinity as $\text{const}\, \tau^{3\alpha-2}$, so that (we recall that α lies between $1/2$ and $2/3$) the self-similar limiting motion has zero momentum and infinite energy. Further, the self-similar solution is limiting to the solution of the original non-self-similar problem with finite p_0 and τ, and t tending to infinity. However, as has already been mentioned, the convergence to the limiting solution is non-uniform in the domain $-\infty < x \leq x_f$. The momentum of the region of compression $x_0(t) \leq x \leq x_f(t)$ grows infinitely with time. The momentum of the region of expansion $-\infty < x \leq x_0(t)$ has a negative sign and its absolute value also grows infinitely with time. Their algebraic sum, equal to $\beta p_0 \tau$, becomes ever smaller compared with the momentum of each of the regions mentioned; it is different from zero only because of the departure of the motion from self-similarity.

We now consider the energy $\mathcal{E}$ of any region $x_1(t) = \xi_1 A t^\alpha \leq x \leq x_f(t)$ in which the motion becomes close to self-similar starting from some instant of time:

$$\mathcal{E} = \int_{x_1}^{x_f} \rho \left[\frac{v^2}{2} + \frac{p}{(\gamma - 1)\rho} \right] dx$$

$$= \rho_0 A^3 t^{3\alpha - 2} \int_{\xi_1}^{1} R \left[\frac{V^2}{2} + \frac{P}{(\gamma - 1)R} \right] \xi^2 d\xi .$$

(4.52)

It is evident that the energy $\mathcal{E}$ tends to zero with increasing t, so that the contribution of the self-similar region to the bulk energy becomes ever less in time, and the basic contribution to the energy is determined by the motion close to the free boundary, where it always remains non-self-similar no matter how much time has passed since the start of the motion.

4.2.5 Explosion at a plane interface – transition from one self-similar intermediate asymptotics to another

The statement of the problem considered in this subsection differs from that for the impulsive loading problem only in that the half-space to the left of the impermeable wall contains not a vacuum but the same gas as the half-space to the right. Assume that the gas density ρ_1 in the left half-space, $x < 0$, is constant at the beginning and much less than the gas density on the right. The gas pressure is assumed to be negligibly small both in the left and right half-spaces.

Obviously the asymptotics for sufficiently large times of the solution to this problem should be the solution to the problem of an instantaneous concentrated explosion at the plane boundary of two half-spaces filled initially by the same gas but at different densities under zero pressure. The latter solution is a self-similar solution of the first kind, it was constructed by Nigmatulin (1965) as a combination of two symmetric 'plane' very intense explosion solutions, obtained by Sedov (1946, 1959). These solutions correspond to initial gas densities ρ_0 and ρ_1 and to certain energies E_1 and E_2 to be determined in the course of solution.

Indeed, let the initial energy of the explosion per unit area of the interface be equal to E. The dimension of the quantity E is equal to MT^{-2}. It is convenient in this case to write the self-similar solution in the form

$$v = v_{f2} f(\zeta, \gamma) , \quad \rho = \rho_{f2} g(\zeta, \gamma) , \quad p = p_{f2} h(\zeta, \gamma) , \quad \zeta = \frac{x}{x_{f2}} , \quad (4.53)$$

where v_{f2}, ρ_{f2}, p_{f2} are the velocity, density and pressure just behind the 'right-hand' shock wave $x = x_{f2}$, which is propagating in the high-density gas (in the positive x-direction). Furthermore, according to dimensional analysis,

$$x_{f2} = \xi_0 \left(\frac{Et^2}{\rho_0} \right)^{1/3} \qquad (4.54)$$

where ξ_0 is a constant that depends on γ only. For the 'left-hand' shock wave propagating in the low-density gas in the negative x-direction we obtain from the dimensional analysis

$$x_{f1} = \zeta_1 x_{f2}, \qquad (4.55)$$

so that the left-hand shock-wave front corresponds to a certain constant value $\zeta = \zeta_1 < 0$. From the conditions applying at a very intense shock wave front, going from the plane of explosion to the left or to the right we obtain respectively

$$p_{f1} = \frac{2}{\gamma + 1} \rho_1 D_1^2, \quad p_{f2} = \frac{2}{\gamma + 1} \rho_0 D_2^2, \qquad (4.56)$$

where D_1 and D_2 are the velocities of the propagation of the left-hand and right-hand shock waves:

$$D_1 = \frac{2}{3} \zeta_1 \xi_0 \left(\frac{E}{\rho_0} \right)^{1/3} \frac{1}{t^{1/3}} = \zeta_1 D_2;$$

$$D_2 = \frac{2}{3} \xi_0 \left(\frac{E}{\rho_0} \right)^{1/3} \frac{1}{t^{1/3}}. \qquad (4.57)$$

The motion established in each of the half-spaces $x \lessgtr 0$ corresponds to a symmetric very intense explosion with densities ρ_1 and ρ_0. However, for a symmetric explosion the pressure $p(0, t)$ at the plane $x = 0$ is a certain fixed part of the pressure at the front, depending on γ only and not on the time. This follows from dimensional analysis in an elementary way. Therefore, from pressure continuity at $x = 0$ it follows that $p_{f1} = p_{f2}$, and we obtain, using relations (4.56), (4.57),

$$\zeta_1 = -\sqrt{\frac{\rho_0}{\rho_1}}. \qquad (4.58)$$

A simple calculation shows that $E_2/E_1 = \sqrt{\rho_1/\rho_0}$. Therefore for a concentrated explosion in the limit $\rho_1/\rho_0 \to 0$ we obtain that the whole energy goes instantaneously into the vacuum.

It is evident, however, that the asymptotics that corresponds to a concentrated explosion at the boundary of the half-spaces does not settle down instantaneously. Indeed, we have assumed, as in subsection 4.2.1, that at the moment $t = \tau$ the whole energy is concentrated in a layer of finite width of the more dense gas. For small ρ_1/ρ_0, in the initial period when the velocity of propagation of the left-hand wave, travelling over less dense gas, is much larger than the characteristic velocity $\sqrt{p_0/\rho_0}$, it is natural to expect that the motion near to the right-hand wave front will resemble the gas motion after impulsive loading considered in the preceding subsections. It was shown above that this motion settles down to a self-similar asymptotics during a time interval of order τ after the beginning. This is demonstrated additionally by numerical solution of the non-self-similar problem with initial conditions

$$u = 0; \quad \rho = \rho_1 = 0.1; \quad p = 0 \quad (x < 0);$$
$$u = 0; \quad \rho = \rho_0 = 1; \quad p = 1 \quad (0 < x < 1); \qquad (4.59)$$
$$u = 0; \quad \rho = \rho_1 = 1; \quad p = 0 \quad (x > 0).$$

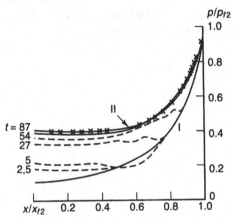

Figure 4.7. The solution to the problem of an explosion at an interface settles down to an intermediate asymptotics, at first that of an impulsive loading (I), and later that of a concentrated very intense explosion (II). The crosses correspond to $t = 100$.

The results of numerical computations are shown in Figure 4.7. There the pressure distribution in the reduced coordinates p/p_{f2}, $\zeta = x/x_{f2}$ is presented for various times t. The curves I and II correspond to the self-similar asymptotics for an impulsive loading and for a concentrated very intense explosion. It is seen that there exists a certain time interval $t_1 < t < t_2$ (for the example under calculation $t_1 = 2.5$, $t_2 = 40$) large enough for the motion to become self-similar with sufficient accuracy, and, at the same time, small enough for the influence of the finite gas density at the left-hand side to be negligible. In this time interval, for a range of substantial pressure variation near the right-hand wave the self-similar asymptotics of an impulsive loading is valid. At $t > t_3$ (for the example considered $t_3 = 100$) over the range of essential pressure variation the self-similar asymptotics of a very intense explosion is valid until the initial pressure of the undisturbed gas (cf. chapter 2) starts to become important.

The analysis of intermediate asymptotics for the problem of an explosion near an interface presented above, and the numerical experiments, were performed by O.S. Ryzhov and his associates (Vlasov, Derzhavina and Ryzhov, 1974). Parkhomenko, Popov and Ryzhov (1977a, b) performed an analysis of the corresponding problems having cylindrical and spherical symmetry. Instead of the impulsive loading solution, which does not exist for such symmetries, in these problems different intermediate asymptotics appear – self-similar solutions of the second kind corresponding to flows with sources or sinks of variable intensity at the axis or at the centre of symmetry.

5

Classification of similarity rules and self-similar solutions. A recipe for the application of similarity analysis

5.1 Complete and incomplete similarity

In chapters 2–4 several instructive and fundamentally different self-similar problems were considered. In the problem of an instantaneous concentrated heat source and the problems of the propagation of very intense thermal and blast waves the situation turned out to be relatively simple. Namely, for them there exists some completely schematized idealized statement of the problem (energy release at a point, initial temperature and pressure equal to zero). Considering this statement of the idealized problem and applying the procedures of dimensional analysis to it in the standard way, we can reveal the self-similarity of the solution, construct the self-similar variables, and obtain the solution in finite form owing to the existence of certain integrals.

Deeper consideration shows, however, that this simplicity is illusory and that, for example, in making the assumption of pointwise release of energy we have, as it is said, gone to the brink of an abyss. In fact, by changing the formulation of the problem, apparently only slightly, in such a way that it would seem that all similarity considerations must preserve their validity, we arrived at a contradiction; it turned out that in the modified problems the required solutions simply do not exist. More detailed analysis showed that in trying to find solutions of the modified problems by the same standard procedure, and starting from the formulation of an idealized problem, it turned out that the very statement of the question was improper. What we actually needed was not exact solutions of the simply formulated idealized, degenerate problem, corresponding to the instantaneous removal at a plane of a finite mass of fluid or the instantaneous release at a point of a finite amount of energy. We were interested, rather, in the asymptotics of solutions of the corresponding non-idealized problems, the existence and uniqueness of whose solutions are either strictly proved or evoke no doubt; the non-idealized problems naturally turned out to be non-self-similar. The passage to the limit as the supplementary parameter that made the problem non-self-similar tended to zero, led to a seemingly empty result: in one case the limit is equal to zero or infinity, depending on the other constraints in the problem; in the other case a limit simply does not exist. Nevertheless, in both cases meaningful intermediate asymptotics exist, and moreover they are self-similar. It was revealed that these asymptotics and not the limits are precisely what we actually need. It turns out that the removal of a fluid mass in the problem of filtration in an elasto-plastic porous medium and the release of energy by a very intense explosion with loss or deposition of energy at the front cannot be considered pointwise or concentrated at a plane. On the contrary, when reducing the size of the region within which the initial energy release or removal of fluid mass occurs we must, in order to obtain a proper asymptotics of the solution to the original non-idealized problem for large times, increase or decrease the amount of mass removed or energy released in such a way that a certain 'moment' of the initial distribution of mass or energy remains constant. It is an essential point that the power to which the length appears in the expression for this moment cannot be given in advance and that in principle it is impossible to determine it from dimensional considerations; it must be found in the course of determining the self-similar asymptotics.

Thus, we have encountered the existence of self-similar solutions of two types. It might seem that their difference is connected with the availability or absence in the problem under consideration of an integral conservation law that is valid also in the non-self-similar stage. However, the impulsive loading problem considered in chapter 4 shows that this is not so; the reason, rather, lies in the character of the transition from the non-self-similar solution to the self-similar asymptotics. It is especially clear for this example that the transition is non-uniform over the region and that this non-uniformity prevents our using conservation laws.

We now give a formal classification of similarity rules, and of self-similar solutions to problems of mathematical physics. We recall (see chapter 1) that, according to dimensional analysis, any relationship among dimensional quantities of the form

$$a = f(a_1, \ldots, a_k, b_1, \ldots, b_m) \tag{5.1}$$

can be represented in the form

$$\Pi = \Phi(\Pi_1, \ldots, \Pi_m) \tag{5.2}$$

where the dimensionless parameters Π, Π_1, ..., Π_m are defined by the relations

$$\Pi = \frac{a}{a_1^p \ldots a_k^r}, \quad \Pi_1 = \frac{b_1}{a_1^{p_1} \ldots a_k^{r_1}}, \quad \ldots, \quad \Pi_m = \frac{b_m}{a_1^{p_m} \ldots a_k^{r_m}}. \tag{5.3}$$

From (5.2) and (5.3) it follows that the function f has the property of generalized homogeneity:

$$f(a_1, \ldots, a_k, b_1, \ldots, b_m) = a_1^p \ldots a_k^r \Phi\left(\frac{b_1}{a_1^{p_1} \ldots a_k^{r_1}}, \ldots, \frac{b_m}{a_1^{p_m} \ldots a_k^{r_m}}\right) \tag{5.4}$$

It is assumed here that the governing parameters a_1, ..., a_k have independent dimensions and that the dimensions of the remaining governing parameters b_1, ..., b_m as well as the dimension of a can be expressed in terms of the dimensions of the parameters a_1, ..., a_k.

Consider now one of the governing parameters, for example, b_m. In traditional discussions 'on a physical level' such a parameter is considered as essential, i.e., as actually governing the phenomenon under consideration, if the value of the corresponding dimensionless governing parameter Π_m is not too large and not too small; let us say, to be specific, that it lies between 1/10 and 10.

If a dimensionless governing parameter, in this case Π_m, is small or large it is conventionally assumed that the influence of this parameter and of the corresponding dimensional parameter b_m can be neglected.

This argument is actually valid if there exists a finite non-zero limit of the function Φ in (5.2), as the parameter Π_m tends to zero or infinity with the other similarity parameters remaining constant, which is certainly not necessarily the case in general. In fact, even more is required: the function Φ must converge sufficiently rapidly to a limit as Π_m tends to zero or infinity, that for $\Pi_m < 0.1$ or $\Pi_m > 10$ it will assume values sufficiently close to that limit. If these conditions are actually satisfied, then for sufficiently small or sufficiently large Π_m the function Φ in (5.2) can, to the required accuracy, be replaced by a function of one less argument:

$$\Pi = \Phi_1 \underbrace{(\Pi_1, \ldots, \Pi_{m-1})}_{m-1 \text{ arguments}} \tag{5.5}$$

where $\Phi_1(\Pi_1, \ldots, \Pi_{m-1})$ is the limit of the function $\Phi(\Pi_1, \ldots, \Pi_m)$ as

Π_m tends to zero (or infinity). In such cases we speak of *complete similarity, or similarity of the first kind, of the phenomenon in the parameter* Π_m.

However, it is clear, and moreover trivial, that, generally speaking, if a similarity parameter Π_m tends to zero or infinity the function Φ does not need to tend to a limit, let alone a finite and non-zero one. Therefore, in general the parameter b_m could remain essential no matter how small or large the value of the corresponding dimensionless parameter Π_m may be.

Suppose, for example, that as $\Pi_m \to 0$ or $\Pi_m \to \infty$ the function Φ tends to zero or infinity. It is clear that in this case the quantity Π_m remains essential, no matter how small or large it becomes: replacing the function Φ in (5.2) by its limiting values as $\Pi_m \to 0$ or ∞ leads to the empty relation $\Pi = 0$ or $\Pi = \infty$. Hence, in this case it is in general impossible simply to delete Π_m from the governing dimensionless parameters and to replace the functions f and Φ in (5.1) and (5.2) by functions with one less argument; one does not have complete similarity in the parameter Π_m. Nevertheless, here too there exists an important exceptional situation where one can decrease the number of arguments and obtain a relation of the form (5.5). Namely, suppose at first that as Π_m tends to zero or infinity the function Φ has the power-type asymptotic representation

$$\Phi = \Pi_m^\alpha \Phi_1(\Pi_1, \ldots, \Pi_{m-1}) + o(\Pi_m^\alpha) \tag{5.6}$$

where α is a constant, the function Φ_1 again depends on $m-1$ arguments, and the second summand is arbitrarily small compared with the first for sufficiently small (or large) Π_m. In this case for sufficiently small or sufficiently large Π_m we have with the required accuracy, a relation of the form (5.5),

$$\Pi^* = \Phi_1(\Pi_1, \ldots, \Pi_{m-1})$$

where

$$\Pi^* = \frac{\Pi}{\Pi_m^\alpha} = \frac{a}{a_1^{p-\alpha p_m} \ldots a_k^{r-\alpha r_m} b_m^\alpha} . \tag{5.7}$$

Thus in this case the relation (5.1) can again be written in terms of a function of a lesser number of arguments, just as in the case of complete similarity. However, firstly, the number α and consequently the form of the dimensionless parameter Π^* can no longer be obtained from considerations of dimensional analysis alone, and, secondly, the argument b_m appears in it and hence does not cease to influence the phenomenon.

However, the power-type asymptotics of the type (5.6) is a special case only. A more general case of a power-type asymptotics is

$$\Phi = \Pi_m^{\alpha_m} \Phi_1 \left(\frac{\Pi_1}{\Pi_m^{\alpha_1}}, \dots, \frac{\Pi_{m-1}}{\Pi_m^{\alpha_{m-1}}} \right) + o\left(\Pi_m^{\alpha_m}\right) \qquad (5.8)$$

where α_1, α_2, ..., α_m are certain constants (naturally, some of them could be equal to zero) and the last summand is arbitrarily small compared with the first for sufficiently small (or large) Π_m. In this case, for sufficiently small (or large) Π_m we arrive again at a relation of the form (5.5),

$$\Pi^* = \Phi_1 \left(\Pi_1^*, \dots, \Pi_{m-1}^* \right) \qquad (5.9)$$

where as before Π^* is defined by the relation (5.7) (with α_m instead of α), and the Π_i^* are defined by the relations

$$\Pi_i^* = \frac{b_i}{a_1^{p_i - \alpha_i p_m} \dots a_k^{r_i - \alpha_i r_m} b_m^{\alpha_i}} . \qquad (5.10)$$

It is evident that in this case, again as in the case of complete similarity, the basic relation (5.1) can be described by a function of a lesser number of arguments; now, however, the forms of the dimensionless parameters Π^*, Π_1^*, ..., Π_{m-1}^* can no longer be obtained from considerations of dimensional analysis alone because these considerations cannot in principle give the quantities α_1, ..., α_m. Moreover, the argument b_m appears in Π^* and along with b_1, ..., b_{m-1}, in $\Pi_1^*, \dots, \Pi_{m-1}^*$, so these parameters b_i do not cease to be essential. The most general exceptional case for power-type asymptotics occurs when several parameters Π_{l+1}, ..., Π_m are small or large, and the function Φ has for small or large Π_{l+1}, ..., Π_m the same property of generalized homogeneity:

$$\Phi = \Pi_{l+1}^{\alpha_{l+1}} \dots \Pi_m^{\alpha_m} \Phi_1 \left(\frac{\Pi_1}{\Pi_{l+1}^{\beta_1} \dots \Pi_m^{\delta_1}}, \dots, \frac{\Pi_l}{\Pi_{l+1}^{\beta_l} \dots \Pi_m^{\delta_l}} \right), \qquad (5.11)$$

which is of exactly the same form as the function f in the relation (5.4). There is, however, one essential difference: the generalized homogeneity of the function f in (5.4) followed from the general physical covariance principle, and the constants p, ..., r_m were obtained by simple rules of dimensional analysis, whereas the generalized homogeneity of the function Φ in (5.11) is a special property of the problem under consideration, and the numbers $\alpha_{l+1}, \dots \delta_l$ cannot be obtained from general principles. Thus in such exceptional cases, despite the fact that there is no complete similarity in the similarity parameters and that all governing parameters b_1, ..., b_m remain essential no matter how small or large the similarity parameters, there is a decrease in the number of arguments of the function Φ that defines the relationship in which we are interested, and we get a relation of the form (5.5) just as in the case of complete similarity. We speak in such cases of *incomplete similarity* or *similarity of the second kind* of a phenomenon in the relevant parameter (or parameters).

The conclusion at which we have arrived is entirely natural: if the values of certain dimensionless parameters Π_i are small or large, then there are three possibilities:

(1) The limits of the corresponding functions Φ as the Π_i tend to zero or infinity exist and are finite and non-zero. The corresponding governing parameters, the dimensional b_i or dimensionless Π_i, can be excluded from consideration, and the number of arguments of the functions Φ decreases. All the similarity parameters can be determined by means of the regular procedures of dimensional analysis. This case corresponds to complete similarity of the phenomenon in the similarity parameters Π_i.

(2) No finite limits exist for the functions Φ as the Π_i tend to zero or infinity, but one of the exceptional cases indicated above holds[†]. If so, the number of arguments of the functions Φ can be decreased, but not all the parameters Π, Π_i can be obtained from dimensional analysis; and the governing parameters b_i remain essential no matter how small (or large) the corresponding similarity parameters. This case corresponds to incomplete similarity in the parameters Π_i.

(3) No finite limits exist for the functions Φ as the $\Pi_i \to 0$ or ∞, and the indicated exceptions do not hold. This case corresponds to lack of similarity of the phenomenon in the parameters Π_i. It has already been remarked that no matter how large (or small) the values of the parameters Π_i, in this case we cannot obtain a relation of the form (5.5) between power-type combinations of the governing and determined parameters that has a smaller number of arguments for the functions Φ.

[†] The question can arise of why one regards as exceptional only asymptotic representations of the power-type forms (5.6), (5.8) and (5.11); is it impossible to factorize the function Φ by another function of Π_i, for example, $\log \Pi_i$? In fact, in this case one no longer gets relations among power-type combinations of dimensional parameters, yet only products of their powers give upon multiplication power-type combinations of the same form. As was proved in chapter 1, dimensions are always power-type monomials. It can be obtained, by exactly the same argument, that a power-type asymptotics follows from the lack of characteristic distinguished values of the parameters $b_{l+1}, \ldots, b_m$.

The difficulty is that *a priori*, until we obtain a non-self-similar so-lution of the complete non-idealized problem[‡], we do not know with which of these three cases we are dealing, irrespective of whether or not we have an explicit mathematical formulation of the problem. Hence one can only recommend assuming in succession each of these possible situations for small (or large) similarity parameters – complete similar-ity, incomplete similarity, lack of similarity – and then comparing the relations obtained under each assumption with data from numerical cal-culations, experiments, or the results of analytic investigations.

5.1.1 Self-similar solutions of the first and second kind

We now consider some problem in mathematical physics that describes a certain phenomenon; let the quantity a be an unknown in this problem and let the quantities $a_1, \ldots, a_k, b_1, \ldots, b_m$ be the independent variables and parameters appearing in the equations and in the boundary, initial, and other conditions that determine solutions.

Self-similar solutions are always solutions of idealized (degenerate) problems, which are obtained if certain parameters b_i and the dimen-sionless parameters Π_i corresponding to them assume zero or infinite values. They are simultaneously exact solutions of degenerate problems and asymptotic (generally intermediate-asymptotic) representations of solutions of wider classes of non-idealized non-self-similar problems as the parameters b_i tend to zero or infinity.

It is clear that if an asymptotics is self-similar, and if the self-similar variables are power-law monomials, then one of the first two cases listed the last subsection must hold; correspondingly, self-similar solutions are divided into solutions of the first and second kind.

Self-similar solutions of the first kind are obtained when in the passage to the limit from the non-self-similar non-idealized problem to the self-similar idealized problem there is complete similarity in the parameters that make the problem non-idealized and its solution non-self-similar. Expressions for all the self-similar variables, independent as well as de-pendent, can be obtained here by applying dimensional analysis.

Self-similar solutions of the second kind are obtained in the case where the idealization of the original problem is such that there is incomplete similarity in the similarity parameters. Then expressions for the self-

---

[‡] If the complete solution of the problem were known, there would be no need to apply the methods of dimensional analysis.

similar variables cannot in general be obtained from dimensional considerations. The parameters that make the problem non-idealized, and its solution non-self-similar, remain in the expressions for the self-similar variables.

In the direct construction of self-similar solutions of the second kind, determination of the exponents of the self-similar variables leads to a nonlinear eigenvalue problem. The constant multiplier appearing in the self-similar variables is left undetermined in the direct construction of self-similar solutions of the second kind. This constant can be found by following, for example by means of numerical calculations, the entire process of evolution of a solution of the non-idealized problem into a self-similar asymptotics.

If the constant can be found from integral conservation laws, this means that for an appropriate choice of governing parameters the problem can be reformulated and reduced to a problem of the first kind. For example, the classical problems of a heat source, a thermal dipole, and a very intense explosion can be represented as self-similar solutions of the second kind if one chooses the governing parameters of the non-idealized pre-self-similar problem inappropriately. The possibility of obtaining solutions to these problems as self-similar solutions of the first kind is connected with the choice, as governing parameters, of the energy of the explosion, the bulk heat, or the dipole moment, which by virtue of the corresponding integral conservation laws do not vary with time[†].

Self-similar solutions of non-power-type are also possible. For such solutions the self-similar variables are no longer represented by power monomials. Instructive examples of such solutions will be considered in chapter 6, where a general idea of their origin will also be given.

[†] On the contrary, an integral conservation law valid also for the pre-self-similar stage is not a neccessary property of the self-similar solution of the first kind. An elegant example illustrating this point was given by Entov (1994). The self-similar asymptotics to the solution to the equation of heat conduction with absorption,

$$\partial_t \theta = \kappa \partial_{xx}^2 \theta - \alpha \theta^n ,$$

where $n > 3$ is a constant, describing the decay of an amount of heat concentrated initially in a finite domain, is, as previously, the point-source self-similar solution of the first kind

$$\theta = \frac{Q}{\sqrt{\kappa t}} e^{-x^2/4\kappa t} .$$

For this distribution obviously the integral $M(t) = \int_{-\infty}^{\infty} \theta(x,t)dx = M_0$ is preserved in time. However, it is not preserved at the pre-self-similar stage, so that M_0 cannot be obtained from initial data.

Examples of self-similar solutions of the first kind are given by the solutions considered in chapter 2 to the problem of an instantaneous heat source and to the problems of the propagation of very intense thermal waves and of very intense blast waves. In fact, we shall return first to the solution of the instantaneous heat source problem, which is the self-similar solution of the equation

$$\partial_t \theta = \kappa \partial_{xx}^2 \theta \qquad (5.12)$$

under the conditions

$$\theta(x,0) \equiv 0, \quad x \neq 0; \quad \int_{-\infty}^{\infty} \theta(x,0)dx \doteq Q; \quad \theta(\pm\infty, t) \equiv 0. \qquad (5.13)$$

If we pass from the idealized degenerate conditions (5.13), corresponding to the concentration at the initial instant of a finite amount of heat at a plane, to the conditions

$$\theta(x,0) = \frac{Q}{l} u_0 \left(\frac{x}{l}\right), \quad \int_{-\infty}^{\infty} u_0(\zeta)d\zeta = 1; \quad \theta(\pm\infty, t) \equiv 0 \qquad (5.14)$$

($u_0(\zeta)$ being an even function that decreases rapidly and monotonically with increasing absolute value of the argument) corresponding to the concentration at the initial instant of the same amount of heat in a region of finite size l, then the solution θ will cease to be self-similar. Dimensional analysis gives

$$\theta = f(t, \kappa, Q, x, l), \quad \Pi = \Phi(\Pi_1, \Pi_2), \qquad (5.15)$$

where

$$\Pi_1 = x/\sqrt{\kappa t}, \quad \Pi_2 = l/\sqrt{\kappa t}, \quad \Pi = \theta\sqrt{\kappa t}/Q.$$

For Π_2 tending to zero, i.e., t tending to infinity or, what is the same thing, the region of initial concentration of heat contracting to a plane, the function Φ converges to a finite limit. In fact, by reducing to dimensionless form relations known from the theory of heat conduction (see Carslaw and Jaeger, 1960; Petrovskii, 1967; Carrier and Pearson, 1976) it is easy to prove that

$$\Phi = \frac{1}{2\sqrt{\pi}} \int_{-\infty}^{\infty} u_0(\zeta) e^{-(\Pi_1 - \Pi_2\zeta)^2/4} d\zeta, \quad \zeta = \frac{x}{l}.$$

As $\Pi_2 \to 0$ the function Φ converges to

$$\left(\frac{1}{2\sqrt{\pi}}\right) e^{-\Pi_1^2/4} = \Phi(\Pi_1, 0).$$

Hence the self-similarity is complete with respect to the parameter Π_2

that makes the problem non-idealized and for sufficiently small Π_2 one can, with any accuracy, replace the function Φ in (5.15) by $\Phi(\Pi_1, 0) = \Phi_1(\Pi_1)$. But the function Φ_1 corresponds to the solution of an idealized, degenerate problem, which is already self-similar.

Complete self-similarity makes it possible to obtain expressions not only for the self-similar variables, as was demonstrated in chapter 2, but also for the meaningful similarity rules. Suppose we want to determine the decay law for the maximum temperature. The latter is obviously achieved at $x = 0$, so that for the idealized problem (in which the heat at the initial instant is concentrated at the plane $x = 0$) we find

$$\theta_{\max} = f(t, \kappa, Q), \quad \Pi_{\max} = \frac{\theta_{\max}\sqrt{\kappa t}}{Q} = \text{const},$$

whence

$$\theta_{\max} = \text{const} \, \frac{Q}{\sqrt{\kappa t}}. \tag{5.16}$$

Such an argument is valid in the present case because, for the non-idealized problem,

$$\theta_{\max} = f(t, \kappa, Q, l), \quad \Pi_{\max} = \Phi_{\max}(\Pi_2) = \Phi(0, \Pi_2), \tag{5.17}$$

and as $\Pi_2 \to 0$ the function $\Phi_{\max}$ converges to a finite limit equal to $1/2\pi^{1/2}$.

The situation is entirely analogous for both very intense thermal waves and very intense explosion waves. Thus, in the case of very intense explosion waves, passing from the idealized formulation of the problem, corresponding to the release of energy at a point, to the non-idealized formulation of the problem, corresponding to the release of energy in a sphere of finite radius r_0, we obtain for the pressure, density and radius of the shock wave,

$$p = \left(\frac{E^2\rho_0^3}{t^6}\right)^{1/5} \Phi_p(\Pi_1, \Pi_2), \quad \Pi_1 = r\left(\frac{Et^2}{\rho_0}\right)^{-1/5},$$

$$\Pi_2 = r_0\left(\frac{Et^2}{\rho_0}\right)^{-1/5}, \quad \rho = \rho_0\Phi_\rho(\Pi_1, \Pi_2), \tag{5.18}$$

$$v = \left(\frac{Et^{-3}}{\rho_0}\right)^{1/5} \Phi_v(\Pi_1, \Pi_2), r_f = \left(\frac{Et^2}{\rho_0}\right)^{1/5} \Phi_f(\Pi_2).$$

For Π_2 tending to zero, i.e. the region of initial release of energy contracting to a point, all the functions Φ_p, Φ_p, Φ_r, Φ_f tend to finite non-zero limits. (This fact has not been proved analytically, but it is verified by numerical calculations and there is no reason to doubt it.) Thus similarity is complete with respect to the parameter Π_2 that makes the problem non-idealized, and for sufficiently small Π_2 we can, to any required accuracy, replace the functions Φ_p, Φ_ρ, Φ_v in (5.18) by

$$\Phi_p(\Pi_1, 0) = \Phi_{p_1}(\Pi_1), \quad \Phi_\rho(\Pi_1, 0) = \Phi_{\rho_1}(\Pi_1), \quad \Phi_v(\Pi_1, 0) = \Phi_{v_1}(\Pi_1),$$

respectively, and the function $\Phi_f(\Pi_2)$ by the constant $C = \Phi_f(0)$. But the functions Φ_{ρ_1}, Φ_{p_1}, Φ_{v_1} and the constant of proportionality C in the formula for the radius of the shock wave correspond to the self-similar solution of the idealized problem (see chapter 2).

Complete self-similarity, in this case also, allows one to obtain meaningful similarity rules, for example for the properties of the motion at the shock front:

$$p_f = \text{const}_1 \left(\frac{E^2 \rho_0^3}{t^6} \right)^{1/5},$$

$$v_f = \text{const}_2 \left(\frac{Et^{-3}}{\rho_0} \right)^{1/5}, \qquad (5.19)$$

$$r_f = \text{const}_3 \left(\frac{Et^2}{\rho_0} \right)^{1/5}.$$

Examples of self-similar solutions of the second kind are the solutions for the modified problems of an instantaneous heat source and of a very intense explosion, and the problem of an impulsive loading, that were considered in chapters 3 and 4.

The solution for the modified instantaneous heat source problem is a self-similar asymptotic solution for large t of the problem with the same initial conditions, (5.14), but with the modified equation

$$\partial_t \theta = \kappa \partial_{xx}^2 \theta \quad (\partial_t \theta \geq 0),$$
$$\partial_t \theta = \kappa_1 \partial_{xx}^2 \theta \quad (\partial_t \theta \leq 0), \qquad (5.20)$$

which is a nonlinear heat conduction equation for $\kappa_1 \neq \kappa$. For this solution dimensional analysis gives

$$\theta = f(t, Q, l, \kappa, \kappa_1, x), \quad \Pi = \Phi(\Pi_1, \Pi_2, \Pi_3), \qquad (5.21)$$

where

$$\Pi_1 = x/\sqrt{\kappa t}, \quad \Pi_2 = l/\sqrt{\kappa t}, \quad \Pi_3 = \kappa_1/\kappa, \quad \Pi = \theta \sqrt{\kappa t}/Q.$$

The general case with $\kappa_1 \neq \kappa$ and $\Pi_3 \neq 1$ differs from the classical case with $\kappa_1 = \kappa$ and $\Pi_3 = 1$ in that for $\Pi_3 \neq 1$ and $\Pi_2 \to 0$ the function Φ in (5.21) no longer tends to a finite non-zero limit, but converges to zero or infinity, depending on whether the similarity parameter Π_3 is larger or smaller than unity. Here we have the first type of incomplete similarity with respect to a parameter: as $\Pi_2 \to 0$,

$$\Phi \to \Pi_2^\alpha \Phi_1(\Pi_1, \Pi_3),$$

where α is some number depending on Π_3 and equal to zero for $\Pi_3 = 1$, positive for $\Pi_3 > 1$, and negative for $\Pi_3 < 1$.

In accordance with what was said above, the relation (5.21) can be written for small Π_2 in the self-similar form

$$\Pi^* = \Phi_1(\Pi_1, \Pi_3), \quad \Pi^* = \frac{\theta(\kappa t)^{(1+\alpha)/2}}{Q l^\alpha}, \qquad (5.22)$$

but now the dependent self-similar variable Π^* can no longer be found by dimensional considerations; the constant α is found by solving a nonlinear eigenvalue problem, and the full solution is found to within a multiplicative constant. Moreover the length l that makes the original problem non-self-similar appears explicitly in this self-similar variable. From (5.22) we get in particular the decay law for the maximum of the quantity θ:

$$\Pi^*_{max} = \Phi_1(0, \Pi_3) = F(\Pi_3); \quad \theta_{max} = \frac{Q l^\alpha}{(\kappa t)^{(1+\alpha)/2}} F(\Pi_3). \qquad (5.23)$$

It is an essential point that although this law also is of power form, it is impossible to obtain it by applying dimensional analysis. The situation is that the decay law for the quantity θ_{max} is determined by the dimension of the constant $Q l^\alpha$. This is unknown in advance, and is determined after the construction of a global self-similar solution, i.e., from the solution of a nonlinear eigenvalue problem. However, since we are dealing in the present case with incomplete similarity of the first type, the independent self-similar variable, in this case Π_1, is found from dimensional analysis. Hence, in particular, for the law of propagation of the boundary between the regions with different κ, one obtains the similarity law

$$x_0(t) = \xi_0(\Pi_3)\sqrt{\kappa t}, \qquad (5.24)$$

which can be established from dimensional considerations.

We turn to the modified very intense blast wave problem. The desired self-similar solution is a self-similar asymptotics for large t of the solution to the equations of adiabatic motion of a gas with adiabatic exponent γ under conditions on the very intense shock wave in which the effective adiabatic exponent $\gamma_1 \neq \gamma$ enters, and initial conditions corresponding to the release at the initial instant of energy E within a sphere of radius r_0.

The application of dimensional analysis to this non-idealized problem again leads us to (5.18) where it is understood that among the arguments of the functions $\Phi_p, \Phi_\rho, \Phi_v$ and Φ_f, the constant parameters $\Pi_3 = \gamma$ and $\Pi_4 = \gamma_1$ also appear.

However, in the case $\gamma_1 \neq \gamma$ there is no complete similarity, as $\Pi_2 \to 0$

in the similarity parameter Π_2 that makes the problem non-idealized; the functions $\Phi_p, \Phi_\rho, \Phi_v,$ and Φ_f do not tend to finite nonzero limits as $\Pi_2 \to 0$; they converge to zero or infinity as depending on whether γ_1 is smaller or larger than γ, and the function Φ_ρ in general does not converge to any limit. Actually, we have here incomplete similarity of the second type: as $\Pi_1 \to 0$ and $\Pi_2 \to 0$

$$\Phi_p = \Pi_2^{2\beta} \Phi_{p2}\left(\frac{\Pi_1}{\Pi_2^\beta}\right), \quad \Phi_\rho = \Phi_{\rho2}\left(\frac{\Pi_1}{\Pi_2^\beta}\right),$$

$$\Phi_v = \Pi_2^\beta \Phi_{v2}\left(\frac{\Pi_1}{\Pi_2^\beta}\right), \quad \Phi_f = \text{const}\, \Pi_2^\beta. \tag{5.25}$$

Hence, in accordance with what was said above, for small Π_1 and Π_2, (5.18) can again be written in the self-similar form

$$\Pi_p^* = \Phi_{p2}(\Pi_1^*), \quad \Pi_\rho^* = \Phi_{\rho2}(\Pi_1^*), \quad \Pi_v^* = \Phi_{v2}(\Pi_1^*), \quad \Pi_f^* = \text{const}, \tag{5.26}$$

where

$$\Pi_p^* = \frac{p}{(E^2\rho_0^3/t^6)^{1/5}\Pi_2^{2\beta}} = \frac{p}{\left[E^{2(1-\beta)}\rho_0^{(3-2\beta)}/t^{2(3+2\beta)}\right]^{1/5} r_0^{2\beta}},$$

$$\Pi_\rho^* = \frac{\rho}{\rho_0},$$

$$\Pi_v^* = \frac{v}{(Et^{-3}/\rho_0)^{1/5}\Pi_2^\beta} = \frac{v}{\left[E^{(1-\beta)}/\rho_0^{(1-\beta)}t^{3+2\beta}\right]^{1/5} r_0^\beta}, \tag{5.27}$$

$$\Pi_f^* = \frac{r_f}{(Et^2/\rho_0)^{1/5}\Pi_2^\beta} = \frac{r_f}{(Et^2/\rho_0)^{(1-\beta)/5} r_0^\beta},$$

$$\Pi_1^* = \frac{\Pi_1}{\Pi_2^\beta} = \frac{r}{(Et^2/\rho_0)^{(1-\beta)/5} r_0^\beta}.$$

However, now neither the dependent self-similar variables Π_ρ^*, Π_v^*, nor the independent self-similar variable Π_1^* can be determined from dimensional considerations, since the constant β is unknown in advance and is found by solving a nonlinear eigenvalue problem. Moreover, the radius r_0 of the sphere within which the release of energy takes place at the initial moment appears explicitly in all the self-similar variables. From (5.27) we get scaling laws for the pressure and velocity at the front of the shock wave and for the radius of the shock wave:

$$p_f = \text{const}_1 \left[\frac{E^{2(1-\beta)} \rho_0^{(3-2\beta)}}{t^{2(3+2\beta)}} \right]^{1/5} r_0^{2\beta},$$

$$v_f = \text{const}_2 \left[\frac{E^{(1-\beta)} t^{-(3+2\beta)}}{\rho_0^{(1-\beta)}} \right]^{1/5} r_0^{\beta}, \qquad (5.28)$$

$$r_f = \text{const}_3 \left(\frac{Et^2}{\rho_0} \right)^{(1-\beta)/5} r_0^{\beta},$$

As is evident, despite the fact that these laws have a scaling, power-law form, it is impossible to derive them by dimensional analysis. The situation is that the dimensions of the constant $A = \sigma E r_0^{5\beta/(1-\beta)}$ that determines these laws can be found only after solving a nonlinear eigen-value problem, to which the construction of a global self-similar solution reduces.

The examples considered are instructive ones. When we turn to the solution of a certain problem, and in particular to the search for its self-similar solutions, we do not know in advance to which type the solu-tions of the idealized formulation of the problems belong. Comparison of the original and modified formulations of the problems considered above shows that the situation can be rather deceiving: from the point of view of whether it is possible to apply dimensional analysis these problems do not differ from one another superficially. Hence, for example, it is ex-tremely tempting to begin by obtaining similarity laws without recourse to solution of the equations. Arguing in the usual fashion, we might assume for the modified problems that, since the initial removal of mass or release of energy occurs in a small region, the size of that region is inessential, i.e., we might assume complete similarity in the similarity parameter that corresponds to the initial length. From this would follow automatically the scaling laws (5.16) and (5.19) corresponding to com-plete similarity. But as a matter of fact, as we have seen, the scaling laws here are quite different. Therefore it is necessary to keep in mind that it is a very strong hypothesis to assume the unimportance of cer-tain parameters that make the problem non-idealized (in the examples considered, the lengths l and r_0). These governing parameters may be essential and yet self-similarity may nevertheless hold. Distinguishing between possible cases of self-similarity requires, in fact, a sufficiently deep mathematical investigation, which is unattainable in seriously non-linear problems. Therefore in obtaining self-similar solutions or similar-ity laws on the basis of dimensional analysis one should take care to verify, if only by means of numerical calculations, that the solutions or similarity laws found actually reflect the required asymptotic behaviour of the analytical solutions to the problem considered. The situation is much more complicated if a mathematical formulation of the problem is lacking; in this case, to verify the basic assumptions one must turn to experiment. The examples considered in the following chapters confirm the necessity of such precautions.

5.1.2 Recipe for similarity analysis

We have discussed above the fundamentals of dimensional analysis, similarity theory and the theory of self-similar phenomena, and have considered numerous illustrative examples. This discussion now allows us to provide a recipe for similarity analysis, i.e., for applying dimensional analysis and dealing with self-similarities. Suppose that we are interested in a property a of some phenomenon (a may be a vector, i.e. there may be several such properties). We proceed in the following way.

(1) *We specify a system of governing parameters $a_1, \ldots, a_k, b_1, \ldots, b_m$ such that a relation of the form*

$$a = f(a_1, \ldots, a_k, b_1, \ldots, b_m)$$

can be assumed to hold. If the problem has an explicit mathematical formulation, the independent variables in the problem and the constant parameters that appear in the equations, boundary conditions and initial conditions, etc., are adopted as the governing parameters. If an explicit mathematical formulation of the problem is unknown, the governing parameters are chosen on the basis of a qualitative model of the phenomenon, which each investigator constructs using his/her own experience and intuition, as well as an analysis of previous studies.

(2) *We choose an appropriate class of systems of units and determine the dimensions of the quanitites under investigation, and of the governing parameters, in this class.* We choose a system of governing parameters with independent dimensions: it is preferable to select those parameters whose importance to the phenomenon being studied is most firmly established.

(3) *We express the dimensions of the quantities under investigation, and of the governing parameters with dependent dimensions as products of powers of the dimensions of the governing parameters with independent dimensions.* We determine the similarity param-

eters and put the function under study into a dimensionless form, the similarity law

$$\Pi = \Phi(\Pi_1, \dots, \Pi_m).$$

(4)   *We estimate the numerical values of the governing similarity parameters.* We select those that are large or small. In some cases, it turns out to be convenient at this stage to choose new similarity parameters – products of powers of the similarity parameters obtained in the previous step: this sometimes makes it easier to perform these estimates.

(5)   *We try to formulate limiting similarity laws under the assumption of complete similarity in any large (or small) similarity parameters.* This means simply discarding these dimensionless governing parameters and the corresponding dimensional parameters. Then, we compare the limiting similarity laws obtained against the available experimental data and/or numerical calculations. If discrepancies are observed,

(6)   *We try to formulate limiting similarity laws under the assumption of incomplete similarity in the large (or small) similarity parameters.* This means that we assume a generalized-homogeneity representation of the function $\Phi(\Pi_1, \dots, \Pi_m)$ in terms of the small (or large) similarity parameters. Once again, we compare the similarity laws obtained against the available experimental data, numerical calculations, etc. If discrepancies are again observed, we then conclude that the phenomenon is not self-similar in the small (or large) similarity parameters. So,

(7)   *Finally, we formulate similarity laws using as few similarity parameters as possible.*

The use of this recipe will be demonstrated below on numerous examples from various fields.

Scaling and transformation groups. Renormalization group

6.1 Dimensional analysis and transformation groups

6.1.1 General concepts

Dimensional analysis, as already mentioned, has a transparent group-theoretical nature. Group considerations can turn out to be useful also in those cases where dimensional analysis becomes insufficient to establish the self-similarity of a solution and determine self-similar variables. A special place belongs here to the renormalization group, a concept now very popular in theoretical physics.

We recall, first of all, the definition of a transformation group. Suppose we have a set of transformations with k parameters,

$$x'_\nu = f_\nu(x_1, \ldots, x_n; A_1, \ldots, A_k) \quad (\nu = 1, \ldots, n) \qquad (6.1)$$

where the f_ν are smooth functions of their variables in some domain. We say that this set forms a k-parameter group of transformations if the following conditions are satisfied:

(1)   Among the transformations (6.1) there exists the identity transformation.

(2)   For each transformation of (6.1) there exists an inverse transformation that also belongs to the set (6.1).

(3)   For each pair of transformations of the set (6.1), i.e., a transformation A with parameters $A_1, \ldots, A_k$ and a transformation B with parameters $B_1, \ldots, B_k$, a transformation C with parameters $C_1, \ldots, C_k$, which also belongs to the set (6.1), exists, and is uniquely determined such that successive realization of the transfor-

mations A and B is equivalent to the transformation C. The transformation C is called the *product* of transformations A and B.

Dimensional analysis, which was considered in detail in chapter 1 is based on the Π-theorem. This theorem allows one to express a function of $n = k + m$ variables in a relationship between dimensional quantities,

$$a = f(a_1, \ldots, a_k, b_1, \ldots, b_m), \qquad (6.2)$$

in terms of a function of m variables (k being the number of governing parameters with independent dimensions) that represents the relationship (6.2) in the form of a relation among dimensionless quantities:

$$\Pi = \Phi(\Pi_1, \ldots, \Pi_m),$$

where

$$\Pi = \frac{a}{a_1^p \ldots a_k^r}, \quad \Pi_1 = \frac{b_1}{a_1^{p_1} \ldots a_k^{r_1}}, \quad \ldots, \quad \Pi_m = \frac{b_m}{a_1^{p_m} \ldots a_k^{r_m}},$$

so that the function f in (6.2) possesses the property of generalized homogeneity:

$$f(a_1, \ldots, a_k, b_1, \ldots, b_m) = a_1^p \ldots a_k^r \Phi\left(\frac{b_1}{a_1^{p_1} \ldots a_k^{r_1}}, \ldots, \frac{b_m}{a_1^{p_m} \ldots a_k^{r_m}}\right).$$

We note now that for any positive $A_1, \ldots, A_k$ the similarity transformation of the governing parameters with independent dimensions

$$a_1' = A_1 a_1, \quad a_2' = A_2 a_2, \quad \ldots, \quad a_k' = A_k a_k, \qquad (6.3)$$

can be obtained by changing from the original system of units of measurement to some other system of units of measurement belonging to the same class of systems of units. At the same time the values of the remaining parameters $a, b_1, \ldots, b_m$ vary in accordance with their dimensions in the following way:

$$\begin{aligned} a' &= A_1^p \ldots A_k^r a, \\ b_1' &= A_1^{p_1} \ldots A_k^{r_1} b_1, \\ &\qquad \cdots\cdots\cdots\cdots\cdots \\ b_m' &= A_1^{p_m} \ldots A_k^{r_m} b_m. \end{aligned} \qquad (6.4)$$

Direct verification shows easily that the transformations (6.3), (6.4) form a k-parameter group. The quantities $\Pi, \Pi_1, \ldots, \Pi_m$ remain unchanged for all transformations of the group (6.3), (6.4), i.e., they are *invariants* of this group. Thus, the Π-theorem is a simple consequence of the covariance principle: relations with a physical meaning among dimensional quantities of the form (6.2) are invariant with respect to the group of similarity transformations of the governing parameters with independent dimensions (6.3), (6.4), corresponding to transition to a different system of units (within a given class). The number of independent invariants of the group is obviously less than the total number of governing parameters by the number k of parameters of the group.

The invariance of the statement, and hence of the solution, of any physically meaningful problem with respect to the group of transformations (6.3), (6.4) is thus necessary according to the general physical covariance principle. It can turn out, however, that there exists a richer group with respect to which the formulation of the problem considered is invariant. Then the number of arguments of the function Φ in the universal (invariant) relation obtained after applying the Π-theorem in its own right should be reduced by the number of parameters of the supplementary group. Here the solution can turn out to be self-similar and the self-similar variables can be determined as a result of using the invariance with respect to the supplementary group, although this self-similarity is not implied by dimensional analysis (which exploits invariance with respect to the group of similarity transformations of quantities with independent dimensions). We consider below some instructive examples that will clarify this idea.

6.1.2 Example: the boundary layer on a flat plate

The boundary-layer problem for high-Reynolds-number flow past a semi-infinite thin plate placed along a uniform stream (the Prandtl–Blasius problem) leads to the system of equations (Kochin, Kibel' and Roze, 1964; Batchelor, 1967; Germain, 1986b; Landau and Lifschitz, 1987; Schlichting, 1968)

$$u\partial_x u + v\partial_y u = \nu\partial^2_{yy} u, \quad \partial_x u + \partial_y v = 0 \qquad (6.5)$$

(x and y being the longitudinal and transverse coordinates, u and v the corresponding velocity components, and ν the kinematic coefficient of viscosity) under the boundary conditions ($x > 0$, $y > 0$)

$$u(0, y) = U, \quad u(x, \infty) = U, \quad u(x, 0) = v(x, 0) = 0 \qquad (6.6)$$

(U being the constant speed of the exterior flow). By comparison, at arbitrary Reynolds number the problem of viscous flow past a semi-infinite thin plate placed along a uniform stream leads to the full Navier–Stokes equations and equation of continuity:

$$u\partial_x u + v\partial_y u = -\frac{1}{\rho}\partial_x p + \nu(\partial^2_{xx} u + \partial^2_{yy} u),$$

$$u\partial_x v + v\partial_y v = -\frac{1}{\rho}\partial_y p + \nu(\partial^2_{xx} v + \partial^2_{yy} v).$$

$$\partial_x u + \partial_y u = 0,$$

under the conditions $(-\infty < x < \infty, \ -\infty < y < \infty)$,

$$u(x,0) = v(x,0) = 0 \quad (x \geq 0),$$

$u(x,y) \to U$, $v(x,y) \to 0$, for any x with $y^2 \to \infty$, and for any y with $x \to -\infty$. Among the governing parameters ν, x, U and y in the boundary-layer problem, only two have independent dimensions: $[\nu] = L^2 T^{-1}, [x] = L, [U] = LT^{-1}, [y] = L$. Hence a direct application of dimensional analysis gives

$$u = U\Phi_1(\Pi_1, \Pi_2), \quad v = U\Phi_2(\Pi_1, \Pi_2),$$
$$\Pi_1 = \xi = Ux/\nu, \quad \Pi_2 = \eta = Uy/\nu. \tag{6.7}$$

Thus (6.5), (6.6) reduce to the form

$$\Phi_1 \partial_\xi \Phi_1 + \Phi_2 \partial_\eta \Phi_1 = \partial_{\eta\eta}^2 \Phi_1,$$
$$\partial_\xi \Phi_1 + \partial_\eta \Phi_2 = 0, \tag{6.8}$$
$$\Phi_1(0, \eta) = \Phi_1(\xi, \infty) = 1, \quad \Phi_1(\xi, 0) = \Phi_2(\xi, 0) = 0,$$

so that this direct application of dimensional analysis does not give any simplification of the problem. Now let $\Phi_1(\xi, \eta)$, $\Phi_2(\xi, \eta)$ be a solution of (6.8), which exists and is unique. Simple verification shows that for any positive α the functions $\Phi_1(\alpha^2 \xi, \alpha\eta)$, $\alpha\Phi_2(\alpha^2 \xi, \alpha\eta)$ also satisfy the equations and all the conditions of the boundary-layer problem, although not all those of the full Navier–Stokes problem. Thus the formulation of the boundary-layer problem has turned out to be invariant with respect to the one-parameter group of transformations

$$\Phi_1'(\xi', \eta') = \Phi_1(\xi, \eta), \quad \Phi_2'(\xi', \eta') = \alpha^{-1}\Phi_2(\xi, \eta), \quad \xi' = \alpha^2\xi, \quad \eta' = \alpha\eta,$$

so that substituting these relations into (6.8) we obtain the same problem, but now in the variables Φ_1', Φ_2', ξ', η' for arbitrary positive α. In view of the uniqueness, the solution too must be invariant with respect to the same group of transformations, i.e. for any α the functions Φ_1 and Φ_2 must satisfy the relations

$$\Phi_1(\xi, \eta) = \Phi_1(\alpha^2\xi, \alpha\eta), \quad \Phi_2(\xi, \eta) = \alpha\Phi_2(\alpha^2\xi, \alpha\eta). \tag{6.9}$$

Since in (6.9) α can be taken equal to any positive number, we obtain, setting $\alpha = 1/\xi^{1/2}$,

$$\Phi_1(\xi, \eta) = \Phi_1\left(1, \frac{\eta}{\sqrt{\xi}}\right) = f_1\left(\frac{\eta}{\sqrt{\xi}}\right) = f_1\left(\frac{y}{\sqrt{\nu x/U}}\right),$$

$$\Phi_2(\xi, \eta) = \frac{1}{\sqrt{\xi}}\Phi_2\left(1, \frac{\eta}{\sqrt{\xi}}\right) = \frac{1}{\sqrt{\xi}}f_2\left(\frac{\eta}{\sqrt{\xi}}\right) = \sqrt{\frac{\nu}{Ux}}f_2\left(\frac{y}{\sqrt{\nu x/U}}\right).$$
$$\tag{6.10}$$

Thus, the self-similarity of the solution to the problem is established and expressions for the self-similar variables are obtained: however, this has been achieved as a result of not only dimensional considerations but also invariance of the problem statement with respect to a group of transformations that is broader than the group of similarity transformations of the quantities with independent dimensions.

The example just considered is instructive in that the application of more general groups of transformations can here be given the form of a use of dimensional analysis, and this device turns out to be useful in many cases. Namely, we shall use different units to measure length in the x-direction and length in the y-direction, i.e., we introduce two dimensions of length, L_x and L_y. This is possible for the boundary-layer equations, in contrast with the full Navier–Stokes equations. (In the latter the term $\nu\partial^2_{yy}u$ appears in sum with the term $\nu\partial^2_{xx}u$, and if we measure x and y in different units these terms will have different dimensions.) Here, then, it is necessary to take

$$[u] = [U] = L_x T^{-1}, \quad [\nu] = L_y^2 T^{-1},$$
$$[v] = L_y T^{-1}, \quad [x] = L_x, \quad [y] = L_y, \tag{6.11}$$

so that both in the boundary-layer equations and in the boundary conditions of the problem all terms will have identical dimensions. Thus among the governing parameters no longer two but rather three have independent dimensions, and the single independent dimensionless similarity parameter will be

$$\Pi'_1 = \frac{y}{\sqrt{\nu x/U}} = \zeta, \tag{6.12}$$

whence follows also the self-similarity of the solution to the problem:

$$u = U f_1(\zeta), \quad v = \sqrt{\frac{\nu U}{x}} f_2(\zeta).$$

Introducing the new function $\varphi(\zeta) = \int_0^\zeta f_1(\zeta)d\zeta$, we easily obtain, from (6.5) and (6.6),

$$f_2 = \tfrac{1}{2}(\zeta\varphi' - \varphi),$$
$$\varphi\varphi'' + 2\varphi''' = 0, \quad \varphi(0) = \varphi'(0) = 0, \quad \varphi'(\infty) = 1,$$

i.e., a nonlinear boundary-value problem for a third-order ordinary differential equation. For the drag F on a section of unit width and length l of the flat plate in a uniform stream of velocity U we obtain from the previous relations, using the results of numerical calculation of the function φ,

$$F = 2 \int_0^l (\sigma_{xy})_{y=0} dx = 2U\sqrt{\frac{U}{\nu}}\rho\nu \int_0^l f_1'(0)\frac{dx}{\sqrt{x}}$$

$$= 4\sqrt{\frac{U^3 l}{\nu}}\rho\nu\phi''(0) = 1.328\rho\sqrt{U^3 l\nu}\,.$$

Here $(\sigma_{xy})_{y=0}$ is the shear stress on the plate. For more details see Kochin, Kibel' and Roze (1964), Batchelor (1967), Germain (1986b), Landau and Lifschitz (1987) or Schlichting (1968).

Introducing the dimensionless parameter $\Pi = F/\rho U^2 l$ corresponding to F, we get

$$\Pi = \frac{1.328}{\sqrt{\mathrm{Re}}}, \quad \mathrm{Re} = \frac{Ul}{\nu}\,.$$

We note in passing that one can also look at this well-known relation as incomplete similarity in the Reynolds number. In fact, the drag F is determined by the following quantities: the length l of the plate, the viscosity ν and density ρ of the fluid, and the velocity U of the stream. Application of the standard procedure of dimensional analysis gives

$$\Pi = \Phi(\mathrm{Re})\,.$$

For the high Reynolds numbers characteristic of the boundary layer there is no complete similarity with respect to Reynolds number, since there does not exist a non-zero limit of the function $\Phi = 1.328\mathrm{Re}^{-1/2}$ as $\mathrm{Re} \to \infty$. Hence the relations

$$\Pi = \mathrm{const}, \quad F = \mathrm{const}\,\rho U^2 l$$

that would have to hold in the case of complete similarity in the Reynolds number must not be expected to be true, no matter how high the Reynolds number. Nevertheless, one has the relation

$$\Pi^* = \frac{F}{\rho\sqrt{U^3 l\nu}} = \mathrm{const} = 1.328\,,$$

corresponding to incomplete self-similarity: the parameter Π^* cannot be obtained from conventional dimensional analysis and contains the dimensional parameter ν whose explicit introduction into the problem violates self-similarity. Using the generalized dimensional analysis (6.11) the reader can easily obtain for the boundary-layer flow the scaling law

$$F = \mathrm{const}\,\rho\sqrt{U^3 l\nu}$$

from dimensional considerations.

6.1.3 Example: limiting self-similar solutions

So-called limiting self-similar solutions[†], i.e., solutions of the form $e^{\alpha t} f(xe^{\beta t})$, in which both the length scale and the scale of the quantity to be determined depend exponentially on the time, give other interesting examples of the use of more general group considerations. We consider such solutions for the equation of nonlinear heat conduction,

$$\partial_t u = \kappa \partial^2_{xx} u^{n+1}, \quad (n > 0). \tag{6.13}$$

An appropriate idealized problem is considered in the semi-infinite domain $x \geq 0$ for $t > -\infty$. We seek a solution of (6.13) that satisfies the conditions

$$u(x, -\infty) = 0, \quad u(0,t) = u_0 e^{\sigma t}. \tag{6.14}$$

The application of dimensional analysis gives, as is easy to prove using the standard procedure,

$$u = u_0 \Phi[x/(\kappa \sigma^{-1} u_0^n)^{1/2}, \sigma t]. \tag{6.15}$$

We now observe that the problem formulated is invariant also with respect to the group of transformations of translation in time. This means that if $u(x,t,u_0,\sigma)$ is a solution of (6.13), (6.14), then $u(x,t - \tau, u_0 e^{\sigma \tau}, \sigma)$ is also a solution of the same equations for any real τ. In fact, substituting $t' = t + \tau$ in (6.13) and (6.14), we get the same problem for the determination of u as a function of the variables x and t', but in place of u_0 will appear $u_0' = u_0 e^{\sigma \tau}$. Hence, from the uniqueness of the solution and from (6.15) it follows that for any τ,

$$u(x,t) = u_0 \Phi[x/(\kappa \sigma^{-1} u_0^n)^{1/2}, \quad \sigma t] = u(x,t')$$
$$= u_0 e^{\alpha \tau} \Phi[x/(\kappa \sigma^{-1} u_0^n e^{n\sigma \tau})^{1/2}, \quad \sigma t - \sigma \tau].$$

But this means that the function Φ satisfies the invariance relation

$$\Phi(\xi, \eta) = e^{\sigma \tau} \Phi(\xi e^{-n\sigma \tau/2}, \quad \eta - \sigma \tau) \tag{6.16}$$

for any τ. Setting $\tau = \eta/\sigma$ we obtain

$$\Phi(\xi, \eta) = e^{\eta} \Phi(\xi e^{-n\eta/2}, 0) = e^{\eta} f(\xi e^{-n\eta/2}), \tag{6.17}$$

from which now follows the self-similarity of the solution of (6.13), (6.14):

$$u = u_0 e^{\sigma t} f[x/(\kappa \sigma^{-1} u_0^n e^{n\sigma t})^{1/2}]. \tag{6.18}$$

The designation of these solutions as *limiting self-similar* is explained in the following way. Equation (6.13) has a family of self-similar solutions of ordinary power-law form satisfying the boundary and initial

---

[†] Sometimes these solutions are called limiting to self-similar, which is unfortunate in our view, since they themselves are also self-similar.

conditions

$$u(x, t_0) = 0, \quad u(0, t) = \mu(t - t_0)^\alpha \quad (t > t_0). \tag{6.19}$$

It is easy to show, using the standard procedure of dimensional analysis, that these solutions can be expressed in the form

$$u = \mu(t - t_0)^\alpha f_\alpha\{x/[\kappa\mu^n(t - t_0)^{\alpha n+1}(\alpha + 1)]^{1/2}\} \tag{6.20}$$

(the factor $(\alpha + 1)$ having been introduced in the self-similar variable for convenience), where the function f_α is a solution of the equation

$$\frac{d^2 f_\alpha^n}{d\xi^2} + \frac{\xi}{2}\frac{df_\alpha}{d\xi} - \frac{\alpha}{\alpha + 1}f_\alpha = 0, \tag{6.21}$$

which satisfies the conditions $f_\alpha(0) = 1$, $f_\alpha(\infty) = 0$ and is continuous, has a continuous derivative $df_\alpha^n/d\xi$, and is in fact identically equal to zero for ξ greater than some $\xi_0(\alpha) < \infty$. We now choose $t_0 = -\alpha/\sigma$, where σ is a constant having the dimensions of inverse time, and let α tend to infinity while keeping $\mu(\alpha/\sigma)^\alpha$ constant and equal to u_0. It is easy to see that the factor $(t - t_0)^\alpha = (\alpha/\sigma)^\alpha(1 + \sigma t/\alpha)^\alpha$ here tends to an exponential, and the solution (6.20) tends to the solution (6.18) as its limit.

Solutions of the form $e^{\alpha t}f(xe^{\beta t})$ have appeared in various problems of fluid mechanics, starting with the paper of Goldstein (1939) devoted to the theory of the boundary layer. The group analysis of the solutions given above and also the explanation of their limiting character for boundary layer theory is given in Barenblatt (1954).

6.1.4 Example: rotation of fluid in a cylindrical container

An instructive example of a self-similar solution for which considerations of dimensional analysis are insufficient for establishing the self-similarity is provided by the remarkable Sobolev problem of small perturbations of a rotating fluid in a cylindrical container (Sobolev, 1954). The equation for pressure perturbation in this problem has, as Sobolev showed, the form

$$\partial_{tt}^2 \Delta p + \omega^2 \partial_{zz}^2 p = 0. \tag{6.22}$$

Here t is the time, z is the coordinate measured along the axis of rotation, $\Delta = \partial_{\rho\rho}^2 + (1/\rho)\partial_\rho + \partial_{zz}^2$ is the Laplace operator, $\rho = (x^2 + y^2)^{1/2}$, x and y are rectangular coordinates in the plane perpendicular to the axis of rotation, and ω is the angular velocity of the rotation.

By the first fundamental solution of (6.22) is meant the solution satisfying the initial conditions

$$p(x, y, z, 0) = \frac{Q}{r}, \quad \partial_t p(x, y, z, 0) = 0, \qquad (6.23)$$

where Q is a constant and $r = (\rho^2 + z^2)^{1/2}$. The desired solution depends on the governing parameters ρ, Q, z, ω and t, whose dimensions are

$$[Q] = [p]L, \quad [\rho] = [z] = L, \quad [\omega] = T^{-1}, \quad [t] = T. \qquad (6.24)$$

Furthermore, dimensional analysis gives, as is easy to show,

$$p = \frac{Q}{\rho}\Phi(\xi, \eta), \quad \xi = \frac{\rho}{z}, \quad \eta = \omega t. \qquad (6.25)$$

Substituting (6.25) into (6.22) and (6.23) and integrating, taking into account the regularity of the solution at the axis of rotation[†], we reduce these relations to the form

$$\xi^2 \partial^3_{\xi\eta\eta}\Phi + \xi^2 \partial_\xi \Phi + \xi\partial_\xi \left(\frac{1}{\xi}\partial^2_{\eta\eta}\Phi\right) = 0,$$
$$\Phi(\xi, 0) = \frac{\xi}{\sqrt{1 + \xi^2}}, \quad \partial_\eta \Phi(\xi, 0) = 0. \qquad (6.26)$$

Since the combination $\zeta = \xi/(1 + \xi^2)^{1/2}$ appears on the right-hand side of one of the conditions, it is convenient to take it as an independent variable (it would appear automatically if we introduced r instead of z as the governing parameter), and to denote $\Phi(\xi, \eta)$ by $\Psi(\zeta, \eta)$. Then (6.26) assumes the form

$$\zeta^2 \partial_\zeta \Psi + \partial^2_{\eta\eta}[\zeta\partial_\zeta(\Psi/\zeta)] = 0,$$
$$\Psi(\zeta, 0) = \zeta, \quad \partial_\eta \Psi(\zeta, 0) = 0. \qquad (6.27)$$

If $\Psi(\zeta, \eta)$ is a solution of the problem then, as is easily verified, $\alpha^{-1}\Psi(\alpha\zeta, \alpha^{-1}\eta)$ also satisfies the equation and all the conditions of the problem for arbitrary positive α. By virtue of the uniqueness of the solution, it follows from this that the function $\Psi(\zeta, \eta)$ satisfies the invariance relation

$$\Psi(\zeta, \eta) = \alpha^{-1}\Psi(\alpha\zeta, \alpha^{-1}\eta) \qquad (6.28)$$

for any $\alpha > 0$. We now set $\alpha = 1/\zeta$ and obtain

$$\Psi(\zeta, \eta) = \zeta\,\Psi(1, \zeta\eta) = \zeta\,\Xi(\zeta\eta), \qquad (6.29)$$

i.e., the function $\Psi(\zeta, \eta)$ can be represented by a function of one variable.

[†] The requirement of regularity also includes the vanishing of $\partial^2_{\eta\eta}\Phi(0, \eta)$, the coefficient of the 'cylindrical' generalized function $\Delta(1/\rho)$, which is obtained when (6.25) is substituted into (6.22). The solution (6.31) that we will obtain satisfies this condition.

Substituting (6.29) into (6.25) and returning to the original variables, we obtain

$$p = \frac{Q}{r} \Xi \left(\frac{\omega \rho t}{r} \right), \tag{6.30}$$

so that the first fundamental solution of Sobolev's equation (6.22) actually turns out to be self-similar. The substitution of (6.30) into (6.22) and (6.23) easily allows one to determine an expression for the function Ξ in terms of Bessel functions of order zero:

$$\Xi = J_0 \left(\frac{\omega \rho t}{r} \right), \quad p = \frac{Q}{r} J_0 \left(\frac{\omega \rho t}{r} \right). \tag{6.31}$$

The use of invariance with respect to a wider group for the proof of self-similarity and the determination of self-similar variables can here too be given the form of an application of dimensional analysis; the simple approach that we shall now apply is also often useful. Namely, we write (6.27) in the form

$$\zeta^2 \partial_\zeta \Psi + \lambda^2 \partial_{\eta\eta}^2 [\zeta \partial_\zeta (\Psi/\zeta)] = 0,$$
$$\Psi(\zeta, 0) = \mu \zeta, \quad \partial_\eta \Psi(\zeta, 0) = 0, \tag{6.32}$$

and temporarily forget that the quantities Ψ, ζ, η are dimensionless and that λ and μ are equal to unity. On the contrary, we assume that ζ has dimension Z, η has dimension H, and Ψ has dimension $[\Psi]$. Then, in order that all terms of (6.32) have identical dimensions, it is necessary that the dimensions of λ and μ be as follows:

$$[\mu] = [\Psi] Z^{-1}, \quad [\lambda] = ZH. \tag{6.33}$$

The solutions Ψ, as follows from (6.32), can depend only on ζ, η, λ and μ, whence we obtain, by means of the standard procedure of dimensional analysis,

$$\Psi = \mu \zeta \Phi \left(\frac{\lambda}{\zeta \eta} \right) = \mu \zeta \Xi \left(\frac{\zeta \eta}{\lambda} \right). \tag{6.34}$$

Setting $\mu = \lambda = 1$, we again get (6.29). The examples given above show how establishing the invariance of a problem with respect to a certain group of continuous transformations allows one to decrease the number of arguments of the function, just as do considerations of dimensional analysis, which are based on invariance with respect to a subgroup of the group of similarity transformations. Therefore of fundamental value is the idea developed by Birkhoff[†] (1960) of generalized inspectional analysis of the equations of mathematical physics, i.e., the idea of looking

---

[†] In Birkhoff's book (1960) a careful citation is given of his predecessors, in particular, T.A. Ehrenfest–Afanassjewa.

for groups with respect to which the equations of some physical phenomenon are invariant, and also for solutions that are invariant with respect to those groups. There naturally arises the question of an algorithm for seeking a maximally broad group of transformations with respect to which a given system of differential equations is invariant. The basic ideas here belong to S. Lie; in recent times a series of general results, and applications to particular systems of equations encountered in various problems of mechanics and physics, have been obtained (see the books by Birkhoff, 1960; Bluman and Cole, 1974; Ovsyannikov, 1978). We refer to the books cited for an account of the general approach and numerous examples. Our account, given above, had as its goal to demonstrate by instructive examples the general idea of using wider groups in the search for self-similar solutions, and to indicate the use, in a series of cases, of formal application of the standard technique of dimensional analysis in working with more general groups. It is clear that dimensional analysis can be applied even without knowing the mathematical formulation of the problem. It would appear that invariance with respect to a more general group than the group of similarity transformations of quantities with independent dimensions can be used only if one has a mathematical formulation of the problem. As a matter of fact this is not so, and invariance with respect to wider groups can also be suggested by physical considerations.

In conclusion we note that the consideration of self-similar solutions as intermediate-asymptotic representations is closely connected with singular perturbation methods, which have been widely developed and applied in the last few decades (Van Dyke, 1975; Cole, 1968; Lagerstrom and Casten, 1972; Kevorkian and Cole, 1980; Hinch, 1991). Namely, self-similar solutions are inner or outer asymptotics of the solutions of the complete problem, depending on which of the scales of the independent variable is taken as the basis for analysis of the intermediate asymptotics. Therefore the determination of the constants appearing in a self-similar solution of the second kind can, in a number of cases, be achieved by matching the self-similar solution with a supplementary asymptotics.

6.2 The renormalisation group and incomplete similarity

6.2.1 The renormalisation group and intermediate asymptotics

Among the additional groups to the groups of similarity transformations of quantities with independent dimensions that lead to self-similarity, a special and very important place belongs to the renormalization group. The renormalization group approach, following the ideas of Stückelberg and Peterman (1953), Gell-Mann and Low (1954), Bogolyubov and Shirkov (1955), Kadanoff (1966), Patashinsky and Pokrovsky (1964) and Wilson (1971), has found extensive applications in modern theoretical physics. Recently N. Goldenfeld, O. Martin and Y. Oono demonstrated a deep relation between the renormalization group method as traditionally used by physicists and the intermediate asymptotics approach developed independently and presented in this book. An important step was that they solved, by the renormalization group method in the form in which it is usually applied by physicists, some typical problems whose solution was previously performed by the method of intermediate asymptotics and, vice versa, they solved by the method of intermediate asymptotics some classical problems in statistical physics solved earlier by the renormalization group approach (Goldenfeld (1989); Goldenfeld, Martin and Oono (1989, 1991); Goldenfeld *et al.*, (1990); Goldenfeld and Oono (1991), Chen, Goldenfeld and Oono (1991); the book by Goldenfeld (1992) is especially recommended). Later, further applications of their approach and its development were performed in the papers by Ginzburg, Entov and Theodorovich (1992) and by Bricmont and Kupiainen (1992).

In this section we present the renormalization group approach including the very definition of a renormalization group from the viewpoint of our intermediate-asymptotics technique, presented above in this book. In the next section we will demonstrate the renormalization group approach in the form commonly used by physicists, following exactly the presentation by Goldenfeld *et al.* (1991).

We now return to the alternative considered in chapter 5.

The basic relation in which we are interested,

$$a = f(a_1, \dots, a_k, b_1, \dots, b_m, \epsilon) \qquad (6.35)$$

can be written in dimensionless form:

$$\Pi = \Phi(\Pi_1, \Pi_2, \dots, \Pi_m, \epsilon). \qquad (6.36)$$

Here we have added an additional constant dimensionless parameter ϵ on which the phenomenon under consideration is assumed to depend. Let the parameter Π_m be small, and incomplete similarity asymptotics be valid for the function Φ, so that for small Π_m

$$\Pi = \Pi_m^{\alpha_m} \Phi_1\left(\frac{\Pi_1}{\Pi_m^{\alpha_1}}, \dots, \frac{\Pi_{m-1}}{\Pi_m^{\alpha_{m-1}}}, \epsilon\right). \qquad (6.37)$$

We begin from a simpler case; the general case of incomplete similarity will be considered later. Generally speaking, the powers $\alpha_1, \ldots, \alpha_m$ do depend on the parameter ϵ. Let, further, $\alpha_1 = \alpha_2 = \ldots = \alpha_m = 0$, i.e., the complete similarity case for $\epsilon = 0$. In this case, for sufficiently small Π_m the function Φ in (6.36) can for $\epsilon = 0$ be replaced by its finite limit $\Phi_1(\Pi_1, \ldots, \Pi_{m-1}, 0)$. So, the dimensional parameter b_m responsible for the dimensionless parameter Π_m disappears from consideration, i.e., from relation (6.35), as does the dimensionless parameter Π_m from (6.36).

Therefore, we can say that at sufficiently small Π_m nothing in the quantitative description of the phenomenon is changed if the transformation group

$$b_1' = b_1, \quad b_2' = b_2, \quad \ldots, \quad b_m' = Bb_m,$$
$$a' = a, \quad a_1' = a_1, \quad \ldots, \quad a_k' = a_k \tag{6.38}$$

is applied, where B, the group parameter, is an arbitrary positive number. Thus, asymptotically the problem statement is invariant with respect to the transformation group (6.38).

For $\epsilon > 0$, the parameters $\alpha_1, \alpha_2, \ldots, \alpha_m$ are generally speaking different from zero; therefore the parameter b_m does not disappear from the resulting relation even for arbitrarily small Π_m. However, the resulting asymptotic form of the relation (6.37) is evidently invariant with respect to the more general transformation group

$$b_1' = B^{\alpha_1} b_1, \quad b_2' = B^{\alpha_2} b_2, \quad \ldots, \quad b_m' = Bb_m, \, a' = B^{\alpha_m} a,$$
$$a_1' = a_1, \quad a_2' = a_2, \quad \ldots, \quad a_k' = a_k. \tag{6.39}$$

We have arrived at a special case of the *renormalization group*. The term reflects the renormalization of the governing parameters with dependent dimensions b_i and of the quantity to be determined, a, entering the basic relation (6.35).

It is essential to note that the statement of the asymptotic invariance of a relation with respect to a renormalization group is equivalent to the statement of incomplete similarity. Indeed, if there exists incomplete similarity in Π_m at small or large Π_m, the asymptotic invariance of relation (6.36) with respect to the transformation group (6.39) is evident, because

$$\Pi' = B^{\alpha_m} \Pi, \quad \Pi_1' = B^{\alpha_1} \Pi_1, \quad \Pi_2' = B^{\alpha_2} \Pi_2, \quad \ldots, \quad \Pi_m' = B\Pi_m. \tag{6.40}$$

and the function Φ in (6.36) has the asymptotic form (6.37), which remains invariant under the transformations (6.40).

Furthermore, without loss of generality relation (6.36) can be represented in the form

$$\Pi = \Phi(\Pi_1, \dots, \Pi_m, \epsilon) = \Pi_m^{\alpha_m} \Psi \left(\frac{\Pi_1}{\Pi_m^{\alpha_1}}, \dots, \frac{\Pi_{m-1}}{\Pi_m^{\alpha_{m-1}}}, \Pi_m, \epsilon \right) \qquad (6.41)$$

or

$$\frac{\Pi}{\Pi_m^{\alpha_m}} = \Psi \left(\frac{\Pi_1}{\Pi_m^{\alpha_1}}, \dots, \frac{\Pi_{m-1}}{\Pi_m^{\alpha_{m-1}}}, \Pi_m, \epsilon \right) . \qquad (6.42)$$

Now we apply the transformation group (6.39) and obtain

$$\frac{\Pi}{\Pi_m^{\alpha_m}} = \Psi \left(\frac{\Pi_1}{\Pi_m^{\alpha_1}}, \dots, \frac{\Pi_{m-1}}{\Pi_m^{\alpha_{m-1}}}, B\Pi_m, \epsilon \right) \qquad (6.43)$$

because the quantities $\Pi_i/\Pi_m^{\alpha_i}$, $\Pi/\Pi_m^{\alpha_m}$ are invariants of the transformation group (6.39). Invariance with respect to the group (6.39) means that relation (6.43) holds for arbitrary values of the group parameter B, i.e. that the function Ψ is independent of its argument Π_m,

$$\Psi \left(\frac{\Pi_1}{\Pi_m^{\alpha_1}}, \dots, \frac{\Pi_{m-1}}{\Pi_m^{\alpha_{m-1}}}, \Pi_m, \epsilon \right) = \Phi_1 \left(\frac{\Pi_1}{\Pi_m^{\alpha_1}}, \dots, \frac{\Pi_{m-1}}{\Pi_m^{\alpha_{m-1}}}, \epsilon \right)$$

whence and from (6.41) follows the validity of the relation

$$\Phi(\Pi_1, \dots, \Pi_m, \epsilon) = \Pi_m^{\alpha_m} \Phi_1 \left(\frac{\Pi_1}{\Pi_m^{\alpha_1}}, \dots, \frac{\Pi_{m-1}}{\Pi_m^{\alpha_{m-1}}}, \epsilon \right) \qquad (6.44)$$

at small Π_m, i.e. incomplete similarity. In the more general case of incomplete similarity, when several parameters $\Pi_{l+1}, \dots, \Pi_m$ are large or small, and at small (for definiteness sake) $\Pi_{l+1}, \dots, \Pi_m$ the function Φ in (6.36) has the same property of generalized homogeneity,

$$\Phi = \Pi_{l+1}^{\alpha_{l+1}} \dots \Pi_m^{\alpha_m} \Phi_1 \left(\frac{\Pi_1}{\Pi_{l+1}^{\beta_1} \dots \Pi_m^{\delta_1}}, \dots, \frac{\Pi_l}{\Pi_{l+1}^{\beta_l} \dots \Pi_m^{\delta_l}} \right) \qquad (6.45)$$

the renormalization group is more complicated:

$$
\begin{aligned}
a_1' &= a_1, \quad \dots, \quad a_k' = a_k ; \\
b_1' &= B_{l+1}^{\beta_1} \dots B_m^{\delta_1} b_1, \quad \dots, \quad b_l' = B_{l+1}^{\beta_l} \dots B_m^{\delta_l} b_l \\
b_{l+1}' &= B_{l+1} b_{l+1}, \quad \dots, \quad b_m' = B_m b_m \\
a' &= B_{l+1}^{\alpha_{l+1}} \dots B_m^{\alpha_m} a
\end{aligned}
\qquad (6.46)
$$

($B_{l+1}, \dots, B_m$ are positive numbers, the parameters of the group). Again, the property of incomplete similarity, (6.45), is equivalent to the property for asymptotic invariance with respect to the renormalization group (6.46).

6.2.2 Perturbation expansion

The next step, which, we emphasize, is an independent one, is a perturbation expansion for small ϵ; generally speaking, this expansion is non-convergent[†]. It enables us to determine the parameters of incomplete similarity $\alpha_1, \ldots, \alpha_m$, which lead to 'anomalous dimensions', as expansions in ϵ, using some quantitative information concerning the phenomenon, for instance, non-integrable conservation laws. The latter point is crucial: if no further information concerning the phenomenon under consideration is available, the parameters $\alpha_1, \ldots, \alpha_m$ entering the renormalization group and incomplete similarity relation cannot be determined.

As a first example, the Cauchy problem for the modified heat conduction equation considered in detail in chapter 3 will be presented here: the equation

$$\partial_t \theta = \begin{cases} \kappa \partial^2_{xx} \theta & (\partial_t \theta \geq 0) \\ \kappa(1 + \epsilon)\partial^2_{xx}\theta & (\partial_t \theta \leq 0) \end{cases} \tag{6.47}$$

where the parameter ϵ is considered as small. The initial condition is assumed in the form

$$\theta(x,0) = \frac{Q}{\sqrt{2\pi}l}e^{-x^2/2l^2} \tag{6.48}$$

where Q is, as before, the initial 'heat charge' and l is a certain length scale. The solution to the problem (6.47), (6.48)

$$\theta = f(t, Q, \kappa, x, l, \epsilon) \tag{6.49}$$

can be represented in the form (cf. chapter 3)

$$\Pi = \Phi(\Pi_1, \Pi_2, \epsilon) \tag{6.50}$$

$$\Pi = \frac{\theta\sqrt{\kappa t}}{Q}, \quad \Pi_1 = \xi = \frac{x}{\sqrt{\kappa t}}, \quad \Pi_2 = \frac{l}{\sqrt{\kappa t}} \tag{6.51}$$

At $\epsilon = 0$ there exists complete similarity with respect to the parameter Π_2: the finite limit of the function Φ at $\Pi_2 \to 0$ is

$$\Phi(\Pi_1, 0, 0) = \frac{1}{2\sqrt{\pi}}e^{-\xi^2/4}, \quad \theta = \frac{Q}{2\sqrt{\pi \kappa t}}e^{-x^2/4\kappa t}. \tag{6.52}$$

The coordinate of the inflection point, where $\partial_t u = 0$ for $\epsilon = 0$, $x = \pm x_0(t)$, is given by the relations

$$\xi = \xi_0 = \sqrt{2}, \quad x_0(t) = \sqrt{2\kappa t}. \tag{6.53}$$

At $\epsilon > 0$ there exists incomplete similarity with respect to Π_2, at small Π_2:

$$\Phi = \Pi_2^\alpha \Phi_1(\Pi_1, \epsilon), \quad \theta = \frac{A}{(\kappa t)^{(1+\alpha)/2}}\Phi_1(\xi, \epsilon), \tag{6.54}$$

$$A = \sigma Q l^\alpha, \quad \sigma = \text{const}.$$

[†] However, in the first example considered below, the expansion does converge.

Let us perform an ϵ-expansion, and restrict ourselves to the first term of this expansion:

$$\alpha = c_1\epsilon + O(\epsilon^2), \quad \Phi_1(\xi, \epsilon) = \frac{1}{2\sqrt{\pi}} e^{-\xi^2/4} + O(\epsilon)$$

$$x_0(t) = \sqrt{2\kappa t} + O(\epsilon) \tag{6.55}$$

The relation (3.47), a non-integrable conservation law, assumes the form

$$\frac{d}{dt} \int_{-\infty}^{\infty} \theta(x, t)dx = 2\kappa\epsilon(\partial_x\theta)_{x=x_0(t)} . \tag{6.56}$$

Using this law gives a convenient way for obtaining the relation for α (Barenblatt, 1993b; Cole and Wagner, 1995). So, we substitute into (6.56) the ϵ-expansion (6.55). We obtain

$$\frac{d}{dt} \int_{-\infty}^{\infty} \theta(x, t)dx = \frac{d}{dt} \left\{ \frac{A}{(\kappa t)^{(1+\alpha)/2}} \left[\int_{-\infty}^{\infty} \frac{e^{-x^2/4\kappa t}}{2\sqrt{\pi}} dx + O(\epsilon) \right] \right\}$$

$$= \frac{d}{dt} \left\{ \frac{A}{(\kappa t)^{\alpha/2}} \left[\int_{-\infty}^{\infty} \frac{e^{-z^2}}{\sqrt{\pi}} dz + O(\epsilon) \right] \right\}$$

$$= -\frac{\alpha}{2} \left[\frac{A\kappa}{(\kappa t)^{1+\alpha/2}} + O(\epsilon) \right]$$

$$= -2\kappa\epsilon \left[\frac{A}{(\kappa t)^{1+\alpha/2}} \left(\frac{1}{2\sqrt{\pi}} \right) \right] \left[\frac{\sqrt{2}}{2} e^{-1/2} + O(\epsilon^2) \right],$$

whence

$$\alpha = \sqrt{\frac{2}{\pi e}}\epsilon + O(\epsilon^2), \tag{6.57}.$$

the result obtained by Goldenfeld, Martin, Oono and Liu (1990) by the traditional renormalization group method, presented in the next section.

The higher approximations were obtained by Cole and Wagner (1995) using the perturbation of the transformation group of the linear problem corresponding to $\epsilon = 0$.

Another example, the modified very intense explosion problem considered in chapter 4, can be treated in the same way, as follows (Barenblatt, 1994).

The assumption of invariance with respect to the renormalization group gives the same self-similar representation for its solution,

$$p = \rho_0 \frac{r^2}{t^2} P(\zeta, \epsilon), \quad \rho = \rho_0 R(\zeta, \epsilon), \quad u = \frac{r}{t} V(\zeta, \epsilon), \quad \zeta = \frac{r}{r_f}$$

$$r_f = \left(\frac{A}{\rho_0}\right)^{\alpha/2} t^{\alpha}, \quad \epsilon = \gamma_1 - \gamma. \tag{6.58}$$

For $\epsilon = 0$ there exists complete similarity and the solution takes the form presented in chapter 2:

$$p = \rho_0 \frac{r^2}{t^2} P(\zeta, 0), \quad \rho = \rho_0 R(\zeta, 0), \quad v = \frac{r}{t} V(\zeta, 0), \quad \zeta = \frac{r}{r_f}$$

$$r_f = \xi_f \left(\frac{E}{\rho_0}\right)^{1/5} t^{2/5}$$

where ξ_f is a function of the adiabatic exponent γ.

Now we use the ϵ-expansion for the functions P, V, R and the exponent α:

$$P = P(\zeta, 0) + O(\epsilon), \quad R = R(\zeta, 0) + O(\epsilon), \quad V = V(\zeta, 0) + O(\epsilon)$$

$$r_f = \xi_f \left(\frac{E}{\rho_0}\right)^{1/5} t^{2/5} + O(\epsilon), \quad \alpha = \frac{2}{5} + a_1 \epsilon + O(\epsilon^2). \tag{6.59}$$

The non-integrable energy conservation law has the form (4.9):

$$\frac{d\mathcal{E}}{dt} = 4\pi r_f^2 D \rho_0 \epsilon \frac{p_f}{(\gamma - 1)(\gamma_1 - 1)\rho_f}, \tag{6.60}$$

where, according to (4.4),

$$p_f = \frac{2}{\gamma_1 + 1} \rho_0 D^2, \quad u_f = \frac{2}{\gamma_1 + 1} D, \quad \rho_f = \frac{\gamma_1 + 1}{\gamma_1 - 1} \rho_0$$

are the values of the pressure, velocity and density behind the very intense shock wave,

$$D = dr_f/dt = \alpha r_f/t$$

is the velocity of the shock wave, and

$$\mathcal{E} = 4\pi \int_0^{r_f} \left[\rho \frac{u^2}{2} + \frac{p}{(\gamma - 1)}\right] r^2 dr$$

$$= 4\pi \rho_0 \frac{r_f^5}{t^2} \int_0^1 \left[R \frac{V^2}{2} + \frac{P}{(\gamma - 1)}\right] \zeta^4 d\zeta$$

is the total energy in the region inside the shock wave. Substituting the last three relations into (6.60) we obtain

$$\frac{d\mathcal{E}}{dt} = (5\alpha - 2)t^{5\alpha - 3}\left(\frac{A}{\rho_0}\right)^{5\alpha/2}$$

$$\times \left\{ 4\pi \int_0^1 \left[R(\zeta, 0)\frac{V^2(\zeta, 0)}{2} + \frac{1}{\gamma - 1}P(\zeta, 0) \right] \zeta^4 d\zeta + O(\epsilon) \right\}$$

$$= \frac{4\pi \left(\frac{2}{5}\right)^3 2\epsilon}{(\gamma - 1)(\gamma + 1)^2} + O(\epsilon^2)$$

whence

$$\alpha - \frac{2}{5} = \frac{64\pi\epsilon\xi_f^5}{625(\gamma - 1)(\gamma + 1)^2} + O(\epsilon^2) \qquad (6.61)$$

because, according to (2.62),

$$4\pi \int_0^1 \left[R(\zeta, 0)\frac{V^2(\zeta, 0)}{2} + \frac{P(\zeta, 0)}{\gamma - 1} \right] \zeta^4 d\zeta = \frac{1}{\xi_f^5}.$$

6.2.3 Traditional renormalisation group approach

Goldenfeld, Martin and Oono (1991) represent the renormalization group approach for solving the problem (6.47), (6.48) as six steps, each a direct counterpart of the conventional procedure followed in quantum field theory or statistical physics. All necessary references are available in this paper.

The *first step* is to construct a 'naive' perturbation expansion in the small parameter ϵ. For $l^2/2\kappa t \to 0$, l being kept fixed, this expansion has the form (see the details in Goldenfeld, Martin, Oono and Liu, 1990)

$$u(x, t) = u_{\mathrm{B}}(x, t) =$$

$$= \frac{Q}{2\sqrt{\kappa t}}e^{-x^2/4\kappa t}\left[1 - \frac{\epsilon}{\sqrt{2\pi e}}\ln\frac{2\kappa t}{l^2} + O(\epsilon^2) \right] + \text{regular terms.}$$

$$(6.62)$$

The subscript B stands for 'bare', the term used by physicists for such a perturbation expansion.

The *second step* is to 'cure' the logarithmic divergence of the perturbation series by introducing the 'renormalized' solution

$$u_{\mathrm{R}}(x, t) = Z\left(\frac{l}{\Lambda}\right)u_{\mathrm{B}}(x, t) \qquad (6.63)$$

where Λ is an arbitrary length. According to the basic idea of this approach the renormalized solution $u_{\mathrm{R}}(x, t)$ should eventually be the correct asymptotic solution of the problem (6.47), (6.48) rather than $u_{\mathrm{B}}(x, t)$, which is divergent. The function $Z(l/\Lambda)$ is referred to as the *renormalization constant*. The quantity Q, the 'initial charge', cannot be obtained from knowledge of $u(x, t)$ at large times, therefore Q is considered to be 'unobservable' at large times in the same way that the 'bare' electric charge is unobservable at long distances according to quantum electrodynamics. The renormalization constant depends on l, so that as $l \to 0$ the divergence in $u_B(x, t)$ should be absorbed into Z to yield a finite $u_R(x, t)$. The removal of the divergence in u_B occurs order by order in the small parameter ϵ, so an expansion of Z in powers of ϵ is assumed. However, Z is dimensionless, so an arbitrary parameter Λ having the dimension of length should be introduced into its argument.

In the *third step* of the renormalization procedure, the expansion

$$Z = 1 + a_1 \left(\frac{l}{\Lambda} \right) \epsilon + a_2 \left(\frac{l}{\Lambda} \right) e^2 + \dots \qquad (6.64)$$

is assumed. The functions a_1, a_2, ... should be chosen to cancel, order by order in ϵ, the divergence in $u_B(x, t)$ as $l \to 0$. It can be shown (see the details in Goldenfeld, Martin, Oono and Liu, 1990) that

$$a_1 = \frac{1}{\sqrt{2\pi e}} \ln \left[C_1 \frac{\Lambda^2}{l^2} \right], \qquad (6.65)$$

where C_1 is an arbitrary dimensionless number; whence

$$u_R = \frac{Q e^{-x^2/4\kappa t}}{2\sqrt{\pi \kappa t}} \left[1 - \frac{\epsilon}{\sqrt{2\pi e}} \ln \left(\frac{2\kappa t}{C_1 \Lambda^2} \right) + O(\epsilon^2) \right]. \qquad (6.66)$$

This expression shows that u_R remains finite as $l \to 0$, because l does not enter it at all. In fact, the relation (6.66) describes a family of solutions, not just a single one.

Step four of the renormalization group procedure is to choose a particular element of the family by requiring that at some time instant t^* the value $u_R(x, t)$ at $x = 0$ is equal to a certain number U:

$$u_R(0, t^*) = U.$$

Then the corresponding solution to order ϵ is

$$u_R = U \left(\frac{t^*}{t} \right)^{1/2} e^{-x^2/4\kappa t} \left[1 - \frac{\epsilon}{\sqrt{2\pi e}} \ln \frac{t}{t^*} + O(\epsilon^2) \right]. \qquad (6.67)$$

This expression will be referred to as the *renormalized perturbation expansion*. The arbitrariness of t^* enables, under some assumptions, the renormalization expansion to be improved.

In the *fifth step* the renormalization-group argument due to Gell-Mann and Low (1954) is used: the $U(t^*)$ dependence is found in such a way as to cancel out the explicit t^*-dependence of the renormalized perturbation expansion:

$$\frac{d u_R}{d t^*} = \frac{\partial u_R}{\partial t^*} + \frac{\partial u_R}{\partial U} \left(\frac{dU}{d t^*} \right) = 0. \qquad (6.68)$$

The partial derivatives $\partial_t u_R$, $\partial_U u_R$ can be explicitly evaluated to $O(\epsilon)$ from the expression (6.66), and the result is

$$t^* \frac{dU}{dt^*} = -U \left[\frac{1}{2} + \frac{\epsilon}{\sqrt{2\pi e}} + O(\epsilon^2) \right] . \tag{6.69}$$

The *sixth step* is to solve this differential equation for U:

$$U = A \left(2\kappa t^*\right)^{-\left[\frac{1}{2} + \frac{\epsilon}{(2\pi e)^{1/2}} + O(\epsilon^2)\right]} \tag{6.70}$$

(A is a constant) and, substituting (6.70) back into (6.67), we obtain

$$u_R = \frac{A}{(2\kappa t^*)^{(1+\alpha)/2}} \left(\frac{t^*}{t}\right)^{1/2} e^{-x^2/4\kappa t} \left[1 - \frac{\epsilon}{\sqrt{2\pi e}} \ln \frac{t}{t^*} + O(\epsilon^2) \right] \tag{6.71}$$

where

$$\alpha = \sqrt{\frac{2}{\pi e}} \, \epsilon + O(\epsilon^2) . \tag{6.72}$$

The value t^* can be selected in an arbitrary way, so we can put $t^* = t$, and the final representation of the renormalized solution is obtained:

$$u_R = \frac{A}{(2\pi\kappa t)^{(1+\alpha)/2}} e^{-x^2/4\kappa t} + O(\epsilon^2) . \tag{6.73}$$

The example considered above gives the general idea of the renormalization group approach in the form in which it is used in statistical physics and quantum field theories. The following basic points should be noted: a scaling law is assumed, depending on a parameter. For the value zero of the parameter the solution is known. An asymptotic expansion is then used to find the solution for small but finite values of the parameter.

If there is no value of the parameter for which there exists complete similarity, ϵ-expansion cannot be performed. In such cases only the ways demonstrated earlier in this book remain for obtaining the exponents, i.e. the 'anomalous dimensions' $\alpha_1, \ldots, \alpha_m$: solving a nonlinear eigenvalue problem, as we have shown in chapter 4 for the problem of impulsive loading, numerical integration (see the examples in chapters 3 and 4), or physical experiment (see chapters 10, 11).

Note, in conclusion, that the difficulties in the traditional renormalization group approach are related essentially to representation of the power function z^α as an exponential and then attempting to use an expansion,

$$z^\alpha = e^{\alpha \ln z} = 1 + \alpha \ln z + \ldots ,$$

which is obviously divergent as $z \to 0$ or $z \to \infty$. The technique of intermediate asymptotics avoids these difficulties.

Self-similar solutions
and travelling waves

7.1 Solutions of travelling-wave type

In various problems in mathematical physics an important role is played by invariant solutions of the *travelling-wave* type. These are solutions for which the distributions of the properties of the motion at different times can be obtained from one another by a translation rather than by a similarity transformation as in the case of self-similar solutions. In other words, one can always choose a moving Cartesian coordinate system such that the distribution of properties of a motion of travelling-wave type is stationary in that system. One can reduce to a consideration of travelling waves the study of the structure of shock-wave fronts in gas dynamics (see, e.g. Kochin, Kibel' and Roze, 1964; Zeldovich and Raizer, 1966, 1967) and in magneto-hydrodynamics (Kulikovsky and Lyubimov, 1965), the structure of flame fronts (Zeldovich, 1948; Zeldovich, Barenblatt, Librovich and Makhviladze, 1985), the investigation of solitary and periodic waves in a plasma and on the surface of a heavy fluid (Jeffrey and Kakutani, 1972; Whitham, 1974; Karpman, 1975; Lighthill, 1978; Eilenberger, 1981; Drazin and Johnson, 1989; Fordy, 1990), and many other problems. In recent years many processes have been studied involving the effects of the propagation of plasma fronts in electrical, electromagnetic, and light (laser) fields, the so-called waves of discharge propagation. These processes also lead to the consideration of solutions of travelling-wave type (Raizer, 1968, 1977).

In accordance with the definition given above, solutions of travelling-wave type can be expressed in the form

$$v = V(x - X(t)) + V_0(t). \tag{7.1}$$

Here v is the property of the phenomenon being considered; x is the spatial Cartesian coordinate, an independent variable of the problem; t is another independent variable, for simplicity identified with time; and $X(t)$ and $V_0(t)$ are time-dependent translations along the x- and v-axes. In particular, if the properties of the process do not depend directly on time, so that the equations governing the process do not contain time explicitly, the travelling-wave propagates uniformly:

$$v = V(x - \lambda t + c) + \mu t. \tag{7.2}$$

Here λ, μ and c are constants; c is the phase shift and λ and μ represent the speeds of translation along the x- and v-axes. For an important class of steady travelling waves the distribution of properties in the wave remains unchanged in time, so that $\mu = 0$, and

$$v = V(x - \lambda t + c). \tag{7.3}$$

In particular, steady travelling waves describe the structure of shock waves and flames.

Travelling waves are closely connected with self-similarities. Indeed, if in (7.1) we set

$$\begin{aligned}
v &= \ln u, \quad t = \ln \tau, \quad V_0(t) = \ln u_0(\tau), \\
V &= \ln U, \quad x = \ln \xi, \quad X(t) = \ln \xi_0(\tau),
\end{aligned} \tag{7.4}$$

we obtain a representation of the travelling wave in the self-similar form

$$u = u_0(\tau) U(\xi/\xi_0(\tau)). \tag{7.5}$$

In particular, the relation (7.2) for a uniformly propagating travelling wave reduces to a self-similar form with power-type self-similar variables,

$$u = B\tau^\mu U(\xi/A\tau^\lambda); \tag{7.6}$$

where A and B are constants.

The simple connection noted here between self-similar solutions and travelling waves is well known; it has been used to simplify the study of some self-similar solutions (see, e.g., Staniukovich, 1960). Surprisingly, however, the connection between the classification of self-similar solutions and the well-known classification of steady travelling waves has long remained unnoticed.

In fact, as already remarked, steady travelling waves describe the structure of the fronts of shock waves, flames, and analogous regions of rapidly changing density, speed, and other properties of the motion which are described by surfaces of discontinuity when dissipative processes are neglected. Another example of such regions is the upper thermocline in the ocean.

One distinguishes two types of such fronts; see, e.g., Sedov (1971). For fronts of the first kind (shock waves, detonation waves, etc.) the speed of propagation of the front is found from the conservation laws of mass, momentum and energy only. The structure of such a front is adapted to the conservation laws in the sense that for one and the same speed of propagation of the front, dictated by the conservation laws, its thickness can be different depending on the character of the dissipative processes in the transition region and the magnitudes of the dissipative coefficients. Of course, analysis of the structure of shock waves allows one to reject unrealizable situations such as shock waves of rarefaction, for which it is impossible to construct the structure but basically the speed of propagation of the front is determined independently of the structure of the transition process.

For fronts of the second kind (a flame, gaseous discharge, etc.) the conservation laws become insufficient for the determination of the speed of the front: this is found as some eigenvalue in the course of determining the structure of the front, that is, a solution of travelling-wave type of the equations describing the dissipative processes in the transition region.

It turns out that this classification of travelling waves corresponds exactly to the classification of self-similar solutions discussed above. Here we consider the simplest examples of steady travelling waves of both types, after which we shall see how the two classifications correspond.

7.2 Burgers shock wave – steady travelling wave of the first kind

The Burgers equation

$$\partial_t v + v \partial_x v = \nu \partial_{xx}^2 v \qquad (7.7)$$

is a successful, though rather simplified, mathematical model of the motion of a viscous, compressible gas. Here v is the speed, ν the kinematic viscosity, x the spatial coordinate, and t the time. If the viscous term is neglected, (7.7) assumes the form of the simplest model equation of gas dynamics,

$$\partial_t v + v \partial_x v = 0 . \qquad (7.8)$$

This last equation has a solution of the type of a uniformly propagating shock wave: $v = V(\zeta)$, $\zeta = x - \lambda t + c$, where $V(\zeta)$ is a step function equal to v_1 for $\zeta > 0$ and to v_2 for $\zeta \leq 0$, with $v_1 < v_2$. The value of the speed of propagation $\lambda = \lambda_0$ is obtained from the law of conservation of momentum at the front of the discontinuity, which corresponds to (7.8):

$$-\lambda_0(v_1 - v_2) + \frac{v_1^2 - v_2^2}{2} = 0, \tag{7.9}$$

whence we find

$$\lambda_0 = \frac{v_1 + v_2}{2}. \tag{7.10}$$

We now take into account the dissipative process, that is, the viscosity, and return to (7.7). We construct a solution of the Burgers equation of travelling-wave type: $v = V(\zeta)$, $\zeta = x - \lambda t + c$. Substituting this expression for v into (7.7), we have

$$-\lambda\frac{dV}{d\zeta} + V\frac{dV}{d\zeta} = \nu\frac{d^2V}{d\zeta^2}, \tag{7.11}$$

whence, integrating and using the condition $V = v_1$ at $\zeta = \infty$, we find

$$\nu\frac{dV}{d\zeta} = -\lambda(V - v_1) + \frac{V^2 - v_1^2}{2}. \tag{7.12}$$

To satisfy the condition at the left endpoint, $V(-\infty) = v_2$, it is necessary to take

$$\lambda = \frac{v_1 + v_2}{2} = \lambda_0, \tag{7.13}$$

after which a solution is obtained in the form

$$\frac{\zeta}{\nu} = \frac{2}{v_2 - v_1} \ln\frac{v_2 - V}{V - v_1}. \tag{7.14}$$

This solution describes the structure of the transition region on the length scale $\nu/(v_2 - v_1)$ that is characteristic for this region. We see that the condition $v_2 > v_1$ imposed above is essential, since a solution describing the structure of the transition region of a wave with $V(-\infty) = v_2 < v_1 = V(\infty)$ does not exist. In fact with (7.13) taken into account, (7.12) assumes the form

$$\nu\frac{dV}{d\zeta} = -\frac{(v_2 - V)(V - v_1)}{2}. \tag{7.15}$$

Since V lies between v_1 and v_2, the right-hand side of (7.15) is always negative, and the left-hand side is negative only for $v_2 > v_1$. The analysis just presented was performed (for the complete set of the equations of gas dynamics) by Taylor (1910, 1963). In fact, it was the first scientific paper of the future giant of applied mathematics and mechanics of the twentieth century.

A solution of travelling-wave type with $\lambda = \lambda_0$ serves as an asymptotic representation of a solution of an initial-value problem for the Burgers equation with initial data of transitional type,

$$v(x,0) \equiv v_2 \ (x \le a); \ v_1 < v(x,0) < v_2 \ (a < x < b); \ v(x,0) \equiv v_1 \ (x \ge b),$$
(7.16)

where a and b $(a < b)$ are arbitrary real numbers, and the function $v(x,0)$ is monotonically non-increasing: $\partial_x v(x,0) \le 0$. This was rigorously proved by Oleinik (1957). As is evident, in the present case the value of the speed of propagation λ_0 is obtained from a conservation law and is independent of the structure of the wave, that is of the viscosity ν. As (7.14) shows, the viscosity determines only the spatial scale of the transition region, that is, the 'width' of the front.

The situation is completely analogous for shock waves in gases and detonation waves: the speed of propagation of these waves is determined from the laws of conservation of mass, momentum and energy alone, and does not require for its determination any consideration of the wave structure. The latter determines only the width of the transition region.

7.3 Flame: steady travelling wave of the second kind

We now consider travelling waves of the second kind, for which the speed of propagation cannot be found from conservation laws alone but is determined by analysis of the structure.

A rigorous mathematical investigation of travelling waves in nonlinear problems with dissipation was first undertaken in the fundamental work of Kolmogorov, Petrovskii and Piskunov (1937), carried out in connection with a biological problem concerning the speed of propagation of a gene that has an advantage in the struggle for life. A remarkable study of this phenomenon was developed independently and simultaneously by Fisher (1937) (see also the book: Murray, 1977). To describe the structure of the transition zone near the boundary of the domains of habitation of genes of both types (advantaged and disadvantaged) they obtained the nonlinear diffusion equation

$$\partial_t v - \kappa \partial_{xx}^2 v = F(v) \,, \tag{7.17}$$

where v is the gene concentration, and $F(v)$ is a continuous function that is differentiable the necessary number of times, defined in the interval $0 \le v \le 1$ and having, in accordance with the physical meaning of the problem, the following properties:

$$F(0) = F(1) = 0; \quad F(v) > 0 \quad (0 < v < 1);$$
$$F'(0) = \alpha > 0; \quad F'(v) < \alpha \quad (0 < v < 1). \tag{7.18}$$

Under these conditions (7.17) has a solution of travelling-wave type, $v = V(\zeta)$, $\zeta = x - \lambda t + c$, satisfying the conditions $v(-\infty) = 1$, $v(\infty) = 0$ for all speeds of propagation λ greater than or equal to $\lambda_0 = 2(\kappa\alpha)^{1/2}$ and for arbitrary c. It is of prime importance that among these solutions only that corresponding to the lowest speed of propagation can be an asymptotic representation as $t \to \infty$ of solutions of the initial-value problem with conditions, as in (7.16), of transitional type:

$$v(x,0) \equiv 1 \quad (x \leq a), \quad 0 < v(x,0) < 1 \quad (a < x < b),$$
$$v(x,0) \equiv 0 \quad (x \geq b). \tag{7.19}$$

In other words, it turns out that the direct consideration of solutions of travelling-wave type gives a continuous 'spectrum' of possible speeds of propagation $\lambda \geq \lambda_0 = 2(\kappa\alpha)^{1/2}$, but only the solution corresponding to the lowest point $\lambda = \lambda_0$ of this spectrum can be an asymptotic solution as $t \to \infty$ of the initial-value problem with conditions of transitional type; the remaining travelling waves are unstable. The quantity λ_0 determines the required speed of propagation of the gene that has an advantage in the struggle for life.

Note that the condition $F'(v) < \alpha$ is not necessary for the establishment of a wave having propagation velocity $\lambda_0 = 2(\kappa\alpha)^{1/2}$. This was shown using analytical investigation and numerical experiment by Aldushin, Zeldovich and Khudyaev (1979).

We now consider in more detail the rather similar problem of thermal flame propagation in gaseous mixtures (Taffanel, 1913, 1914; Daniell, 1930; Zeldovich and Frank–Kamenetskii, 1938a, b; Zeldovich, 1948). We shall formulate the simplest schematization of the problem. Suppose that in the course of a reaction a component of a gaseous mixture, whose concentration we denote by n, is annihilated. The reaction rate q, that is, the mass of combustible matter annihilated in unit volume in unit time, depends on the concentration n and the temperature θ. We introduce the notation

$$q = \frac{1}{\tau}\Phi(n, \theta), \tag{7.20}$$

where Φ is a function having the dimensions of density, and τ is a constant with the dimension of time – the characteristic time of the reaction – a quantity that is ordinarily very small in comparison with a characteristic time for the large-scale motion of the gas. It is known from physical chemistry that the temperature dependence of reaction rates is very strong: a small change in temperature greatly changes the reaction rate. We shall assume that this reaction is irreversible, so that $\Phi \geq 0$. Furthermore the original state of the gaseous mixture, $n = 1$, $\theta = \theta_1$, is assumed to be uniform and stable. For this, it is *sufficient* that the function $\Phi(n, \theta)$ be equal to zero not only for the initial temperature $\theta = \theta_1$ but also in some interval of temperature $\theta_1 \le \theta \le \theta_1 + \Delta$ close to it (the meaning of this condition will be elucidated below). It is obvious also that the reaction does not take place in the absence of combustible matter. Thus, it is assumed that the function $\Phi(n, \theta)$ satisfies the conditions

$$\Phi(n,\theta) \ge 0; \quad \Phi(n,\theta) \equiv 0, \quad (0 \le n \le 1, \ \theta_1 \le \theta \le \theta_1 + \Delta); \quad \Phi(0,\theta) = 0.$$
$$(7.21)$$

The velocity of the gas motion due to the spreading of the flame is small compared with the speed of sound; therefore we can neglect the compressibility of the gas and assume that the density of the gaseous mixture depends only on the temperature[t]: $\rho = \rho(\theta)$. Finally, the reaction is assumed to be exothermic: combustion yields a release of heat. We denote by Q the thermal effect of the reaction, that is, the amount of heat released upon combustion of a unit mass of combustible gas. In accordance with what has been said, the system of basic equations of motion for the mixture of combustible gas and the products of combustion formed in the course of the reaction can be written in the form

$$\partial_t(\rho v_i) + \partial_\alpha(\rho v_i v_\alpha) = -\partial_i p,$$
$$\partial_t \rho + \partial_\alpha(\rho v_\alpha) = 0,$$
$$\rho = \rho(\theta),$$
$$\partial_t(\rho n) + \partial_\alpha(\rho n v_\alpha) = \partial_\alpha(\rho D \partial_\alpha n) - \frac{1}{\tau}\Phi(n,\theta),$$
$$\partial_t(\rho \sigma \theta) + \partial_\alpha(\rho \sigma \theta v_\alpha) = \partial_\alpha(k \partial_\alpha \theta) + \frac{Q}{\tau}\Phi(n,\theta).$$
$$(7.22)$$

The first three equations are the usual equations for the motion of an incompressible fluid whose density depends on temperature (where repeated Greek indices α indicate summation from $\alpha = 1$ to $\alpha = 3$). In addition we have the equations for the balance of mass of combustible gas and for the conservation of energy. In these equations the v_i are the components of the velocity vector of the mixture, p is the gas pressure, k is the coefficient of thermal conductivity, D is the diffusion coefficient,

---

[t] Without essential loss of generality we can neglect the difference between the density of the combustible gas and the density of the combustion products at the same temperature, so that the density of the mixture does not depend on the concentration. This assumption is by no means crucial and is made only to simplify the calulations.

and σ is the specific heat at constant pressure; we will assume the last
two coefficients to be constant.

The problem under consideration has two length scales that differ substantially in magnitude: the inner scale $L_1 = (D\tau)^{1/2}$ characterizing the
size of the region in which the processes of chemical reaction, diffusion,
and heat transfer occur, and the outer scale $L_2 = L$ characterizing the
size of the container or combustion chamber, the diameter of the burner,
etc. In view of the great disparity in these two scales it is natural to
apply to this problem the method of matched asymptotic expansions
(Van Dyke, 1975; Cole, 1968; Lagerstrom and Casten, 1972; Kevorkian
and Cole, 1980; Hinch, 1991). We first consider the 'outer' asymptotic
expansion of the solution, that is, we change to dimensionless variables
in which we take the outer scale L as the length scale and $L/(D/\tau)^{1/2}$
as the time scale. Then the equations for the balance of mass of combustible gas and for the conservation of energy, the last two equations
of (7.22), assume the form

$$\epsilon[\partial_\Omega(\rho n) + \partial_\alpha(\rho n V_\alpha)] = \epsilon^2 \partial_\alpha(\rho \,\partial_\alpha n) - \Phi(n, \theta)\,,$$

$$\epsilon[\partial_\Omega(\rho \sigma \theta) + \partial_\alpha(\rho \sigma \theta V_\alpha)] = \epsilon^2 \partial_\alpha \left(\frac{k}{D}\partial_\alpha \theta\right) + Q\Phi(n, \theta)\,; \qquad (7.23)$$

here Ω is the dimensionless 'slow' time $t(D/\tau)^{1/2}/L$; $V_\alpha = v_\alpha/(D/\tau)^{1/2}$,
the operator ∂_α is taken into the dimensionless spatial variables associated with L, and $\epsilon^2 = L_1^2/L^2 \ll 1$. Thus everywhere except in the
narrow region in which the gradients of temperature and concentration
are large (of order $1/\epsilon$) we can assume the reaction rate to be equal to
zero. The intermediate transition region must necessarily be narrow, of
relative width not greater than ϵ, since the changes of temperature and
concentration in it are bounded and the gradient is of order $1/\epsilon$. Hence
it follows that the whole region occupied by the gas splits (Figure 7.1)
into (1) a region of cold unburnt gas, where the reaction has not yet
started since the gas has not yet warmed up; (2) a region occupied by
the hot products of combustion, where the reaction no longer continues
since all the combustible matter there has burned; and a narrow transition region where the combustion reaction is going on and transport
processes, diffusion and heat transfer, are taking place. If we pass to
the limit $\epsilon \to 0$, that is, to the first order approximation, the transition
region becomes a discontinuity surface on which occur jumps in speed,
density, temperature, and concentration, but not in pressure. The speed
of propagation through the gas of the discontinuity surface – the flame –
(the 'normal' speed of the flame) is not determined by the equations of

motion and the conditions of balance of mass, momentum and energy on
the discontinuity surface. One could regard this quantity as a physico-
chemical constant that is defined independently; for example, it might
be determined from experiments. Thus one obtains a closed system of
relations, the so-called gas-dynamic theory of combustion (cf. Landau
and Lifschitz, 1987). For an analytic determination of the normal speed
of propagation of the flame we must turn to the 'inner' asymptotic expansion of the solution and consider the phenomena in the transition
zone, taking $L_1 = (D\tau)^{1/2}$ as the characteristic length scale and τ as
the characteristic time. We choose the direction of the normal to the
median of the transition zone (Figure 7.1), as the direction of the coordinate x, measured from the median, and change to a dimensionless
variable $\xi = x/L_1$. Because of the narrowness of the transition zone,
only derivatives with respect to ξ are of order unity. Derivatives with
respect to the other spatial variables that would be of order unity in the
outer scale L are negligibly small in the new scale.

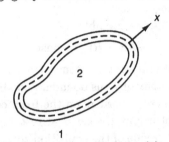

Figure 7.1. The region of motion splits into (1) a region occupied by
the cold combustible mixture, (2) a region occupied by the products of
combustion, and a narrow transition region in which the chemical reaction
and the processes of diffusion and thermal conduction are occurring.

Keeping only leading terms in the equations for the balance of combustible matter and energy, we write these equations in the form

$$\rho\partial_\omega n + \rho V \partial_\xi n = \partial_\xi(\rho\partial_\xi n) - \Phi(n,\theta)\,,$$

$$\rho\sigma\partial_\omega\theta + \rho\sigma V \partial_\xi\theta = \partial_\xi\left(\frac{k}{D}\partial_\xi\theta\right) + Q\Phi(n,\theta)\,. \qquad (7.24)$$

Here $V = v_x/(D/\tau)^{1/2}$, v_x being the component of the velocity of the
mixture along the x-axis, and $\omega = t/\tau$ is the 'faster' time, i.e., the
dimensionless time referred to the scale τ. Furthermore, to the same
approximation the equation for the conservation of mass is written in
the form

$$\partial_\omega\rho + \partial_\xi(\rho V) = 0\,. \qquad (7.25)$$

We shall seek a solution of (7.24), (7.25) of steady travelling-wave type:

$$n = N(\zeta), \quad \theta = \Theta(\zeta), \quad V = V(\zeta), \quad \zeta = \xi - \lambda\omega + c, \qquad (7.26)$$

where c is a constant that cannot be determined in the course of construction of the travelling-wave solution (it may be obtained by matching with the 'outer' expansion) and λ is the speed of propagation of the travelling wave, which is unknown and subject to determination. Substituting (7.26) into (7.24) and (7.25), and keeping in mind that $\rho = \rho(U(\zeta)) = R(\zeta)$, we obtain for the determination of the unknown functions N, Θ, and V the system of ordinary differential equations

$$-\lambda R\frac{dN}{d\zeta} + RV\frac{dN}{d\zeta} = \frac{d}{d\zeta}\left(R\frac{dN}{d\zeta}\right) - \Phi(N,\Theta),$$

$$-\lambda R\sigma\frac{d\Theta}{d\zeta} + RV\sigma\frac{d\Theta}{d\zeta} = \frac{d}{d\zeta}\left(\frac{k}{D}\frac{d\Theta}{d\zeta}\right) + Q\Phi(N,\Theta), \qquad (7.27)$$

$$-\lambda\frac{dR}{d\zeta} + \frac{d}{d\zeta}RV = 0.$$

Integrating the last equation, we obtain

$$-\lambda R + RV = \text{const.} \qquad (7.28)$$

The distributions of temperature, concentration, and velocity in the transition zone must satisfy obvious boundary conditions: on one side of the transition zone, where it borders on the fresh combustible mixture, combustion has not yet begun, the gas is at rest, and its temperature is prescribed. On the other side of the transition zone, combustible matter is fully burnt. According to standard asymptotic procedure, in view of the smallness of the inner scale L_1 compared with the outer L, the first boundary condition should be imposed at $\zeta = \infty$ and the second at $\zeta = -\infty$:

$$N(\infty) = 1, \quad \Theta(\infty) = \theta_1, \quad N(-\infty) = 0, \quad V(\infty) = 0. \qquad (7.29)$$

Substituting these conditions into (7.28), we reduce this relation to the form

$$\lambda(\rho_0 - R) + RV = 0. \qquad (7.30)$$

Here ρ_0 is the density of the fresh combustible mixture. Substituting (7.30) into the first two equations of (7.27), we reduce them to the form

$$-\lambda\rho_0\frac{dN}{d\zeta} = \frac{d}{d\zeta}\left(\rho\frac{dN}{d\zeta}\right) - \Phi(N,\Theta),$$

$$-\lambda\rho_0\frac{d\Theta}{d\zeta} = \frac{d}{d\zeta}\left(\frac{k}{D\sigma}\frac{d\Theta}{d\zeta}\right) + \frac{Q}{\sigma}\Phi(N,\Theta). \qquad (7.31)$$

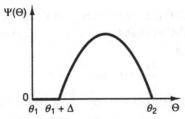

Figure 7.2. The function $\Psi(\Theta)$ vanishes in some interval close to $\Theta = \theta_1$.

It is known from physical chemistry that if the combustible matter and the products of combustion have close molecular weights, one can assume the quantity $k/\rho\sigma D$ to be equal to unity. Under this assumption, multiplying the first equation of (7.31) by Q/σ and adding the two equations we find that the system (7.31) has the integral

$$\frac{QN(\zeta)}{\sigma} + \Theta(\zeta) = \text{const}, \qquad (7.32)$$

which is called in the theory of combustion the Lewis–von Elbe similarity law for the fields of concentration and temperature. From (7.29) we find const $= \theta_1 + Q/\sigma$, and from this and (7.32) we obtain

$$\Theta(-\infty) = \theta_1 + Q/\sigma = \theta_2. \qquad (7.33)$$

Using the similarity law (7.32), one can decompose (7.31) and reduce it to a single equation for the temperature,

$$\lambda\frac{d\Theta}{d\zeta} + \frac{d}{d\zeta}\left[\frac{\rho(\Theta)}{\rho_0}\frac{d\Theta}{d\zeta}\right] + \Psi(\Theta)\frac{\rho_0}{\rho(\Theta)} = 0. \qquad (7.34)$$

Here

$$\Psi(\Theta) = \frac{Q}{\rho_0\sigma}\Phi\left(1 - \frac{\sigma}{Q}(\Theta - \theta_1), \Theta\right)\frac{\rho(\Theta)}{\rho_0}. \qquad (7.35)$$

By assumption, $\Phi(n, \Theta) \equiv 0$ for $\theta_1 \leq \Theta \leq \theta_1 + \Delta$[†]. Hence, and since $\rho(\Theta)$ is positive and bounded, $\Psi(\Theta)$ is identically equal to zero in the interval $\theta_1 \leq \Theta \leq \theta_1 + \Delta$, vanishes for $\Theta = \theta_2$, and is positive for $\theta_1 + \Delta < \Theta < \theta_2$ (Figure 7.2). Setting $p = [\rho(\Theta)/\rho_0]d\Theta/d\zeta$ and taking Θ as independent variable, we reduce (7.34) to the form

$$p\frac{dp}{d\Theta} + \lambda p + \Psi(\Theta) = 0. \qquad (7.36)$$

[†] As remarked above, this condition guarantees the stability of the original state. In fact, let us set $\partial_\xi n$ and $\partial_\xi\theta$ identically equal to zero in (7.24), so that n and θ depend only on time. Then, if the assumed condition is satisfied, a small change in the temperature of the gas mixture does not cause a reaction to start.

It follows from (7.29) that the solution of (7.36) of interest to us satisfies the obvious condition that the heat flux vanishes on the boundaries of the transition zone:

$$\frac{d\Theta}{d\zeta} = 0 \quad \text{for} \quad \zeta = \pm\infty. \tag{7.37}$$

From this and (7.29) we obtain boundary conditions for the function $p(\Theta, \lambda)$:

$$p = 0, \quad \Theta = \theta_1; \quad p = 0, \quad \Theta = \theta_2. \tag{7.38}$$

7.4 Nonlinear eigenvalue problem

We have again, as in the case of self-similar solutions of the second kind, a nonlinear eigenvalue problem: (7.36) is an equation of the first order and (7.38) gives us two boundary conditions. We shall show, following Zeldovich (1948), that there exists a unique eigenvalue λ for which the desired solution exists. We consider the phase portrait of (7.36) in the region of interest to us in the Θp-plane (Figure 7.3). At $\Theta = \theta_2$ and $p = 0$, (7.36) has a singular point of saddle type. Through this singular point pass two separatrices with slopes $-\lambda/2 \pm [\lambda^2/4 - \Psi'(\theta_2)]^{1/2}$; since $\Psi'(\theta_2) < 0$, the slope of one of the separatrices is positive, the other negative. It is clear that only the separatrices can satisfy the second condition of (7.38). Furthermore, for $\lambda = 0$, (7.36) can be integrated in finite form: the solutions satisfying (7.38) for $\Theta = \theta_2$ have the form

$$p = \pm \left\{ 2 \int_{\Theta}^{\theta_2} \Psi(\Theta) d\Theta \right\}^{1/2}, \tag{7.39}$$

so that the ordinates of the points of intersection of the corresponding integral curves with the vertical axis are

$$p_1 = \left\{ 2 \int_{\theta_1}^{\theta_2} \Psi(\Theta) d\Theta \right\}^{1/2} > 0, \; p_2 = - \left\{ 2 \int_{\theta_1}^{\theta_2} \Psi(\Theta) d\Theta \right\}^{1/2} < 0, \tag{7.40}$$

We now consider the function $q(\Theta, \lambda) = \partial_\lambda p$ for all solutions of (7.36) satisfying the second condition of (7.38). It is clear that $q(\theta_2, \lambda) \equiv 0$ since $p(\theta_2, \lambda) \equiv 0$. Differentiating (7.36), we obtain for the function q the equation

$$\frac{dq}{d\Theta} = -1 + \frac{\Psi(\Theta)}{p^2} q. \tag{7.41}$$

Close to the point $\Theta = \theta_2$ the separatrices behave, according to the

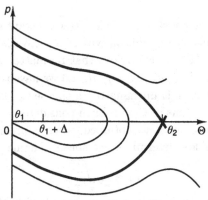

Figure 7.3. Phase portrait: picture of the integral curves of the first-order equation (7.36).

above, like $p = (\Theta - \theta_2)\{-\lambda/2 \pm [\lambda^2/4) - \Psi'(\theta_2)]^{1/2}\}$. Differentiating with respect to λ, we find that the corresponding curves $q(\Theta, \lambda)$ behave near to $\Theta = \theta_2$ like $q = K(\Theta - \theta_2)$, where

$$K = \left\{ -\frac{1}{2} \pm \frac{\lambda}{2\sqrt{\lambda^2 - 4\Psi'(\theta_2)}} \right\}$$

is negative for both separatrices, that is, $q > 0$ for $\Theta < \theta_2$. Furthermore there cannot be an intersection of the curve $q(\Theta, \lambda)$ with the axis $q = 0$ at some point intermediate between θ_1 and θ_2, because at a point of intersection one would have $dq/d\Theta = -1$, which is geometrically impossible. Thus, $q(\theta_1 + \Delta, \lambda) > 0$. But for $\theta_1 \leq \Theta \leq \theta_1 + \Delta$ we have $\Psi(\Theta) \equiv 0$, and from this and (7.41) we get $q(\theta_1, \lambda) = q(\theta_1 + \Delta, \lambda) + \Delta > \Delta$. Since

$$p(\theta_1, \lambda) = p(\theta_1, 0) + \int_0^\lambda q(\theta_1, \lambda) d\lambda > -\left\{ 2 \int_{\theta_1}^{\theta_2} \Psi(\Theta) d\Theta \right\}^{1/2} + \lambda\Delta, \quad (7.42)$$

it follows that one can find a value $\lambda = \lambda_0$, and moreover only one, such that the lower separatrix reaches the point $p = 0$, $\Theta = \theta_1$, that is, satisfies all the conditions of the problem.

Thus, the existence and uniqueness of the solution of the nonlinear eigenvalue problem is proved. Using the methods developed by Kolmogorov, Petrovsky and Piskunov (1937), Kanel' (1962) proved that this solution is an asymptotic representation as $t \to \infty$ of the solutions of a certain naturally defined class of initial-value problems with transitional-type conditions. We note that in the gene propagation problem as well as in the flame propagation problem, direct construction of a solution of travelling-wave type, $\theta = \Theta(\xi - \lambda\omega + c)$ determines the solution to within the constant c. This latter constant can be found only by matching an invariant solution with a non-invariant solution of the original problem. Here it is obvious that no matter what intermediate state of the system $V(\xi, \omega)$, $\theta(\xi, \omega)$, $n(\xi, \omega)$ we have taken as initial, the value of the constant c is unchanged. In this sense the constant c is an integral of the equations of the problem considered (see Lax, 1968).

The eigenvalue λ_0 we have obtained, when expressed in the original dimensional variables, determines the speed of flame propagation:

$$m = \lambda_0 \sqrt{\frac{D}{\tau}}. \tag{7.43}$$

Furthermore, since the 'faster' time was involved in the problem of determining the speed of the travelling wave, it is clear, that in the natural outer time scale, passage to the asymptotics occurs in fact very quickly, and the 'pre-asymptotic' evolution of the solution is of no value.

7.5 Flame propagation in a reacting mixture: an intermediate asymptotics

In the above presentation it was assumed that the chemical reaction does not occur at the initial temperature of the mixture: $\Phi(n, \theta_1) = 0$. Moreover, a strong and apparently rather artificial condition was additionally imposed on the reaction rate: that it also vanishes in a certain temperature interval close to the initial temperature, $\Phi(n, \theta) \equiv 0$, $\theta_1 \leq \theta \leq \theta_1 + \Delta$. However, in physical chemistry Arrhenius' law for the reaction rate is commonly assumed,

$$\Phi(n, \theta) = An^p \exp\left(-\frac{E}{R\theta}\right), \tag{7.44}$$

where the constants A and p are the pre-exponent and the reaction order, E is the activation energy of the chemical reaction, and R is the universal gas constant. The relation (7.44) gives for practically occurring values of the initial temperature of the mixture, values of the reaction rate that although rather small are, however, finite. It is of interest to find out whether this means that the solution to the nonlinear eigenvalue problem for flame propagation does not exist, in which case there would be no intermediate asymptotics of travelling-wave type in this problem.

In fact neither of the conditions imposed above on the reaction rate is necessary. It is sufficient to have the reaction rate at the initial temperature much less than maximal for the process under consideration. If so, an intermediate asymptotics for the distributions of temperature, concentration, etc. will remain a solution of travelling-wave type – a propagating flame solution.

Let us demonstrate this. To simplify let us make some additional assumptions that are by no means crucial. We assume, at first, that the Lewis–von Elbe similarity law for the temperature and concentration fields,

$$\frac{Qn}{\sigma} + \theta = \theta_2 = \text{const}, \tag{7.45}$$

is valid also at the stage of non-steady flame propagation; here θ_2 is, as before, the constant temperature of the burnt mixture. Furthermore we neglect the gas extension in combustion and the gas motion arising due to extension. Therefore we assume that the density of the gas mixture does not depend on temperature. Finally, we assume that all thermophysical properties of the mixture are constant and that $k/\rho\sigma D = 1$.

Under these assumptions the energy equation (the second equation of (7.24)) can be rewritten in the form

$$\partial_\omega \theta = \partial^2_{\xi\xi}\theta + f(\theta) \tag{7.46}$$

where

$$f(\theta) = \frac{Q}{\rho\sigma}\Phi\left(\frac{(\theta_2 - \theta)\sigma}{Q}, \theta\right). \tag{7.47}$$

Let us consider for equation (7.46) the 'ignition problem' with the following initial condition at $\omega = 0$:

$$\theta(\xi, 0) = \theta_2 \quad (|\xi| \le \xi_0), \quad \theta(\xi, 0) = \theta_1^0 \quad (|\xi| \ge \xi_0). \tag{7.48}$$

where θ_1^0 is a constant. The physical meaning of the initial-value problem (7.48) is obvious: it is a model of flame initiation by a certain mass of burnt gas. The constant ξ_0 defines the value of this mass as equal (per unit area of the flame surface) to $2\rho\xi_0\sqrt{D\tau}$. The constant θ_1^0 is the initial temperature of the cold gas mixture. From symmetry considerations we can construct the solution at $\xi \ge 0$ simply by assuming, at $\xi = 0$, the condition of thermal insulation,

$$\partial_\xi\theta = 0. \tag{7.49}$$

Let the temperature θ_1^0 be so small that the reaction rate at this temperature and the quantity $\delta = f(\theta_1^0)/\theta_1^0$ are also small. Then we can assume that in a certain temperature interval near $\theta = \theta_1^0$ the function f can be represented in the form $f(\theta) = \delta F(\theta)$, where $\delta \ll 1$ and the quantity $F(\theta)$ is of order θ.

Taking into account the reaction in the region ahead of the flame front means that the condition $\Phi(n, \theta_1) = 0$ is replaced by the relation

$$\frac{d\theta_1}{d\omega} = f(\theta_1) = \delta F(\theta_1) \qquad (7.50)$$

which is obtained from the basic equation of energy conservation in the reacting mixture (see (7.22)) if the temperature gradients are negligibly small. Let us introduce now a new dimensionless 'slow' time $\bar{\Omega} = \omega\delta$ based on the time scale τ/δ of the reaction in the region ahead of the flame front. Then the relation (7.50) can be written in the form of a system,

$$\frac{d\theta_1}{d\bar{\Omega}} = F(\theta_1), \quad \frac{d\bar{\Omega}}{d\omega} = \delta. \qquad (7.51)$$

Prescribing the initial temperature of the mixture ahead of the front, θ_1^0 and integrating (7.51) we obtain the dependence of the temperature ahead of the front on the slow time $\bar{\Omega}$,: $\theta_1(\bar{\Omega})$. It is convenient to introduce the relative variation of temperature inside the reaction zone according to the relation

$$\mu = \frac{\theta(\xi, \omega) - \theta_1(\bar{\Omega})}{\theta_2 - \theta_1(\bar{\Omega})}. \qquad (7.52)$$

Substituting (7.52) into equation (7.46) we obtain an equation for μ in the form

$$\partial_\omega \mu = \partial_{\xi\xi}^2 \mu + f_1(\mu, \bar{\Omega}) \qquad (7.53)$$

where

$$f_1(\mu, \bar{\Omega}) = \frac{[f(\theta_1 + (\theta_2 - \theta_1)\mu) - f(\theta_1)(1 - \mu)]}{\theta_2 - \theta_1(\bar{\Omega})}. \qquad (7.54)$$

The function $f_1(\mu, \bar{\Omega})$, in contrast with $f(\theta)$, vanishes both at $\mu = 0$, $\theta = \theta_1$ and $\mu = 1$, $\theta = \theta_2$.

To zeroth order in δ the second relation of the system (7.51) gives $d\bar{\Omega}/d\omega = 0$, so that the function $f_1(\mu, \bar{\Omega})$ ceases to depend on the time ω explicitly: $\bar{\Omega}$ becomes a parameter. Therefore we can again construct a solution to equation (7.53) of travelling-wave type,

$$\mu = \mu(\zeta), \quad \zeta = \xi - \lambda(\bar{\Omega})\omega + c(\bar{\Omega}), \qquad (7.55)$$

which represents an intermediate asymptotics of the solution to the non-invariant problem under consideration, (7.46)–(7.48), in the transition region. Substituting (7.55) into equation (7.53) and using the boundary conditions $\mu = 1$ at $\xi = -\infty$, $\mu = 0$ at $\xi = \infty$, we obtain the nonlinear eigenvalue problem

$$\frac{d^2\mu}{d\zeta^2} + \lambda(\bar{\Omega})\frac{d\mu}{d\zeta} + f_1(\mu, \bar{\Omega}) = 0, \quad \mu(-\infty) = 1, \quad \mu(\infty) = 0, \qquad (7.56)$$

where the solution determines the functions $\mu(\zeta, \bar{\Omega})$ and $\lambda = \lambda_0(\bar{\Omega})$. The function $c(\bar{\Omega})$ in (7.55) remains undetermined when an invariant solution of travelling-wave type is constructed directly. It can be found numerically by matching with the asymptotics of the original non-invariant solution.

In Figure 7.4 the results are presented of a comparison of the numerical solution to the problem (7.46)–(7.48) with the intermediate asymptotics, i.e., the solution of travelling-wave type (7.55). The function $c(\bar{\Omega})$ in the solution (7.55) was determined by requiring coincidence of the points corresponding to $\mu = 0.5$ in the invariant solution and in the numerical solution of the non-invariant problem. As it is seen, the intermediate asymptotics in the transitional region is close to the solution obtained by numerical calculation.

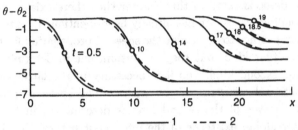

Figue 7.4. The numerical experiment shows good agreement of the numerical solution and the intermediate asymptotics of travelling-wave type, in the transitional region. Solid line, the numerical solution at various time instants; broken line, intermediate asymptotics obtained by solution of a nonlinear eigenvalue problem. The points of forced coincidence are marked.

The analysis presented in this section has followed, with slight modifications, the papers of Ya.B. Zeldovich and his co-workers (Zeldovich, 1978; Aldushin, Zeldovich and Khudyaev, 1979). It is interesting to note that as the numerical calculations showed (Aldushin, Zeldovich and Khudyaev, 1979) the dependence of the propagation velocity on the slow time $\bar{\Omega}$ agrees over a large temperature interval, with the Kolmogorov–Petrovsky–Piskunov formula

$$\lambda_0(\bar{\Omega}) = 2\sqrt{f_1'(0, \bar{\Omega})} \tag{7.57}$$

even though the condition $f_1'(\mu, \bar{\Omega}) < f_1'(0, \bar{\Omega})$ is not satisfied here. An analytical explanation of this was given in the same paper.

Let us now consider in which cases the reaction ahead of the flame must be taken into account. In the problem under consideration we have two small parameters: the ratio of the front thickness to the external length scale, $\epsilon = \sqrt{D\tau}/L$, and the ratio of the dimensionless heat generation rate to the initial temperature $\delta = f(\theta_1^0)/\theta_1^0$. The variation of the mixture's composition far from the flame occurs in times of the order τ/δ; a characteristic time of flame propagation over the vessel is $L/\sqrt{D/\tau}$; the speed of the flame is of order of magnitude $\sqrt{D/\tau}$. The occurrence of reaction in the region ahead of the flame should be taken into account if the second characteristic time has the order of magnitude of the first one or is even larger,

$$\frac{\tau}{\delta} \lesssim \frac{L}{\sqrt{D/\tau}}, \tag{7.58}$$

i.e., if $\delta \gtrsim \epsilon$. If $\delta \ll \epsilon$, the condition $\Phi(n, \theta_1) = 0$ is acceptable and, even more, it can be assumed that $\Phi(n, \theta) \equiv 0$ within some temperature interval near the initial temperature of the mixture.

We have demonstrated in this chapter that there exist two types of steady travelling waves. As has already been mentioned, 'external' conservation laws suffice to determine the speed of propagation for travelling waves of the first kind, but they are insufficient to determine this for waves of the second kind, and it is necessary in the latter case to invoke the internal structure of the waves. The speed of propagation of travelling waves of the second kind is determined, in fact, by the condition for global existence of the internal structure, that is, by the condition for the existence of a solution of travelling-wave type, to the equations of motion in the transition region that satisfies the boundary conditions for this region[†].

This situation corresponds to the classification of self-similar solutions considered above. In fact, for a solution of the type

$$\theta = \Theta(x - \lambda t + c),$$

we again set $x = \ln \xi$, $t = \ln \tau$, $c = -\ln A$. Then this solution can be written, as we have seen already, in the form

$$\Theta(x - \lambda t + c) = \Theta\left(\ln \frac{\xi}{A\tau^\lambda}\right) = \Theta_1\left(\frac{\xi}{A\tau^\lambda}\right), \tag{7.59}$$

that is, in self-similar form. It is obvious that the classification of solutions of travelling-wave type formulated above carries over into the language of self-similar solutions. In particular, the exponent λ in the expression for the self-similar variable corresponds to the speed of propa-

---

[†] Sometimes (see below) the speed of propagation is determined non-uniquely upon consideration of the structure. This means that it depends on the initial conditions of the original problem, the asymptotic solution of which serves as the travelling wave.

gation in solutions of travelling-wave type. The constant A corresponds to the phase shift. Therefore the classification of such solutions, into solutions for which the speed of propagation can be found from conservation laws at the shock front alone, and solutions for which this speed is obtained from the conditions for global existence of the inner structure, corresponds to the classification of self-similar solutions into those solutions of the first and second kind. The correspondence between self-similar solutions and travelling waves will be used more than once in what follows.

Invariant solutions: asymptotic conservation laws, spectrum of eigenvalues, and stability

8.1 Asymptotic conservation laws

For problems in mathematical physics leading to self-similar solutions of the first kind, such as the heat source problem, and the very intense explosion problem, there exist integral conservation laws, valid even at the initial, pre-self-similar stage of the motion. Such integral conservation laws do not exist for 'modified' problems leading to self-similar solutions of the second kind. They are replaced by certain non-integrable relations that do not allow one to determine the basic constant entering a self-similar solution.

Nevertheless, for these modified problems leading to self-similar solutions of the second kind there exist instead 'asymptotic' conservation laws, valid in the intermediate-asymptotic range only, which reveal the existence of more complicated integrals of motion. The following examples will clarify this general idea. We begin with the modified heat source problem considered in chapters 3 and 6, which is an initial-value problem for the nonlinear equation

$$\partial_t \theta = \kappa \partial_{xx}^2 \theta \quad (\partial_t \theta \geq 0), \quad \partial_t \theta = \kappa (1 + \epsilon) \partial_{xx}^2 \theta \quad (\partial_t \theta \leq 0) \qquad (8.1)$$

under the initial condition

$$\theta(x, 0) = \theta_0(x) = \frac{Q}{\sqrt{2\pi} l} e^{-x^2/2l^2}, \quad (-\infty < x < \infty). \qquad (8.2)$$

At $\epsilon = 0$ we have the case of complete similarity, so that asymptotically, for $t \gg l^2/\kappa$, the solution is represented by a self-similar solution of the first kind,

$$\theta(x,t) = \frac{Q}{2\sqrt{\pi\kappa t}}e^{-x^2/4\kappa t}\,, \tag{8.3}$$

and at any time $t > 0$ the following integral conservation law holds

$$Q(t) = \int_{-\infty}^{\infty} \theta(x,t)dx = Q = \text{const.} \tag{8.4}$$

For $\epsilon > 0$ the asymptotics for $t \gg l^2/\kappa$ takes the different form of a self-similar solution of the second kind:

$$\theta(x,t) = \frac{A}{(\kappa t)^{(1+\alpha)/2}}\Phi_1\left(\frac{x}{\sqrt{\kappa t}},\epsilon\right),\quad A = \sigma Q l^\alpha\,. \tag{8.5}$$

For $|x| < x_0(t) = \xi_0\sqrt{\kappa t}$ the derivative $\partial_t\theta$ is negative; for $|x| > x_0(t)$ this derivative is positive. The dimensionless constants α, ξ_0 are determined from the solution to a nonlinear eigenvalue problem (see chapter 3); the constant σ is determined by matching (8.5) with the solution at the pre-self-similar stage. The integral conservation law (8.4) is replaced by a non-integrable relation,

$$\frac{d}{dt}\int_{-\infty}^{\infty} \theta(x,t)dx = 2\kappa\epsilon(\partial_x\theta)_{x=x_0(t)}\,, \tag{8.6}$$

which does not allow one to determine σ. Nevertheless we can define the quantity

$$Q(t) = \int_{-\infty}^{\infty} \theta(x,t)dx\,, \tag{8.7}$$

which, asymptotically, for $t \gg l^2/\kappa$, is determined by the relation

$$Q(t) = \frac{Ac}{(\kappa t)^{\alpha/2}},\quad c = \int_{-\infty}^{\infty}\Phi_1(\xi,\epsilon)d\xi\,; \tag{8.8}$$

the quantity $x_0(t)$ is asymptotically determined by the relation

$$x_0(t) = \xi_0(\epsilon)\sqrt{\kappa t}\,. \tag{8.9}$$

From (8.8) and (8.9) an asymptotic conservation law is obtained:

$$Q(t)[x_0(t)]^\alpha = Ac\xi_0^\alpha = \sigma c\xi_0^\alpha Q l^\alpha = \text{const}\,Q l^\alpha\,. \tag{8.10}$$

(From the initial condition (8.2) it follows that $Q(0) = Q$.)

The conservation law (8.10) has two important distinctions. Firstly, it holds only asymptotically, for $t \gg l^2/\kappa$, i.e. it does not hold at the pre-self-similar stage. The second property is that the constant in (8.10), which is equal to $\sigma c\xi_0^\alpha$, contains only one constant, σ, depending on the initial conditions, i.e., on pre-self-similar stage; the other constants, ξ_0, c, α, are determined by the solution to the nonlinear eigenvalue problem considered in chapter 3. The constant σ will remain invariant, if instead of the initial condition (8.2) we take $\theta(x,t)$ at an arbitrary time t. Therefore the σ is an 'integral' of the problem. However, this integral is now a more complicated functional of the initial conditions and cannot be determined by some integral relation, as it was for $\epsilon = 0$.

Let us demonstrate briefly analogous asymptotic conservation laws for other problems, considered above, having asymptotic self-similar solutions of the second kind.

For the modified very intense explosion problem considered in chapters 4 and 6 the shock wave radius r_f at large times, when it is much larger than the initial radius r_0, is represented asymptotically by the relation

$$r_f = \left(\frac{A}{\rho_0}\right)^{(1-\beta)/5} t^{2(1-\beta)/5}, \quad A = \sigma E r_0^{5\beta/(1-\beta)} \qquad (8.11)$$

where β is determined by the solution of a nonlinear eigenvalue problem and σ is determined by the initial conditions. For $\gamma_1 \neq \gamma$, in contrast with the case $\gamma_1 = \gamma$, it is impossible to determine σ from an integral conservation law.

The asymptotic relation for the bulk energy of gas motion inside the shock wave is

$$\mathcal{E}(t) = 4\pi \int_0^{r_f} \left[\rho\frac{u^2}{2} + \frac{p}{(\gamma - 1)}\right] r^2 dr = \rho_0 \frac{r_f^5}{t^2} e, \qquad (8.12)$$

where

$$e = 4\pi \int_0^1 \left[R(\zeta, \epsilon)\frac{V^2(\zeta, \epsilon)}{2} + \frac{P(\zeta, \epsilon)}{\gamma - 1}\right] \zeta^4 d\zeta, \quad \zeta = \frac{r}{r_f}. \qquad (8.13)$$

From (8.11), (8.12) the asymptotic conservation law is obtained:

$$\mathcal{E}(t)r_f^{5\beta/(1-\beta)} = Ae = \sigma e E r_0^{5\beta/(1-\beta)}. \qquad (8.14)$$

Again σ is a complicated functional of the initial conditions, determined by the evolution of the solution at the pre-self-similar stage.

For the problem of ideal fluid flow past a wedge with an opening angle α, considered in chapter 3, the intermediate-asymptotic solution for the flow potential has the form (3.15):

$$\phi = \beta U L^{-\lambda} r^{1+\lambda} \cos[(\lambda + 1)\theta + \gamma] \qquad (8.15)$$

where $\lambda = \alpha/(\pi - \alpha)$, $\gamma = -\pi\alpha/(\pi - \alpha)$ are obtained from the solution to an eigenvalue problem and β is obtained from matching (8.15) with the solution to the problem of the flow past a wedge of finite length L. For $\alpha = 0$, we have a flat plate, $\lambda = \gamma = 0$, and (8.15) gives the potential for a uniform flow with velocity U, so that $\beta = 1$. For $\alpha > 0$ the velocity components u_r, u_θ are not constant:

$$u_r = \partial_r\phi = \beta(1 + \lambda)U L^{-\lambda} r^\lambda \cos[(\lambda + 1)\theta + \gamma]$$
$$u_\theta = \partial_\theta\phi/r = -\beta(1 + \lambda)U L^{-\lambda} r^\lambda \sin[(\lambda + 1)\theta + \gamma], \qquad (8.16)$$

so that the velocity magnitude does not depend on the polar angle θ:

$$u(r) = \sqrt{u_r^2 + u_\theta^2} = \beta(1 + \lambda)UL^{-\lambda}r^\lambda. \tag{8.17}$$

The asymptotic conservation law assumes the form

$$u(r)r^{-\lambda} = \beta(1 + \lambda)B, \quad B = UL^{-\lambda}. \tag{8.18}$$

In the same way asymptotic conservation laws can be obtained for the Guderley very intense implosion problem, and the von Weizsäcker–Zeldovich impulsive loading problem considered in chapter 4. Further on in the text we will obtain the asymptotic conservation laws for many examples of interest.

The existence of asymptotic conservation laws for problems leading to self-similar solutions of the second kind was noted by Batchelor and Linden (1992).

8.2 Spectrum of eigenvalues

In determining the exponent of time in the expression for self-similar variables in self-similar solutions of the second kind or, what is the same, the speed of propagation for solutions of travelling-wave type, we have arrived at special eigenvalue problems for nonlinear operators. These problems are by their nature close to classical eigenvalue problems for linear differential operators, and for them too there arises the question of the structure of the spectrum, i.e., of the set of eigenvalues.

We recall the well-known problem of a vibrating string,

$$\partial_{tt}^2 u = \partial_{xx}^2 u + q(x)u \tag{8.19}$$

(u being the displacement, x the coordinate measured along the string, and t the time) under the conditions of fixed ends:

$$u(0, t) = u(l, t) = 0 \tag{8.20}$$

(l being the length of the string). Separating variables, we seek a solution in the form

$$u = \exp(i\sqrt{\lambda}t)\Psi(x, \lambda). \tag{8.21}$$

For the determination of $\Psi(x, \lambda)$ we thus obtain the boundary-value problem

$$\Psi''(x, \lambda) + [\lambda + q(x)]\Psi(x, \lambda) = 0,$$
$$\Psi(0, \lambda) = \Psi(l, \lambda) = 0. \tag{8.22}$$

In general, for arbitrary λ, a non-trivial (not identically equal to zero) solution to this boundary-value problem does not exist. However, there are exceptional values of λ, *eigenvalues*, for which non-trivial solutions of the boundary-value problem do exist. These eigenvalues form a set (the *spectrum*) having a certain structure; this may be discrete, continuous, mixed, etc., depending on the properties of the function $q(x)$.

One can look at all this somewhat differently. Relations (8.19) and (8.20) are invariant with respect to the two-parameter transformation group

$$u' = \alpha u, \quad t' = t + \beta, \quad x' = x. \tag{8.23}$$

This means that substituting (8.23) into (8.19) and (8.20), we again obtain the same problem but in the variables u', x', t' for arbitrary group parameters – the constants α and β. In separating variables we actually look for solutions that are invariant with respect to some one-parameter subgroup of this group. The subgroup corresponds to the following relation between the parameters α and β,

$$\alpha = \exp(-i\sqrt{\lambda}\beta), \tag{8.24}$$

and the invariant solution has the form (8.21). The eigenvalues λ determining the subgroup are found from the condition that there exists an invariant global solution of the form (8.21), i.e., a solution satisfying (8.20).

The situation is completely analogous for solutions of travelling-wave type. In order that such a solution exists, the equations and the boundary conditions must be invariant with respect to the two-parameter group of translational transformations

$$x' = x + \alpha, \quad t' = t + \beta, \quad u' = u. \tag{8.25}$$

Here too in finding a solution of travelling-wave type we seek a one-parameter subgroup of this transformation group corresponding to $\alpha = \lambda\beta + $ const, where λ is an eigenvalue, and a solution, invariant with respect to that subgroup, $u(x', t') = u(x, t)$.

The eigenvalues λ that extract from the basic group a one-parameter subgroup are also determined by the condition that there exists an invariant global solution, i.e., the invariant solution, satisfying the boundary conditions. In this case too, the spectrum of eigenvalues can have various types of structure. Thus, in the problem, considered in the previous chapter of the propagation of a gene, it is continuous and semi-bounded: $\lambda \geq \lambda_0$. In the problem of flame propagation (given the assumption that the reaction rate becomes equal to zero over some temperature interval near to the temperature of the cold mixture, as considered in section 7.3) the spectrum consists of one point. There is the peculiar situation of the remarkable Korteweg-de Vries equation, which arose initially in the theory of surface waves on shallow water (see Whitham, 1974; Lighthill, 1978), and was later encountered in numerous other problems (see Jeffrey and Kakutani, 1972; Karpman, 1975; Eilenberger, 1981; Novikov et $al.$, 1984; Drazin and Johnson, 1989; Fordy, 1990; Ablowitz and Clarkson, 1991):

$$\partial_t u + u\partial_x u + \beta \partial_{xxx}^3 u = 0 \,. \tag{8.26}$$

In the theory of surface waves, u is, to within a constant factor, the horizontal velocity component, which is constant, in the present approximation, over the channel depth; $\beta = c_0 h^2/6$, $c_0 = (gh)^{1/2}$, g is the acceleration of gravity, h the undisturbed depth of the fluid layer, t the time and x the horizontal coordinate in a system moving with speed c_0 relative to fluid at rest at infinity. An analogous equation is valid also in the corresponding approximation for the elevation of the free surface over its undisturbed level. Equation (8.26) has solutions of travelling-wave type, the so-called solitons (Figure 8.1),

$$u = \frac{u_0}{\cosh^2(\sqrt{u_0/12\beta}\,\zeta)} \tag{8.27}$$

where $\zeta = x - \lambda t + c$ and $u_0 = 3\lambda$. (The name 'soliton' reflects the particle-like behaviour of such solutions: after a 'collision' they remain the same except that 'phases' c become, generally speaking different.)

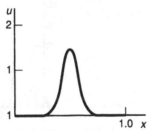

Figure 8.1 Solitary wave-soliton.

The solution (8.27) satisfies the conditions

$$u(\infty) = u(-\infty) = 0 \tag{8.28}$$

for any $\lambda > 0$; the spectrum of eigenvalues λ is continuous and semi-bounded: $\lambda \geq 0$. There is, however, an essential difference between the continuous spectrum in the problem of gene propagation and in this problem. In the former problem only the lowest point $\lambda = \lambda_0$ of the spectrum satisfies the requirement that the solution of the initial-value problem with initial data of transitional type tends to the given solution of travelling-wave type as $t \to \infty$; for all other λ this is not so, and therefore the corresponding solutions are unstable. For the Korteweg-de Vries equation a remarkable discovery was made by Gardner, Greene, Kruskal and Miura (1967) (see also Lax, 1968[†]): as $t \to \infty$ for large positive x the solution of the initial-value problem, for initial data $u(x,0)$ that decrease sufficiently rapidly at $x = \pm\infty$, is represented asymptotically (Figure 8.2) by a finite sum of solutions of the form (8.27):

$$u \sim \sum_{n=1}^{N} 2|\mu_n| \cosh^{-2} \left\{ \sqrt{\frac{|\mu_n|}{6\beta}} \left(x - \tfrac{2}{3}|\mu_n|t + c_n \right) \right\}, \qquad (8.29)$$

where the μ_n are the discrete eigenvalues of the Schrödinger operator, well-known from quantum mechanics, with potential equal to $-u(x,0)$:

$$\frac{d^2\Psi}{dx^2} + \frac{1}{6\beta}[\mu + u(x,0)]\Psi = 0, \quad \Psi(\pm\infty) = 0, \qquad (8.30)$$

and the 'phases' c_n are certain constants, also determined by the initial condition. Hence any solution of soliton type can be an intermediate asymptotics of the solution of an initial-value problem as $t \to \infty$, but exactly which one it is will be determined by the initial conditions.

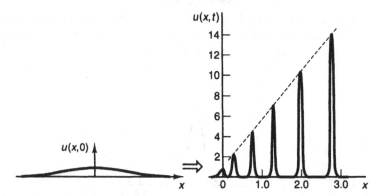

Figure 8.2. Initial elevation of the free surface of a heavy fluid in a shallow channel generates a finite series of solitary waves (solitons).

For self-similar solutions too there is an analogous situation. In fact, for a self-similar solution to exist it is necessary that the equations and boundary conditions of the idealized problem be invariant with respect to some group – a subgroup of the group of similarity transformations of

[†] The L, A-pair technique invented in this paper permitted the result to be understood from a more general point of view and analogous results to be obtained for other equations.

the independent and dependent variables. In searching for self-similar solutions one seeks a subgroup of that group and a solution that this subgroup leaves invariant. Here too, eigenvalues λ that are determined by the condition of global existence of an invariant (this time, self-similar) solution arise naturally. The examples considered above illustrate what has been said. Thus in the problem considered in chapter 3 of the filtration in an elasto-plastic porous medium, the basic equation

$$\partial_t \theta = \kappa \partial_{xx}^2 \theta \quad (\partial_t \theta \geq 0), \quad \partial_t \theta = \kappa_1 \partial_{xx}^2 \theta \quad (\partial_t \theta \leq 0) \tag{8.31}$$

and the condition at infinity

$$\theta(\pm\infty, t) = 0 \quad (t \geq 0), \tag{8.32}$$

and also the condition of continuous matching, with continuous x-derivative, at the points $x = \pm x_0(t)$ where the quantity $\partial_t \theta$ vanishes, are all invariant with respect to the two-parameter group

$$\theta' = \alpha\theta, \quad t' = \beta^2 t, \quad x' = \beta x, \tag{8.33}$$

which is a subgroup of the three-parameter group of similarity transformations $\theta' = A_1\theta$, $t' = A_2 t$, $x' = A_3 x$. We seek a one-parameter subgroup of the group (8.33) for which $\alpha = \beta^{2\lambda}$, and also a self-similar solution $u = t^\lambda f(x/t^{1/2})$ that remains invariant under this one-parameter subgroup. The parameter λ, the eigenvalue, is determined from the condition for the global existence of a self-similar solution; the spectrum turns out to be discrete and, in fact, for $\epsilon = \kappa_1/\kappa - 1$ not too large, consists of one point only.

Analogously, in the problem of a very intense explosion with loss or deposition of energy at the wave front (chapters 4 and 6), the equations for spherically symmetric adiabatic motion of an ideal gas, the conditions at an intense shock wave, the condition $v(0, t) = 0$, and the conditions $\rho(\infty, t) = \rho_0$, $v = 0$, $p = 0$ at infinity are all invariant with respect to the two-parameter group of transformations

$$p' = \alpha^2 p, \quad \rho' = \rho, \quad v' = \alpha v, \quad r' = \alpha\beta r, \quad t' = \beta t. \tag{8.34}$$

We seek a one-parameter subgroup of this group for which $\alpha = \beta^\lambda$ and also a self-similar solution

$$\rho = \rho_0 R\left(\frac{r}{t^{1+\lambda}}\right), \quad p = \rho_0 \frac{r^2}{t^2} P\left(\frac{r}{t^{1+\lambda}}\right), \quad v = \frac{r}{t} V\left(\frac{r}{t^{1+\lambda}}\right) \tag{8.35}$$

that is invariant with respect to this subgroup. As we have seen, the spectrum of eigenvalues λ, determined by the condition for the global existence of a self-similar solution, turns out for $\gamma < 2$ to consist of one point.

In the problem of a converging very intense shock wave, first con-

sidered by Guderley (1942), which also leads to self-similar solutions of
the Bechert–Guderley type (8.35) but with other boundary conditions,
the spectrum for values of the adiabatic exponent $\gamma > \gamma_0 \cong 1.87$ turns
out (cf. Brushlinskii and Kazhdan, 1963) to be continuous and semi-
bounded. There is a conjecture due to I.M. Gelfand according to which
the intermediate asymptotics of the non-self-similar problem as $t \to 0$
(the time of collapse) selects the lowest point of the spectrum, just as in
the problem of propagation of a gene, but the question actually remains
open, since numerical calculations have not yet been able to confirm this
conjecture.

There is an instructive self-similar interpretation of the result presented above (8.29) for the Korteweg-de Vries equation (8.26). If we set
$x = \ln \xi$ $t = \ln \tau$, equation (8.26) can be written in the form

$$\tau \partial_\tau u + \xi u \partial_\xi u + \beta(\xi^3 \partial^3_{\xi\xi\xi} u + 3\xi^3 \partial^2_{\xi\xi} u + \xi \partial_\xi u) = 0. \qquad (8.36)$$

The solution of travelling-wave type (8.27) here assumes the self-
similar form

$$u = \frac{12\lambda}{2 + \eta^{\sqrt{\lambda/\beta}} + \eta^{-\sqrt{\lambda/\beta}}}, \qquad \eta = \frac{\xi}{A\tau^\lambda}. \qquad (8.37)$$

Here, $A = e^{-c}$ is constant. We note that the right-hand side of (8.37)
is not small only for η of order unity; it is small if η is either large
or small. The spectrum of eigenvalues λ, obtained by direct construction of solutions of travelling-wave type, is continuous and semibounded:
$\lambda \geq 0$. The result of Gardner, Greene, Kruskal and Miura (1967) presented above can be expressed in the following way in the self-similar
interpretation: an asymptotic solution of the initial-value problem for
(8.36) as $\tau \to \infty$ and for large ξ can be represented in the form

$$u \sim \sum_{n=1}^N 12\lambda_n \left\{ 2 + \left(\frac{\xi}{A_n \tau^{\lambda_n}}\right)^{\sqrt{\lambda_n/\beta}} + \left(\frac{\xi}{A_n \tau^{\lambda_n}}\right)^{-\sqrt{\lambda_n/\beta}} \right\}^{-1}. \qquad (8.38)$$

Thus the initial distribution $u(\xi, 0)$, which by assumption decreases
sufficiently rapidly as $\xi \to 0$ or ∞, determines N positive constants λ_1,
..., λ_N and N positive constants A_1, ..., A_N, and selects N intervals in
ξ. Here, inside each of the intervals $\xi = O(\tau^{\lambda_n})$, the asymptotics of the
solution is self-similar and has the form

$$u \sim 12\lambda_n \left\{ 2 + \left(\frac{\xi}{A_n \tau^{\lambda_n}}\right)^{\sqrt{\lambda_n/\beta}} + \left(\frac{\xi}{A_n \tau^{\lambda_n}}\right)^{-\sqrt{\lambda_n/\beta}} \right\}^{-1}. \qquad (8.39)$$

Outside the intervals mentioned the solution u is small: $u = o(1)$. Here
it is significant that in the self-similar asymptotics not only do the con-

stants A_n depend as usual on the initial conditions of the original non-idealized problem, but so also do the exponents λ_n in the expressions for the self-similar variables. We meet an analogous situation later in considering the self-similar decay of isotropic turbulence. This example has once again emphasized the insufficiency in the general case of dimensional analysis for determining the exponents of the self-similar variables.

The examples given above demonstrate the variety of possible structures of the spectrum of a nonlinear eigenvalue problem arising in the construction of a self-similar solution.

8.3 Stability of invariant solutions

8.3.1 Stability of travelling waves

The statement of the stability problem for invariant and, in particular, self-similar solutions is distinguished by certain peculiarities. In the present and following sections a general approach to the investigation of the stability of self-similar and other invariant solutions is outlined and illustrated by several instructive examples.

A simple example will immediately take us to the heart of the matter. The equation

$$\partial_t \theta = \kappa \partial^2_{xx} \theta + f(\theta)\,, \tag{8.40}$$

where $f(\theta)$ is bounded together with its first derivative and satisfies the conditions

$$f(\theta) \equiv 0 \quad (\theta_1 \le \theta \le \theta_1 + \Delta)\,, \quad f(\theta_2) = 0\,,$$
$$f(\theta) > 0 \quad (\theta_1 + \Delta < \theta < \theta_2)\,, \tag{8.41}$$

is a simplified model of thermal flame propagation (θ being the temperature), where the density of the gaseous mixture is assumed constant and the concentration of combustible matter and the temperature at any moment are related by the Lewis–von Elbe similarity law; this follows readily from what was presented in chapter 7. The equation (8.40) has a solution of travelling-wave type

$$\theta = \Theta(\zeta)\,, \quad \zeta = x - \lambda t + c\,, \tag{8.42}$$

where c is an arbitrary constant and the speed of propagation λ is uniquely determined by solving a nonlinear eigenvalue problem, the equation

$$\lambda \frac{d\Theta}{d\zeta} + \kappa \frac{d^2\Theta}{d\zeta^2} + f(\Theta) = 0\,, \tag{8.43}$$

obtained by substituting (8.42) into (8.40) with the conditions

$$\Theta(-\infty) = \theta_2, \quad \Theta(\infty) = \theta_1. \qquad (8.44)$$

It is easy to show that the solution (8.42) is a monotonically decreasing function (Figure 8.3). The function $\Theta(\zeta)$ cannot have a minimum lying between $\Theta = \theta_1$ and $\Theta = \theta_2$, because at this point one would have to have $d\Theta/d\zeta = 0$, $d^2\Theta/d\zeta^2 > 0$, $f(\Theta) > 0$, which is impossible by virtue of (8.43). Neither can the function have a maximum within these same limits, because there would then have to be a minimum between $\Theta = \theta_1$ and $\Theta = \theta_2$, which is impossible by the preceding argument.

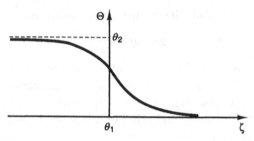

Figure 8.3. The temperature distribution in a travelling wave is monotonic.

The stability of the solution (8.42) is of paramount importance. In fact, as has already been remarked more than once, the invariant solution (8.42) is of physical interest first of all as an asymptotic representation of a certain class of solutions to the non-idealized initial-value problem for equation (8.40) with initial data of transitional type. If this solution were unstable, so that a small perturbation imposed on the temperature distribution at some moment led to a large deviation in the temperature distribution at later moments, then the solution would be physically meaningless.

Here it is necessary, however, to define stability and instability precisely. Suppose that at some moment $t = t_0$ the temperature distribution is determined by the relation

$$\theta(x, t_0) = \Theta(x - \lambda t_0 + c) + \delta\phi(x), \qquad (8.45)$$

where δ is a small parameter and $\phi(x)$ is a finite function, i.e., one equal to zero outside some finite interval; such a temperature distribution corresponds to the solution of travelling-wave type already considered plus a small local addition. At first glance, a natural definition of the

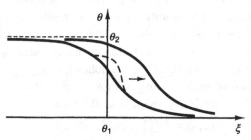

Figure 8.4. A perturbed solution (broken line) tends to a shifted unperturbed one and there is no reason to consider this as an instability.

stability of the travelling wave would appear to be the following: if a solution of the type (8.45) of any 'perturbed' initial-value problem can be represented for $t > t_0$ in the form

$$\theta(x,t) = \Theta(\zeta) + w(\zeta,t), \qquad (8.46)$$

where the function $w(\zeta,t)$ tends to zero as $t \to \infty$, then the original solution is stable; otherwise, it is not.

Just such a definition of stability was adopted for the stability of flames by Rosen (1954), who arrived at the conclusion that instability is possible in the problem just considered and indicated some approximate criteria for the stability of solutions, etc. In fact, such a definition of stability is insufficient, and must be replaced by another one; this turns out to be essential for the case of travelling waves and of self-similar and, in general, invariant solutions; it was clarified by Barenblatt and Zeldovich (1957b).

Actually, a solution of travelling-wave type is invariant under a one-parameter group of translations with respect to the coordinate and time. Hence the solution (8.42) is determined by (8.43) and (8.44) up to a constant. Consequently a definition of the stability of a travelling wave must also have the corresponding invariance. If in fact the perturbed solution tends not to the original unperturbed solution as $t \to \infty$, but to a shifted one (Figure 8.4), then there is no reason to consider this transition as an instability. Thus an invariant definition of the stability of the travelling wave (8.42) consists of the following: *a solution is stable if one can find constant a such that the solution of the perturbed problem can be represented for $t > t_0$ in the form*

$$\theta(x,t) = \Theta(\zeta + a) + w(\zeta,t), \qquad (8.47)$$

where $w(\zeta,t)$ tends to zero as $t \to \infty$; otherwise the solution is considered unstable.

In what follows we shall restrict ourselves to the investigation of linear stability. For small δ, the quantity a must be small. Expanding $\Theta(\zeta+a)$ in a series and restricting ourselves, in accord with our adoption of the linear approximation, to the first term of the expansion, we reformulate the definition of stability (8.47) in the following way: *if one can find a constant a, tending to zero along with δ, such that the solution of the perturbed problem can be represented in the form*

$$\theta(x,t) = \Theta(\zeta) + a\Theta'(\zeta) + w(\zeta,t), \tag{8.48}$$

where $w(\zeta,t) \to 0$ as $t \to \infty$, then the unperturbed solution is stable.

We shall prove the stability of the travelling wave (8.42), in the sense indicated, for any $f(\theta)$ satisfying (8.41). In (8.40) we set $\theta(x,t) = \Theta(\zeta) + \delta v(\zeta,t)$. Discarding terms of higher than first order in δ, and using the fact that $\Theta(\zeta)$ satisfies (8.43), we obtain for $v(\zeta,t)$ the equation

$$\partial_t v - \lambda \partial_\zeta v = \kappa \partial^2_{\zeta\zeta} v + f'(\Theta(\zeta))v. \tag{8.49}$$

Applying the method of separation of variables, we construct a solution of the initial-value problem for (8.49), with arbitrary initial distribution of the perturbation $v(\zeta,0)$, vanishing outside some finite interval, in the form of a Fourier series,

$$v = \sum_{n=1}^{\infty} c_n e^{-\mu_n t}\Psi(\zeta,\mu_n), \tag{8.50}$$

where the function $\Psi(\zeta,\mu_n)$ is the nth eigenfunction of the operator defined by the equation

$$\lambda\frac{d\Psi}{d\zeta} + \kappa\frac{d^2\Psi}{d\zeta^2} + [\mu + f'(\Theta(\zeta))]\Psi = 0 \tag{8.51}$$

and by the conditions of tending to zero faster than any power $|\zeta|$ for $\zeta = \pm\infty$. Here μ_n is the nth eigenvalue.

The coefficients c_n in (8.50) are determined by expanding the initial condition in series with respect to the functions $\Psi(\zeta,\mu_n)$. According to (8.50), if it could be shown that all the eigenvalues μ_n are non-negative, then the stability of the travelling wave in the sense of (8.48) would be proved. We thus note that differentiating (8.43) with respect to ζ,

$$\lambda\frac{d}{d\zeta}\left(\frac{d\Theta}{d\zeta}\right) + \kappa\frac{d^2}{d\zeta^2}\left(\frac{d\Theta}{d\zeta}\right) + f'(\Theta)\frac{d\Theta}{d\zeta} = 0, \tag{8.52}$$

we find that $d\Theta/d\zeta$ satisfies (8.51) for $\mu = 0$. Observing further that $d\Theta/d\zeta$ tends exponentially to zero as $\zeta \to \pm\infty$ we see that $d\Theta/d\zeta$ coincides to within a constant factor with the eigenfunction corresponding to $\mu = 0$. From the proof of the monotonicity of the function $\Theta(\zeta)$ given above it follows that $d\Theta/d\zeta$ does not vanish for finite ζ. But an eigenfunction of the operator Ψ has as many zeros as its ordinal number.[†] Therefore $d\Theta/d\zeta$ corresponds to the smallest eigenvalue. Since this is equal to zero, there are no negative eigenvalues in the problem; all $\mu_n \geq 0$. As for $\mu_n = 0$, this eigenvalue does not spoil the stability, since the corresponding eigenfunction is equal to $d\Theta/d\zeta$ to within a constant factor, and this eigenfunction corresponds to just a shift of the travelling wave. Thus the linear stability of the progressive wave (8.42) in the sense formulated is proved[‡]. Generally speaking the proof of linear stability is only part of the problem: the rejected nonlinear terms in principle could make the solution unstable. In this special case, however, it is not so: the global stability of the travelling-wave solution in the sense (8.47) follows from the results by Kanel' (1962) who used essentially the technique of the paper by Kolmogorov, Petrovsky and Piskunov (1937).

It is clear that under assumptions that are sufficiently broad for our purposes the considerations presented have a completely general meaning. In particular, they are easily reformulated to apply to the stability of self-similar solutions, as we now demonstrate.

8.3.2 Stability of self-similar solutions

As we have seen, self-similar solutions are determined by direct construction to within some constant A that is found from conservation laws, or by following the evolution of a non-self-similar solution if the conservation laws assume a non-integrable form.

By definition, *a self-similar solution is stable if the solution of any perturbed problem with sufficiently small perturbations can be represented in the form of a self-similar solution corresponding to a constant A' that is in general different from A plus some additional term whose ratio to the unperturbed solution tends to zero as $t \to \infty$.*[§]

Relying on this definition, we proceed to investigate the stability of the solution to the modified heat source problem that was considered in chapter 3 (Kerchman, 1971). We present this investigation here to

---

[†] In fact (8.51) reduces to self-adjoint form if one sets $\Psi = e^{-(\lambda\zeta/2\kappa)}\phi$ in it. But the factor $e^{-(\lambda\zeta/2\kappa)}$ does not vanish, and for a self-adjoint operator the property formulated holds, as is well known.

[‡] We emphasize that the argument given proves the stability of a flame only under the assumptions indicated. In particular, if similarity of the fields of concentration and of temperature does not hold (for example, for the burning of powder), some instability arises.

[§] For convenience we identify the non-self-similar variable with the time t.

demonstrate the technique of linear stability analysis for self-similar solutions of the second kind.

As shown in chapter 3 a self-similar intermediate asymptotics of the
solution to the initial-value problem given by the equations

$$\partial_t \theta = \kappa \partial_{xx}^2 \theta \quad (\partial_t \theta \geq 0), \quad \partial_t \theta = \kappa_1 \partial_{xx}^2 \theta \quad (\partial_t \theta \leq 0) \qquad (8.53)$$

is obtained, for $\kappa_1 \neq \kappa$, in the form of a self-similar solution of the second
kind,

$$\theta = \frac{A}{(\kappa t)^{(1+\alpha)/2}} f(\xi, \epsilon), \quad \xi = \frac{x}{\sqrt{\kappa t}}, \quad \epsilon = \frac{\kappa_1}{\kappa} - 1, \qquad (8.54)$$

where the function $f(\xi, \epsilon)$ can be expressed in terms of parabolic cylinder
functions. For the self-similar solution (8.54), if $|x| < \xi_0 (\kappa t)^{1/2}$ the
derivative $\partial_t \theta < 0$, but if $|x| > \xi_0 (\kappa t)^{1/2}$, $\partial_t \theta > 0$, so that the change of
coefficient in (8.53) occurs for

$$x = \pm \xi_0 \sqrt{\kappa t}.$$

The constants α and ξ_0 are found from the set of equations

$$D_{\alpha+2} \left(\frac{\xi_0}{\sqrt{2}} \right) = 0,$$

$$M \left(-1 - \frac{\alpha}{2}, \frac{1}{2}, \frac{\xi_0^2}{4(1+\epsilon)} \right) = 0. \qquad (8.55)$$

We now consider, in accordance with the general procedure outlined
above for analytical investigations of the linear stability of a solution,
the perturbed initial-value problem, for which the initial condition at
$t = t_0$ can, without loss of generality, be written in the form

$$\theta(x, t_0) = \left[A f \left(\frac{x}{\sqrt{\kappa t_0}}, \epsilon \right) + \delta v_0 \left(\frac{x}{\sqrt{\kappa t_0}} \right) \right] (\kappa t_0)^{-(1+\alpha)/2}.$$

Here δ is a small parameter, and the function $v_0(\xi)$ vanishes outside
some finite interval in ξ. In the linear approximation, for $t > t_0$,

$$\theta(x, t) = \frac{1}{(\kappa t)^{(1+\alpha)/2}} [A f(\xi, \epsilon) + \delta v(\xi, \tau)], \qquad (8.56)$$

where $v(\xi, \tau)$ is the perturbation. (Instead of the time t it is convenient
to take $\tau = \ln(t/t_0)$ as independent variable for the perturbation.) The
surfaces $x_1(t)$ and $x_2(t)$ on which $\partial_t u$ vanishes are also shifted, so that

$$x_1 = [\xi_0 + \beta_1(\tau)]\sqrt{\kappa t}, \quad x_2 = -[\xi_0 + \beta_2(\tau)]\sqrt{\kappa t}. \qquad (8.57)$$

The perturbation is not necessarily symmetric, so $\beta_1(\tau) \neq \beta_2(\tau)$. Substituting the perturbed solution (8.56), (8.57) into the basic equation,

we obtain an equation for the perturbation when $\xi > 0$ in the form

$$\partial_\tau v = (1 + \epsilon)\partial^2_{\xi\xi} v + \frac{\xi}{2}\partial_\xi v + \frac{1+\alpha}{2} v, \ (0 \le \xi \le \xi_0),$$

$$\partial_\tau v = (1 + \epsilon)\partial^2_{\xi\xi} v + \frac{\xi}{2}\partial_\xi v + \frac{1+\alpha}{2} v + \frac{\epsilon}{\delta} A \frac{d^2 f}{d\xi^2}, \ (\xi_0 \le \xi \le \xi_0 + \beta_1(\tau)),$$

$$\partial_\tau v = \partial^2_{\xi\xi} v + \frac{\xi}{2}\partial_\xi v + \frac{1+\alpha}{2} v, \ (\xi_0 + \beta_1(\tau) \le \xi < \infty),$$

$$(8.58)$$

and an analogous equation for $\xi < 0$.

Furthermore, from (8.56) we obtain an expression for the derivative $\partial_t \theta$ of the perturbed solution:

$$\partial_t \theta = \frac{1}{(\kappa t)^{(1+\alpha)/2}} \frac{1}{t} \left\{ -\frac{1+\alpha}{2} [Af(\xi,\epsilon) + \delta v(\xi,\tau)] \right.$$
$$\left. - \frac{\xi}{2}[Af'(\xi,\epsilon) + \delta\partial_\xi v] + \delta\partial_\tau v \right\}.$$

Setting $\xi = \xi_0 + \beta_1(\tau)$ in this relation, linearizing, and keeping in mind that $f''(\xi_0) = 0$ and $\partial_t \theta = 0$ for $\xi = \xi_0 + \beta_1(\tau)$, we obtain

$$\delta \left[\partial_\tau v - \frac{1+\alpha}{2} v - \frac{\xi}{2}\partial_\xi v \right] - \left(1 + \frac{\alpha}{2} \right) Af'(\xi_0)\beta_1(\tau) = 0,$$

whence it follows that the displacement of the boundary is proportional to the small parameter δ. Linearizing (8.58), we obtain for $v(\xi,\tau)$ the linear equation

$$\partial_\tau v = (1 + \epsilon)\partial^2_{\xi\xi} v + \frac{1+\alpha}{2} v + \frac{\xi}{2}\partial_\xi v \quad (|\xi| \le \xi_0),$$
$$\partial_\tau v = \partial^2_{\xi\xi} v + \frac{1+\alpha}{2} v + \frac{\xi}{2}\partial_\xi v \quad (|\xi| \ge \xi_0).$$

$$(8.59)$$

At $\xi = \xi_0$ the functions v and $\partial_\xi v$ must be both continuous. In fact, from the second equation of (8.58) we get, integrating from $\xi = \xi_0$ to $\xi = \xi_0 + \beta_1(\tau)$,

$$(1 + \epsilon)\partial_\xi v \bigg|_{\xi=\xi_0}^{\xi=\xi_0+\beta_1(\tau)}$$

$$= \int_{\xi_0}^{\xi_0+\beta_1(\tau)} \left[\partial_\tau v - \frac{\xi}{2}\partial_\xi v - \frac{1+\alpha}{2} v \right] d\xi \ - \ \frac{\epsilon}{\delta} A \frac{df}{d\xi}\bigg|_{\xi=\xi_0}^{\xi=\xi_0+\beta_1(\tau)}.$$

The quantities under the integral sign on the right-hand side are bounded and, by the preceding argument, $\beta_1(\tau)$ is of order δ, so that the entire integral is of order δ. Furthermore, for $\xi = \xi_0$ the quantity $f''(\xi,\epsilon)$ vanishes, so the second term on the right-hand side is also of order δ; and from this and the linearity of the approximation follows the continuity of $\partial_\xi v$ at $|\xi| = |\xi_0|$. The continuity of v is proved by multiplying it by ξ and using the same kind of integration and subsequent estimates.

A solution to the initial-value problem for the perturbation is sought in the form

$$v(\xi, \tau) = \sum_{n=0}^{\infty} c_n e^{-\mu_n \tau} \Psi(\xi, \mu_n), \qquad (8.60)$$

where the function $\Psi(\xi, \mu_n)$ is an eigenfunction of the operator determined by the equations

$$(1 + \epsilon) \frac{d^2 \Psi}{d\xi^2} + \frac{\xi}{2} \frac{d\Psi}{d\xi} + \frac{1 + \alpha + 2\mu}{2} \Psi = 0 \quad (|\xi| \leq \xi_0),$$

$$\frac{d^2 \Psi}{d\xi^2} + \frac{\xi}{2} \frac{d\Psi}{d\xi} + \frac{1 + \alpha + 2\mu}{2} \Psi = 0 \quad (|\xi| \geq \xi_0),$$

$$(8.61)$$

and by the condition of vanishing more rapidly than any power of $|\xi|$ at $\xi = \pm\infty$, so that

$$\Psi(\pm\infty, \mu_n) = 0 \qquad (8.62)$$

(μ_n being the nth eigenvalue of this operator). Furthermore, the functions $\Psi(\xi, \mu_n)$ together with their first derivatives with respect to ξ are continuous at $\xi = \xi_0$.

It is convenient to consider separately the symmetric (Ψ_1) and antisymmetric (Ψ_2) eigenfunctions of the operator (8.61), (8.62). A symmetric solution of (8.61) satisfying (8.62) must be representable in the form (cf. subsection 3.2.4).

$$\Psi_1 = C_1 \left\{ \exp\left[-\frac{\xi^2}{8(1+\epsilon)}\right] \right\} \left[D_{\alpha+2\mu}\left(\frac{\xi}{\sqrt{2(1+\epsilon)}}\right) \right.$$

$$\left. + D_{\alpha+2\mu}\left(-\frac{\xi}{\sqrt{2(1+\epsilon)}}\right) \right], \quad (|\xi| \leq \xi_0), \qquad (8.63)$$

$$\Psi_1 = C_2 \left\{ \exp\left(-\frac{\xi^2}{8}\right) \right\} D_{\alpha+2\mu}\left(\frac{|\xi|}{\sqrt{2}}\right) \quad (|\xi| \geq \xi_0).$$

To determine the constants C_1 and C_2 we use the continuity of Ψ and $d\Psi/d\xi$ for $|\xi| = \xi_0$. Thus we get a system of homogeneous linear algebraic equations; the condition that the determinant of this system vanishes gives the characteristic equation

$$\Delta(\mu) \equiv (\alpha + 2\mu + 1) D_{\alpha+2\mu}\left(\frac{\xi_0}{\sqrt{2}}\right) M\left(-1 - \frac{(\alpha + 2\mu)}{2}, \frac{1}{2}; \frac{\xi_0^2}{4(1+\epsilon)}\right)$$

$$+ D_{\alpha+2\mu+2}\left(\frac{\xi_0}{\sqrt{2}}\right) M\left(-\frac{(\alpha + 2\mu)}{2}, \frac{1}{2}; \frac{\xi_0^2}{4(1+\epsilon)}\right)$$

$$= 0. \qquad (8.64)$$

The quantities α and ξ_0 are determined, as before, by the relations (8.55). Using these relations it is easy to show that $\mu_0 = 0$ is a root of (8.64). We now show that the other roots of this equation are positive. Equation (8.64) can be put into the form

$$\Delta(\mu) \equiv \left(\frac{\xi_0}{\sqrt{2}}\right) D_{1+\alpha+2\mu}\left(\frac{\xi_0}{\sqrt{2}}\right) M\left(-1 - \frac{(\alpha + 2\mu)}{2}, \frac{1}{2}; \frac{\xi_0^2}{4(1 + \epsilon)}\right)$$
$$+ D_{\alpha+2\mu+2}\left(\frac{\xi_0}{\sqrt{2}}\right)\left[M\left(-\frac{(\alpha + 2\mu)}{2}, \frac{1}{2}; \frac{\xi_0^2}{4(1 + \epsilon)}\right)\right.$$
$$\left. - M\left(-1 - \frac{(\alpha + 2\mu)}{2}, \frac{1}{2}; \frac{\xi_0^2}{4(1 + \epsilon)}\right)\right]$$
$$= 0. \tag{8.65}$$

It is known (see Abramowitz and Stegun, 1970), that the function $M(\alpha + l, 1/2, x_0)$ is a monotonically increasing function of l for $l > 0$ if x_0 is the smallest positive root of the equation $M(\alpha, 1/2, x) = 0$. If ζ_0 is the smallest positive root of the equation $D_{\alpha+2}(\zeta) = 0$, then

$$D_{\alpha+2\mu+2}(\zeta_0) > 0 \quad \text{for } \mu < 0.$$

Therefore $\Delta(\mu) > 0$ for all negative μ, and there are thus no negative roots of (8.65).

Further, the antisymmetric solution has the form

$$\Psi_2 = C_3 \left\{\exp\left[-\frac{\xi^2}{8(1 + \epsilon)}\right]\right\}\left[D_{\alpha+2\mu}\left(\frac{\xi}{\sqrt{2(1 + \epsilon)}}\right)\right.$$
$$\left. - D_{\alpha+2\mu}\left(-\frac{\xi}{\sqrt{2(1 + \epsilon)}}\right)\right], \quad (0 \le |\xi| \le \xi_0),$$
$$\Psi_2 = C_4\left\{\exp\left(-\frac{\xi^2}{8}\right)\right\} D_{\alpha+2\mu}\left(\frac{\xi}{\sqrt{2}}\right), \quad (\xi_0 \le |\xi| < \infty).$$

Its characteristic equation can be reduced to the relation

$$\Delta_1(\mu) \equiv \frac{\xi_0}{\sqrt{2}} D_{\alpha+2\mu}\left(\frac{\xi_0}{\sqrt{2}}\right) M\left(-1 - \frac{(\alpha + 2\mu - 1)}{2}, \frac{1}{2}; \frac{\xi_0^2}{4(1 + \epsilon)}\right)$$
$$+ (1 + \epsilon) D_{1+\alpha+2\mu}\left(\frac{\xi_0}{\sqrt{2}}\right)\left[M\left(-\frac{(\alpha + 2\mu - 1)}{2}, \frac{1}{2}; \frac{\xi_0^2}{4(1 + \epsilon)}\right)\right.$$
$$\left. - M\left(-1 - \frac{(\alpha + 2\mu - 1)}{2}, \frac{1}{2}; \frac{\xi_0^2}{4(1 + \epsilon)}\right)\right]$$
$$= 0. \tag{8.66}$$

Comparison with (8.65) shows that the smallest root of (8.66) is equal to $\mu_1 = 1/2$. Subsequent investigation reveals that the smallest positive root of (8.65) is equal to $\mu_2 = 1$ and the corresponding root of (8.66) is equal to $\mu_3 = 3/2$. Thus, (8.56) and (8.60) show that a solution of the perturbed initial-value problem can be written in the following form:

$$\theta(x,t) = \frac{1}{(\kappa t)^{(1+\alpha)/2}}\left[(A + \delta c_0)f(\xi,\epsilon) + \delta c_1\left(\frac{t_0}{t}\right)^{1/2}\Psi\left(\xi,\frac{1}{2}\right)\right.$$

$$\left. + \delta c_2\left(\frac{t_0}{t}\right)\Psi(\xi,1) + \delta c_3\left(\frac{t_0}{t}\right)^{3/2}\Psi\left(\xi,\frac{3}{2}\right) + o\left(\left(\frac{t_0}{t}\right)^{3/2}\right)\right]$$

(8.67)

(the c_i being the coefficients of the expansion of the function $v_0(\xi)$ in a Fourier series with respect to the eigenfunctions of the operator (8.61), (8.62)). Thus the self-similar solution constructed in chapter 3 turns out to be stable with respect to small perturbations. It is evident that in the present case the constant A also turns out to have been shifted to: $A' = A(1 + \delta c_0)$, so that the invariance we have incorporated in the definition of the stability of self-similar solutions is used in this case too. Here again we have to emphasize that the proof of linear stability forms only part of the problem. In the special case considered here global stability in the sense proposed at the beginning of this subsection follows from the results by Kamin, Peletier and Vázquez (1991).

In the linear case $\kappa_1 = \kappa$, $\epsilon = 0$, $\alpha = 0$ one gets the expected result for the stability of a self-similar solution, of instantaneous heat-source type, of the classical equation of heat conduction. The representation (8.67) of the solution of the perturbed initial-value problem in this case assumes the form

$$\theta(x,t) = \frac{1}{\sqrt{\kappa t}}\left[(A + \delta c_0)e^{-\xi^2/4} + \delta c_1\left(\frac{t_0}{t}\right)^{1/2}\frac{\xi}{2}e^{-\xi^2/4}\right.$$

$$\left. + \delta c_2\left(\frac{t_0}{t}\right)\frac{\xi^2 - 2}{4}e^{-\xi^2/4} + \ldots\right].$$

(8.68)

The coefficients of (8.61) in the linear case ($\epsilon = 0$) are actually continuous, and that equation can be written in the form

$$\frac{d^2\Psi}{d\xi^2} + \frac{\xi}{2}\frac{d\Psi}{d\xi} + \frac{1+2\mu}{2}\Psi = 0.$$

(8.69)

For $\mu = 0$ a solution to this equation that satisfies the condition (8.62) of rapid convergence to zero at infinity is $e^{-\xi^2/4}$. This function does not vanish for any finite ξ; hence it is the zeroth eigenfunction, and $\mu = \mu_0 = 0$ is the zeroth eigenvalue. Furthermore, the derivative of $e^{-\xi^2/4}$ with respect to ξ, equal to $-(\xi/2)e^{-\xi^2/4}$, vanishes except at infinity only for $\xi = 0$, and it satisfies (8.69) for $\mu = 1/2$, and also the conditions at infinity. This is consequently the eigenfunction with $n = 1$, and $\mu = \mu_1 = 1/2$ is the corresponding eigenvalue. Thus $\Psi(\xi,\mu_1) = -(\xi/2)e^{-\xi^2/4}$. In general the eigenvalues are given by $\mu_n = n/2$, $n = 0, 1, 2, \ldots$, and the eigenfunction corresponding to μn is are equal to the nth derivative of $e^{-\xi^2/4}$.

It is easy to obtain the result for the linear case $\epsilon = 0$ directly (Zeldovich and Barenblatt, 1958; see also Zeldovich and Raizer, 1967), since in this case there exists an explicit representation of the solution to the perturbed initial-value problem.

Scaling in the deformation
and fracture of solids

9.1 Transition from self-similarity of the first kind to self-similarity of the second; a linear elasticity problem

9.1.1 Equilibrium of an elastic wedge under the action of a concentrated couple applied at its tip

The consideration of linear problems is instructive for our purposes: for them one can follow analytically the transition to self-similar asymptotics of the solutions of non-idealized problems, and the transition at certain critical values of the parameter from self-similarities of the first kind to self-similarities of the second kind. For nonlinear problems, as a rule one has not been able to do this. The example given below, as well as the problem of ideal flow past a wedge considered in chapter 3, is simple enough that construction of the complete solution to the non-idealized problem with non-self-similar solution is possible; at the same time it exhibits clearly enough the complexities that can appear in nonlinear problems.

We shall consider some problems in the theory of elasticity for the case of plane strain, when the components of the elastic fields – the stress tensors, deformation tensors, displacement vectors, etc. – are identical in all planes perpendicular to some direction. The equilibrium equations for plane strain have the form (Muskhelishvili, 1963; Germain, 1986b; Landau and Lifshitz, 1986)

$$\partial_r\sigma_{rr}+\frac{1}{r}\partial_\theta\sigma_{r\theta}+\frac{1}{r}(\sigma_{rr}-\sigma_{r\theta})=0, \quad \frac{1}{r}\partial_\theta\sigma_{\theta\theta}+\partial_r\sigma_{r\theta}+\frac{2}{r}\sigma_{r\theta}=0. \quad (9.1)$$

Here r, θ are polar coordinates in the deformation plane and σ_{rr}, $\sigma_{\theta\theta}$,

$\sigma_{r\theta}$ are the corresponding components of the stress tensor. (In what follows, polar coordinates are what we shall need.) These equations are identically satisfied by introduction of the Airy stress function Ψ:

$$\sigma_{rr} = \frac{1}{r}\partial_r\Psi + \frac{1}{r^2}\partial^2_{\theta\theta}\Psi, \quad \sigma_{\theta\theta} = \partial^2_{rr}\Psi, \quad \sigma_{r\theta} = -\partial_r\left(\frac{1}{r}\partial_\theta\Psi\right). \quad (9.2)$$

Hooke's law relates the components of the stress tensor to the first spatial derivatives of a single displacement vector, whence it follows that the three components of the stress tensor satisfy a certain integrability condition, the so-called compatibility relation. If we substitute into this relation the expressions (9.2) for the components of the stress tensor in terms of the stress function, we obtain for this function the biharmonic equation

$$\Delta^2\Psi = \left[\frac{1}{r}\partial_r(r\partial_r) + \frac{1}{r^2}\partial^2_{\theta\theta}\right]^2 \Psi = 0. \quad (9.3)$$

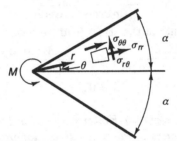

Figure 9.1. A wedge of opening angle 2α under the action of a couple of moment M applied at its tip.

We begin with an instructive problem first considered by Carothers (1912) and Inglis (1922). Namely, we take (Figure 9.1) an infinite wedge of opening angle 2α and at the tip of the wedge we apply a couple of moment M. The stress function Ψ that governs the elastic field depends in this case on four parameters, M, r, θ and α, whose dimensions in the natural class FLT are respectively $F, L, 1$ and 1. (The dimension of the moment of a couple in the plane problems of elasticity coincide with the dimensions of force, since in such a problem one is really dealing with a couple per unit thickness.) By virtue of (9.2) the dimension of the stress function also coincides with the dimension of force. Hence the standard procedure of dimensional analysis leads to a relation

$$\Psi = M\Phi(\theta, \alpha), \quad (9.4)$$

so that the stress function is independent of the radius.

Substituting (9.4) into (9.3), we get for $\Phi(\theta, \alpha)$ the ordinary differential equation

$$\Phi^{IV} + 4\Phi'' = 0. \quad (9.5)$$

Furthermore the lateral faces of the wedge are free of stress over their entire extent:

$$\sigma_{\theta\theta}(r, \pm\alpha) = 0, \quad \sigma_{r\theta}(r, \pm\alpha) = 0,$$

and from this, (9.2), and (9.4) we get the boundary conditions for the function $\Phi(\theta, \alpha)$:

$$\partial_\theta \Phi(\pm\alpha, \alpha) = 0. \tag{9.6}$$

The equation and boundary conditions determine the solution to within a constant factor, which is found from the following consideration: we make a cut along a circle of arbitrary radius, calculate the total moment of the stresses acting on the cut, and equate the result to M, because the cut-off part of the wedge must be in equilibrium. As a result the final expression for the stress function is obtained in the form

$$\Psi = \frac{M(2\theta\cos 2\alpha - \sin 2\theta)}{2(\sin 2\alpha - 2\alpha\cos 2\alpha)}, \tag{9.7}$$

and for the components of the stress field we have

$$\sigma_{rr} = \frac{2M\sin 2\theta}{(\sin 2\alpha - 2\alpha\cos 2\alpha)r^2}, \quad \sigma_{r\theta} = \frac{M(\cos 2\alpha - \cos 2\theta)}{(\sin 2\alpha - 2\alpha\cos 2\alpha)r^2}, \tag{9.8}$$

$$\sigma_{\theta\theta} = 0.$$

9.1.2 The Sternberg–Koiter paradox. Intermediate asymptotics of the non-self-similar problem

Everything seemed to be clear with this problem, and for a long time it occupied a permanent place in the text-books. However, in the remarkable paper of Sternberg and Koiter (1958) attention was drawn for the first time to a strange property of the solution (9.7), (9.8) just obtained: as the angle α approaches the value $\alpha = \alpha_* \approx 0.715\pi$ for which the denominator in (9.7), (9.8) vanishes (which is perfectly admissible from the physical point of view), the stresses at all points of the wedge tend to infinity according to (9.8). In this connection the following question arises: is the self-similar solution of the idealized problem (9.7) an asymptotics of some non-self-similar solution of the non-idealized problem; in other words, does it have a certain physical meaning?

In order to clarify this matter, Sternberg and Koiter considered for the same wedge the following non-idealized (and therefore non-self-similar) problem (Figure 9.2a). On finite segments of the lateral faces of the wedge $\theta = \pm\alpha$, $0 \leq r \leq r_0$, there is distributed, according to some law, a normal loading that is antisymmetric with respect to the axis of the wedge and statically equivalent to a couple with moment M. The tangential stress is as before equal to zero everywhere on the faces of the wedge. Thus one has the conditions

$$\sigma_{\theta\theta}(r,\alpha) = -\sigma_{\theta\theta}(r,-\alpha) = p(r),$$
$$\sigma_{r\theta}(r,\alpha) = \sigma_{r\theta}(r,-\alpha) \equiv 0 \quad (0 < r < \infty),$$

(9.9)

where $p(r)$ is a function, identically equal to zero for $r \geq r_0$, and satisfying the conditions

$$\int_0^{r_0} p(r)dr = 0, \quad \int_0^{r_0} p(r)r\,dr = \frac{M}{2}.$$

(9.10)

Furthermore, to get a unique solution one imposes the additional regularity requirement of boundedness of the resulting force on any radial cut of the wedge:

$$\int_0^\infty \sigma_{\theta\theta}(r,\theta)dr < \infty, \quad \int_0^\infty \sigma_{r\theta}(r,\theta)dr < \infty.$$

(9.11)

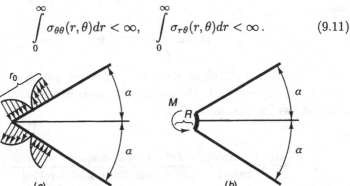

Figure 9.2. Non-idealized problems of the elastic equilibrium of a wedge under the action of a couple of moment M: (a) forces distributed over the lateral faces of the wedge; (b) forces applied to a stiff ring segment of finite radius R.

To get a solution of the problem posed we apply following Sternberg and Koiter, the Mellin integral transformation in the variable r. As is well known (see Sneddon, 1951), the Mellin transform of a function and its inverse are given by the relations

$$\bar{f}(s) = \int_0^\infty f(r)r^{s-1}dr, \quad f(r) = \frac{1}{2\pi i}\int_{c-i\infty}^{c+i\infty} \bar{f}(s)r^{-s}ds.$$

(9.12)

Applying the Mellin transformation to the biharmonic equation (9.3), we get for the transform, $\bar{\Psi}(s,\theta)$, of the stress function the ordinary differential equation

$$\left[\frac{d^2}{d\theta^2} + s^2\right]\left[\frac{d^2}{d\theta^2} + (s+2)^2\right]\bar{\Psi}(s,\theta) = 0.$$

(9.13)

The stress field sought is antisymmetric, thus the stress function must be also so. The general antisymmetric solution of (9.13) has the form

$$\bar{\Psi}(s,\theta) = A(s)\sin s\theta + B(s)\sin(s+2)\theta. \tag{9.14}$$

Further, (9.9) with (9.2) taken into account can be written in the form

$$r^2\partial_{rr}^2\Psi(r,\pm\alpha) = \pm p(r)r^2, \quad \partial_r\left[\frac{1}{r}\partial_\theta\Psi(r,\pm\alpha)\right] = 0. \tag{9.15}$$

Applying the Mellin transformation to these conditions and integrating by parts (in order to do this it is necessary to multiply by r^2), we get the boundary conditions for the function $\bar{\Psi}(s,\theta)$:

$$\bar{\Psi}(s,\pm\alpha) = \pm\frac{\bar{p}(s)}{s(s+1)}, \quad \frac{d\bar{\Psi}(s,\pm\alpha)}{d\theta} = 0. \tag{9.16}$$

Here

$$\bar{p}(s) = \int_0^{r_0} p(r)r^{s+1}dr. \tag{9.17}$$

From (9.14) and (9.16) we determine the constants $A(s)$ and $B(s)$; substituting the result into the inversion formula we obtain the solution for the stress function in the form

$$\Psi(r,\theta) =$$
$$\int_{c-i\infty}^{c+i\infty}\frac{\bar{p}(s)[s\cos s\alpha\sin(s+2)\theta - (s+2)\cos(s+2)\alpha\sin s\theta]r^{-s}ds}{2\pi is(s+1)[(s+1)\sin 2\alpha - \sin 2(s+1)\alpha]}.$$
$$\tag{9.18}$$

The relations for the components of the stress tensor are obtained from this by differentiation:

$$\sigma_{\theta\theta}(r,\theta) =$$
$$\int_{c-i\infty}^{c+i\infty}\frac{\bar{p}(s)[s\cos s\alpha\sin(s+2)\theta - (s+2)\cos(s+2)\alpha\sin s\theta]r^{-s-2}ds}{2\pi i[(s+1)\sin 2\alpha - \sin 2(s+1)\alpha]}$$

$$\sigma_{rr}(r,\theta) =$$
$$\int_{c-i\infty}^{c+i\infty}\frac{\bar{p}(s)[(s+2)\cos(s+2)\alpha\sin s\theta - (s+4)\sin(s+2)\theta\cos s\alpha]r^{-s-2}ds}{2\pi i[(s+1)\sin 2\alpha - \sin 2(s+1)\alpha]}$$

$$\sigma_{r\theta}(r,\theta) =$$
$$\int_{c-i\infty}^{c+i\infty}\frac{\bar{p}(s)[(s+2)\cos s\alpha\cos(s+2)\theta - (s+2)\cos(s+2)\alpha\cos s\theta]r^{-s-2}ds}{2\pi i[(s+1)\sin 2\alpha - \sin 2(s+1)\alpha]}$$
$$\tag{9.19}$$

It is evident that the integrands in (9.9) are meromorphic functions of the complex variable s whose poles correspond to the zeros of the entire function

$$(s+1)\sin 2\alpha - \sin 2(s+2)\alpha = G(s,\alpha). \qquad (9.20)$$

In the integrals (9.18), (9.19) the abscissa of the line of integration $\text{Re}(s) = c$ can be chosen arbitrarily within one band of regularity of the integrand. Which band of regularity to take is determined by the conditions imposed on the stress at infinity. The requirements of vanishing at infinity and of regularity of the stress, i.e., the satisfying of (9.11), allow us to select the band of regularity containing the point $s = -1$. The representation of the solution by (9.18), (9.19) is convenient for calculating asymptotics.

To calculate the integrals we must close the contour of integration, adding to the line $\text{Re}(s) = c$ a semicircle of large radius on the right or left, depending on whether we are interested in the asymptotics of the stress field for $r \to 0$ or $r \to \infty$, and then letting the radius of the circle tend to infinity. Thus the required integral is expressed in terms of the sum of the residues at the poles contained in the contour obtained, i.e., for the stress, at the points corresponding to the roots of the function (9.20). The principal terms in the asymptotic solution for $r \to \infty$ of interest to us are thus determined by the roots of (9.20) that have the smallest real parts.

Investigation of the roots of (9.20) shows that the situation changes at the value $\alpha = \alpha_* \approx 0.715\pi$ that makes the expression $\sin 2\alpha - 2\alpha \cos 2\alpha$ vanish. Namely, for $0 < \alpha < \alpha_*$ the root of (9.20) having the smallest real part is actually the simple root $s = 0$. For $\alpha = \alpha_*$ the root $s = 0$ becomes double: for $s = 0$, not only $G(s, \alpha)$ but also $G'(s, \alpha) = \sin 2\alpha - 2\alpha \cos 2(s+1)\alpha$ vanishes. Finally, for $\alpha_* < \alpha \leq \pi$ there appears a real simple negative root $s = \lambda(\alpha)$, where $\lambda(\alpha)$ varies monotonically from zero for $\alpha = \alpha_*$ to $-1/2$ for $\alpha = \pi$. Hence the principal terms in the expansion for $r \to \infty$ are different in these three cases:

(1) For $0 < \alpha < \alpha_*$ and $r \to \infty$,

$$\Psi = \frac{M(2\theta \cos 2\alpha - \sin 2\theta)}{2(\sin 2\alpha - 2\alpha \cos 2\alpha)} + o(1),$$

$$\sigma_{rr} = \frac{2M \sin 2\theta}{(\sin 2\alpha - 2\alpha \cos 2\alpha)r^2} + o(r^{-2}), \quad \sigma_{\theta\theta} = o(r^{-2}), \qquad (9.21)$$

$$\sigma_{r\theta} = \frac{M(\cos 2\alpha - \cos 2\theta)}{(\sin 2\alpha - 2\alpha \cos 2\alpha)r^2} + o(r^{-2}).$$

(2) For $\alpha = \alpha_*$ and $r \to \infty$,

$$\Psi = \frac{M}{12\alpha_*^2 \sin 2\alpha_*} \left\{ 3 \left[g(r_0) - \ln \frac{r}{r_0} \right] (2\theta \cos 2\alpha_* - \sin 2\theta) \right.$$

$$\left. - 3\theta \cos 2\theta + 4 \sin 2\theta - 5\theta \cos 2\alpha_* - 6\alpha_*\theta \sin 2\alpha_* \right\} + o(1), \quad (9.22a)$$

$$\sigma_{rr} = \frac{M}{12\alpha_*^2 \sin 2\alpha_* r^2} \left\{ \left[12g(r_0) - 12 \ln \frac{r}{r_0} - 1 \right] \sin 2\theta \right.$$

$$\left. + 12\theta \cos 2\theta - 6\theta \cos 2\alpha_* \right\} + o(r^{-2}), \quad (9.22b)$$

$$\sigma_{\theta\theta} = \frac{M(2\theta \cos 2\alpha - \sin 2\theta)}{4\alpha_*^2 \sin 2\alpha_* r^2} + o(r^{-2}), \quad (9.22c)$$

$$\sigma_{r\theta} = \frac{M}{12\alpha_*^2 \sin 2\alpha_* r^2} \left\{ \left[6g(r_0) - 6 \ln \frac{r}{r_0} + 1 \right] (\cos 2\alpha_* - \cos 2\theta) \right.$$

$$\left. + 6\theta \sin 2\theta - 6\alpha_* \sin 2\alpha_* \right\} + o(r^{-2}), \quad (9.22d)$$

where

$$g(r_0) = \frac{2}{M} \int\limits_0^{r_0} p(r)r \ln\left(\frac{r}{r_0}\right) dr \, ;$$

(3) For $\alpha_* < \alpha \leq \pi$ and $r \to \infty$,

$$\Psi = \frac{\bar{p}(\lambda)[(\lambda + 2)\cos(\lambda + 2)\alpha \sin \lambda\theta - \lambda \cos \lambda\alpha \sin(\lambda + 2)\theta]}{\lambda(\lambda + 1)(\sin 2\alpha - 2\alpha \cos 2(\lambda + 1)\alpha)r^\lambda}$$

$$+ \frac{M(2\theta \cos 2\alpha - \sin 2\theta)}{2(\sin 2\alpha - 2\alpha \cos 2\alpha)} + o(1), \quad (9.23a)$$

$$\sigma_{\theta\theta} = \frac{\bar{p}(\lambda)[(\lambda + 2)\cos(\lambda + 2)\alpha \sin \lambda\theta - \lambda \cos \lambda\alpha \sin(\lambda + 2)\theta]}{[\sin 2\alpha - 2\alpha \cos 2(\lambda + 1)\alpha]r^{\lambda+2}}$$

$$+ o(r^{-2}), \quad (9.23b)$$

$$\sigma_{rr} = \frac{\bar{p}(\lambda)[(\lambda + 4)\cos \lambda\alpha \sin(\lambda + 2)\theta - (\lambda + 2)\cos(\lambda + 2)\alpha \sin \lambda\theta]}{[\sin 2\alpha - 2\alpha \cos 2(\lambda + 1)\alpha]r^{\lambda+2}}$$

$$+ \frac{2M \sin 2\theta}{(\sin 2\alpha - 2\alpha \cos 2\alpha)r^2} + o(r^{-2}), \quad (9.23c)$$

$$\sigma_{r\theta} = \frac{\bar{p}(\lambda)(\lambda + 2)[\cos(\lambda + 2)\alpha \cos \lambda\theta - \cos \lambda\alpha \cos(\lambda + 2)\theta]}{(\sin 2\alpha - 2\alpha \cos 2(\lambda + 1)\alpha)r^{\lambda+2}}$$

$$+ \frac{M(\cos 2\alpha - \cos 2\theta)}{(\sin 2\alpha - 2\alpha \cos 2\alpha)r^2} + o(r^{-2}), \quad (9.23d)$$

where $\bar{p}(\lambda)$ is determined by (9.17) as before.

We now apply dimensional analysis to the original non-self-similar problem. Without loss of generality we can write the function $p(r)$ in the form

$$p(r) = \frac{M}{2r_0^2} \varphi \left(\frac{r}{r_0} \right) . \tag{9.24}$$

Here φ is a dimensionless function of its dimensionless argument. It is evident that the solution Ψ is governed by the following quantities: M, r_0, r, θ and α, whose dimensions are respectively F, L, L, 1 and 1. Consequently the standard procedure of dimensional analysis gives

$$\Psi = M\Phi \left(\frac{r}{r_0}, \theta, \alpha \right) . \tag{9.25}$$

The previous considerations by means of which we arrived at (9.4) were based on the implicit assumption that at large distances from the tip of the wedge the parameter r/r_0 is very large, and hence the length r_0 of the part of the lateral face of the wedge on which the loading is distributed is an inessential parameter.

The analysis just performed showed that this is actually so only for $0 < \alpha < \alpha_*$. If $\alpha > \alpha_*$, then the size r_0 remains an essential parameter, no matter how far we go from the tip of the wedge. Nevertheless the asymptotics of the stress function, and hence also of all the components of the stress tensor, are self-similar; but this self-similarity is of the second kind, not determined by dimensional considerations.

Indeed, if as is seen from (9.23) there exists a real number λ such that the function $\Phi(\eta, \theta, \alpha)$, where $\eta = r/r_0$, behaves like $\eta^{-\lambda}\Phi_1(\theta, \alpha)$ as $\eta \to \infty$ (i.e., as $r \to \infty$ or $r_0 \to 0$), then by virtue of (9.25) the limiting solution obtained by shrinking r_0 to zero, i.e., for $\eta \to \infty$, has the form

$$\Psi = \frac{M r_0^{\lambda}}{r^{\lambda}} \Phi_1(\theta, \alpha) . \tag{9.26}$$

It is clear here that if we want to get a correct asymptotics of the solution of the non-self-similar problem as $r/r_0 \to \infty$ by shrinking r_0 to zero it is impossible to keep M constant; it should also tend to zero so that the product $M r_0^{\lambda}$ remains constant.

Let us substitute (9.26) into the biharmonic equation (9.3). We obtain for $\Phi_1(\theta, \alpha)$ the ordinary equation

$$\Phi_1^{IV} + [\lambda^2 + (\lambda + 2)^2]\Phi_1'' + \lambda^2(\lambda + 2)^2\Phi_1 = 0, \tag{9.27}$$

which coincides with (9.13) for $s = \lambda$. The solution of interest to us must be antisymmetric and satisfy the conditions $\Phi_1 = 0$, $d\Phi_1/d\theta = 0$ for $\theta = \pm \alpha$. The latter conditions follow from the fact that both components of the stress on the lateral faces of the wedge are equal to zero. From these conditions we get a relation for Φ_1 to within a dimensionless constant factor β,

$$\Phi_1 = \beta[(\lambda + 2)\cos(\lambda + 2)\alpha \sin \lambda\theta - \lambda \cos \lambda\alpha \sin(\lambda + 2)\theta], \qquad (9.28)$$

and also the characteristic equation for determining λ,

$$(\lambda + 1)\sin 2\alpha - \sin 2(\lambda + 1)\alpha = 0, \qquad (9.29)$$

which coincides with the condition for vanishing of the function (9.20). As was said above, for $\alpha_* < \alpha < \pi$ (9.29) has a real negative root, and at the same time for $0 < \alpha < \alpha_*$ the root of this equation with the smallest real part is zero. Therefore for $0 < \alpha < \alpha_*$ the function $\Phi(\eta, \theta, \alpha)$ in the expression for the solution of the non-self-similar problem tends to a finite non-zero limit as $\eta \to \infty$ (as the region of application of the load shrinks to zero), and the dimensional considerations developed in subsection 9.1.1 turn out to be applicable and to lead to the correct final result. For $\alpha_* < \alpha \leq \pi$, however, the limiting solution can be written in the form

$$\Psi = \frac{A}{r^\lambda}[(\lambda + 2)\cos(\lambda + 2)\alpha \sin \lambda\theta - \lambda \cos \lambda\alpha \sin(\lambda + 2)\theta], \qquad (9.30)$$

$\lambda < 0$, where the constant $A = \beta M r_0^\lambda$ can no longer be determined if we seek a self-similar solution of the second kind directly. It can be found only if we follow the transition from a solution of the non-self-similar problem to a self-similar asymptotics. In fact the asymptotic representation of the solution to the non-self-similar problem for large r/r_0 has, in the case $\alpha_* < \alpha \leq \pi$, a principal term that coincides with (9.30) (cf. (9.23)) if one takes

$$A = \frac{\int_0^{r_0} p(r)r^{\lambda+1}dr}{\lambda(\lambda + 1)[\sin 2\alpha - 2\alpha \cos 2(\lambda + 1)\alpha]}. \qquad (9.31)$$

Thus the asymptotics of the solution obtained by shrinking the region of application of the loading on the lateral faces of the wedge to zero 'remembers' for $\alpha_* < \alpha \leq \pi$ not the ordinary moment of force, i.e., not the integral $\int_0^{r_0} p(r)rdr = M/2$, but a more complicated fractional-power moment of the system of forces acting on the lateral faces of the wedge. Here the exponent to which the radius appears in the moment depends on the opening angle of the wedge, and is determined by solving the eigenvalue problem for the linear equation (9.27) under the conditions that the solution and its derivative vanish at the endpoints of the interval.

The solution just considered is instructive in many respects. It contains the parameter α, the opening angle of the wedge. As is evident from the preceding analysis, for angles less than some critical value we can use the naive arguments of dimensional analysis, considering only the prescribed moment of forces acting on the wedge; we get a self-similar solution of the first kind, which is completely determined by direct construction with the help of dimensional analysis. For wedge angles larger than the critical one, naive considerations of dimensional analysis are not applicable, because it is impossible, for $\alpha > \alpha_*$, to delete r_0 from the list of governing parameters while leaving M in the list. Nevertheless, by shrinking to zero the region of application of the loading on the lateral faces of the wedge, we obtain in this case too a self-similar limiting solution. The attempt to construct this solution directly as a self-similar solution of the second kind determines the limiting solution, just as for any self-similar solution of the second kind, only to within a constant. The value of this constant can be obtained by matching the self-similar solution with a solution of the non-self-similar problem. It can be expressed, as carrying out the matching shows, in terms of some fractional moment. That is, to what exponent r is raised in this moment can be determined only after solving the eigenvalue problem; it is impossible to determine this exponent in advance from dimensional considerations. Finally, for a wedge angle equal to the critical one, dimensional considerations turn out to be meaningless; they do not lead to any simplification of the solution, and arguing about the smallness of the part on which loading is applied, in order to arrive at an idealized problem, is not valid. In other words, similarity in the parameter η does not occur, no matter how large η may be.

Nevertheless, as (9.22) shows, the asymptotics of the solution is self-similar in this case, since the expression for $\Phi = \Psi/M$ can for large $\eta = r/r_0$ be written in the form

$$\Phi = \ln \eta \, \Phi_3(\theta) \, .$$

This self-similarity, however, is not of scaling type and is itself no longer a solution.

It is obvious that it would be impossible to establish what was said above without knowing the non-self-similar solution of the complete non-idealized problem. In nonlinear problems an analysis similar to that presented above is practically never possible; as already mentioned, one of the main reasons that we are generally interested in self-similar solutions of idealized problems is the desire to obtain some idea of the structure of the solutions of complicated non-idealized nonlinear problems. The example presented clearly demonstrates that it is insufficient simply to construct a self-similar solution; it is necessary to verify that this solution is an intermediate asymptotics for at least a certain restricted class of non-idealized problems. After the fundamental paper of Sternberg and Koiter (1958) there appeared other studies of the same kind. An elegant work by Moffatt and Duffy (1980) should be especially noted; it is concerned with the elastic field near the angular points of the bar cross-section's contour in torsion[†]. The papers by Dundurs and Markenscoff (1989), Dempsey (1981) and Ting (1984) contain interesting additional remarks concerning plane-strain problems for an elastic wedge.

Self-similar solutions of the second kind for the flow of rigid-plastic material in a wedge-form channel were constructed recently by Alexandrov and Goldstein (1993a, b). An instructive application of the intermediate-asymptotic approach and solutions for the experimentally obsesrved phenomenon of separation in plastic extrusion are presented in these papers.

9.1.3 The use of self-similar solutions for estimating integral characteristics: the stiffness of a wedge

Budiansky and Carrier (1973) carried out in connection with the same problem of a wedge on which a couple acts, an instructive investigation concerning the application of self-similar solutions to the estimation of the bulk integral characteristics of the solution to non-idealized problems. Namely, they considered (Figure 9.2(b)) an elastic wedge truncated along a circular arc $r = R$ close to its tip and reinforced by an absolutely stiff ring segment at the cut $r = R$. Using the fact that the same equations of the plane theory of elasticity apply to the case of plane stress (thin plates) as to the case of plane strain, Budiansky and Carrier considered a 'generalized wedge' consisting of a tightly wound infinite helicoid. This makes it possible to consider the problem for arbitrary angles α, including those greater than π. A couple is applied to the ring segment with torque M (per unit thickness of the wedge). It is clear that the reinforced boundary will turn by a small angle Ω. This angle Ω is proportional to the applied torque M per unit wedge thickness; it is natural to call the quantity M/Ω the torsional stiffness of the wedge. As considerations of dimensional analysis show, this quantity, which is governed by the shear modulus G, by Poisson's ratio ν, and by the radius R

[†] In fact, Moffatt and Duffy (1980) considered this problem in a different, hydrodynamic interpretation: a Poiseuille flow in a cylindrical pipe with angular points at the cross-section's contour. The analogy between these problems is now well known. Note, in general, that the analysis of mathematically equivalent self-similar problems of plane elasticity for wedge-form regions and slow viscous flows in wedge-form vessels was performed over nearly sixty years without any cross-correlations, so that the difficulties were always got over twice. (Compare two fundamental papers, Williams (1952) and Moffatt (1964).) The hydrodynamic interpretation of the Sternberg–Koiter solution was given by Barenblatt and Zeldovich (1972), and Moffatt and Duffy (1980). In the paper by Anderson and Davis (1993) two-fluid viscous flow in a wedge-form vessel was considered. Effects similar to the Sternberg–Koiter problem were also treated.

of the circle along which the wedge is cut, is equal to $GR^2C(\nu)$, where the quantity $C(\nu)$ is called the *dimensionless stiffness*[†]. The problem consists in obtaining a sufficiently reliable estimate for the dimensionless stiffness. Budiansky and Carrier used self-similar solutions for this in an effective way that was instructive from a general point of view. They started from the principle of minimality of the complementary energy, proved in the theory of elasticity, according to which, among all virtual stress fields that vanish along the lateral faces of the wedge and have on the arc $r = R$ zero resultant force and torque equal to M, the actual stress field minimizes the stress energy (per unit thickness)

$$W = \frac{1}{2} \int\limits_{R}^{\infty} \int\limits_{-\alpha}^{\alpha} \sigma_{\mu\nu}\epsilon_{\mu\nu} r \, dr \, d\theta \qquad (9.32)$$

(with summation over repeated Greek indices). Here the components of the deformation tensor ϵ_{ij} are expressed in terms of the components of the stress tensor σ_{ij} by Hooke's law for a generalized state of plane stress (δ_{ij} being the components of a unit tensor):

$$\epsilon_{ij} = \frac{1}{2G}\left(\sigma_{ij} - \frac{\nu}{1+\nu}\sigma_{\gamma\gamma}\delta_{ij}\right). \qquad (9.33)$$

The exact value of the total stress energy is evidently equal to $M\Omega/2$. Consequently, if $\bar{W}$ is the energy corresponding to a certain virtual stress field, then

$$\bar{W} \geq W = \frac{M\Omega}{2} = \frac{M^2}{2GR^2C(\nu)}, \qquad (9.34)$$

whence

$$C(\nu) \geq \frac{M^2}{2GR^2\bar{W}}. \qquad (9.35)$$

The idea consists in using self-similar stress fields to obtain values of $\bar{W}$ as close as possible to the actual one, and by the same token obtaining good estimates for $C(\nu)$. Here the energy $\bar{W}$ is found from the relation

$$\bar{W} = -\frac{R}{2} \int\limits_{-\alpha}^{\alpha} (\sigma_{rr}u_r + \sigma_{r\theta}u_\theta)_{r=R} \, d\theta, \qquad (9.36)$$

obtained from the energy equation and the condition of rapid decrease of the stress at infinity. If we take as the virtual elastic field the field corresponding to the Carothers–Inglis solution (9.7), (9.8), then a simple

---

[†] Contrary to previously considered cases here the elastic displacements appear in the problem statement; therefore among the governing parameters appear the elastic constants G and ν.

but lengthy calculation according to the indicated recipe gives a lower bound for $C(\nu)$:

$$C(\nu) = \frac{16(\sin 2\alpha - 2\alpha \cos 2\alpha)^2}{4(\kappa + 5)\alpha + 8\alpha \cos 4\alpha - (\kappa + 7)\sin 4\alpha}. \tag{9.37}$$

Here $\kappa = (3 - \nu)/(1 + \nu)$.

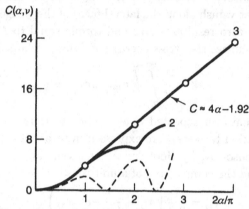

Figure 9.3. Dependence of the dimensionless stiffness $C(\nu)$ of a wedge on its opening angle for $\nu = 0$.

This estimate is represented by the broken line in Figure 9.3 for the case $\nu = 0$. It has obvious absurdities. For example, according to it the wedge loses stiffness at the critical values at which the numerator of (9.37) vanishes, and long before the critical angles the stiffness starts to decrease as the angle α increases. The latter is obviously wrong since for any wedge an admissible stress field is given by that for a wedge of any smaller angle, but extended as identically zero up to the boundary of the wedge, so that the stiffness is a non-decreasing function of the opening angle of the wedge: $dC/d\alpha \geq 0$. In fact, calculating the stiffness for $\alpha = \alpha_*$ on the basis of a non-self-similar solution, given by the first terms of (9.22) and also belonging to the admissible stress fields, leads to a non-zero estimate of the stiffness. Furthermore, for angles not equal to the critical ones, admissible fields were taken to be represented by the sum of the solutions (9.7) and (9.30), where the coefficient A in (9.30) was chosen so as to minimize the stress energy $\bar{W}$. Also, λ was taken as the real root of (9.29) giving minimal stress energy. The corresponding estimate for the stiffness, represented in Figure 9.3 by the lower solid line, goes significantly higher than the previous estimate and, what is most important, passes smoothly through the critical angles. However, this estimate is unsatisfactory for large α. For large α the 'generalized' wedge must in fact behave like a closed elastic ring, for which $\sigma_{r\theta} = M/2\pi r^2$, $\sigma_{rr} = \sigma_{\theta\theta} = 0$ at all points, and the stress energy is equal to

$$\frac{M\Omega}{2} = 2\pi \int\limits_{R}^{\infty} \frac{\sigma_{r\theta}^2}{2G} r \, dr = \frac{M^2}{8\pi R^2 G}; \qquad (9.38)$$

consequently the dimensionless stiffness $C(\nu)$ is $C = M/\Omega G R^2 = 4\pi$. Thus for large α each increase of α by π must increase the dimensionless stiffness by 4π, so that $dC/d\alpha \sim 4$ and we have the asymptotic formula

$$C = 4\alpha - \delta(\nu). \qquad (9.39)$$

To obtain a very accurate estimate of the dimensionless stiffness, Budiansky and Carrier (1973) used the fact that for $\alpha = N\pi/2$, where N is an integer, all the roots of (9.29) are real and are expressed by a simple formula,

$$\lambda = \frac{m}{N} - 1, \qquad (9.40)$$

where m is an integer. For the elastic field one can take an expression in the form of the sum of a large number of solutions of the type (9.30),

$$\Psi = \sum_{m=1(m\neq N)}^{K} \frac{A_m}{r^{(m/N)-1}} \left[\left(\frac{m}{N} + 1 \right) \cos \frac{(m+N)\pi}{2} \sin \left(\frac{m}{N} - 1 \right) \theta \right.$$
$$\left. - \left(\frac{m}{N} - 1 \right) \cos \frac{(m-N)\pi}{2} \sin \left(\frac{m}{N} + 1 \right) \theta \right]$$
$$+ \frac{M[2(-1)^N \theta - \sin 2\theta]}{2N\pi(-1)^{N+1}}, \qquad (9.41)$$

where $K \geq N$, and m is positive. (The last term corresponds to the root that is equal to zero, and is expressed by just the same formulae as in (9.7).) The coefficients A_m are also obtained from the condition of minimality of the stress energy $\bar{W}$. The results of the corresponding calculations are represented in Figure 9.3 by the upper solid line. It is evident that this estimate is considerably higher and, what is important, is compatible with (9.39). The calculations give the following values for $\delta(\nu)$:

$$\delta(0) = 1.92, \quad \delta(0.25) = 1.68, \quad \delta(0.5) = 1.48.$$

This example shows that self-similar solutions can be used successfully to obtain estimates of the bulk characteristics of non-self-similar solutions of the non-idealized problems. The investigation conducted by Budiansky and Carrier shows, however, that such use of self-similar solutions should be made only with great care. In fact, for example, for all $\alpha < \alpha_*$ the solution at large distances from the tip is well approximated by the Carothers–Inglis solution. It would seem that this implies the possibility of using the latter solution to estimate the stiffness for

$\alpha < \alpha_*$. This leads, however, to an unnatural decrease in stiffness, connected with the unsuitability of the self-similar solution for describing the stress field close to $r = R$, which makes an essential contribution to the stress energy[t].

9.2 Similarity laws for brittle and quasi-brittle fracture

9.2.1 Basic concepts of fracture mechanics

The classical theory of elasticity does not allow us to determine the strength of a body or structure directly. The reason is that this theory is a *linear* one. In fact, the components of the stress are related to the components of small deformation (strain) by Hooke's law, which is *linear*. The components of small strain are *linear* combinations of space derivatives of the elastic displacement components. Therefore, after substitution of the relations for stress into static equilibrium equations that are *linear* with respect to stress components a *linear* system of differential equations for the displacement components is obtained. Moreover, in typical problems in elasticity either the actual components of the displacement or the acting loadings i.e., ultimately, *linear* combinations of displacement space derivatives, are prescribed at *known* boundaries.

The linearity of the elasticity theory leads to a result that is in obvious contradiction to everyday experience. Let a solution be obtained for stress σ and elastic displacement $\mathbf{u}$ corresponding to a certain load $\mathbf{F}$. Then, increasing the acting load arbitrarily, say C times, where C can be arbitrarily large, so that $\mathbf{F}_1 = C\mathbf{F}$, we obtain that the solution to an elasticity problem always exists and can be represented in the form $\sigma_1 = C\sigma$, $\mathbf{u}_1 = C\mathbf{u}$. Therefore from the viewpoint of classical elasticity theory the structure can support arbitrarily large loads, so that the very statement of the strength problem is impossible. To obtain the fracture phenomenon within a mathematical model of deformable solid a certain nonlinearity is necessary; though obviously not every sort of nonlinearity makes the fracture phenomenon possible. Meanwhile, for many materials (metals, ceramics, polymers) under ordinary conditions the strains before fracture are small and generally plastic deformations are either completely absent (brittle fracture) or concentrated in a tiny

---

[t] Remember the analogous situation in the Guderley very intense implosion problem, and the von Weizsäcker–Zeldovich impulsive loading problem.

vicinity of the fracture surface (quasi-brittle fracture)[†], so classical elasticity theory should be applicable! Therefore the first strength theories that were proposed complemented the elasticity theory by certain *local* fracture conditions. These conditions bounded the stresses in a structure by a certain constant, which was considered as a material property. It was assumed, more precisely, that the fracture occurs when a certain (positive definite) combination of stress components reaches a limiting value and in the place where the latter is reached.

The crucially important step in the construction of the mathematical model of brittle fracture was performed by the great British scientist and engineer A.A. Griffith (Griffith, 1920). He understood that representing the shape of a structure in an ideal form as in the drawing of a designer we define the boundaries of the body incompletely. In fact, along with 'legitimate' boundaries every structure always contains defects – flaws and cracks – whose surfaces also form a part of the boundary of this structure. But the strength theories mentioned above, complementing elasticity theory by a certain bound to stresses, are in principle inappropriate for the calculation of structural strength for bodies with cracks, since at the crack tip the stresses according to elasticity theory are always infinite.

Cracks are capable to extend under increasing loads and this very aspect makes the elasticity problem for a body with cracks essentially nonlinear even for perfectly linear elastic materials. Therefore in the fracture problem a certain additional material property must appear that characterizes the material's resistance to crack propagation in it. As such a property Griffith himself chose the energy of crack formation per unit area. Irwin (1949) and Orowan (1949) extended the Griffith concept to more realistic quasi-brittle fracture, and so enlarged in an essential way its field of applicability.

A different, *force*, approach to the theory of brittle and quasi-brittle fracture, based on explicitly taking into account 'cohesion forces' that complement the basic loads was developed by the author (Barenblatt, 1959a). It was shown (Barenblatt, 1964), that the estimation of a limiting load, i.e., of the structural strength, leads, from a mathematical viewpoint, to a global problem: the determination of the existence domain (for the given loading parameters) of the solution of a nonlinear problem, the elastic equilibrium of a body with cracks. The last problem belongs to a class of complicated *free-boundary* problems (the cracks are

---

[†] Everybody remembers a china cup or vase broken in childhood. The child tries to put the pieces together, and seemingly there is no difference from what was before, but punishment is inevitable.

able to extend with increasing loads), so it is unrealistic to expect an analytic result for more or less general cases. Therefore the value of experiment, numerical and physical, increases and, consequently, so does the role of scaling and similarity rules. We refer for the details of the fracture problem statement to the reviews by Irwin (1958), Barenblatt (1962), Barenblatt (1964), Raizer (1970) and Barenblatt (1993c) and the monographs by Muskhelishvili (1966), Panasyuk (1968), Liebowitz (1968a,b), Landau and Lifshitz (1986), Bui (1977) and Hahn (1976). We will restrict ourselves in this and the following sections to discussing the scaling laws for static brittle and quasi-brittle fracture where time effects are negligible (Barenblatt, 1956, 1962, 1964).

The plastic head of a crack for a classical example of quasi-brittle fracture – a crack in organic glass – is presented in Figure 9.4(a). As we see, the head size is small, of order of a few hundredths of a millimetre. So, we take as the crack surface under such circumstances the sharp boundary between the elastic region (region 2 in Figure 9.4(b)) and the strongly plastically deformed region (the so-called 'crazy', white region, i.e. region 1 in Figure 9.4(a)). In our approach the forces $G(s)$ (Figure 9.4(b)) acting from the side of the plastic head on the elastic region should also be taken into account in addition to the basic loads. In the ideal case of purely brittle fracture such forces also exist: they are the molecular cohesion forces. Therefore the forces $G(s)$ are unified by the common title *cohesion forces*. Thus, some additional dimensional quantities should enter into consideration: the crack head size d, which is, generally speaking determined by the load and by the microstructure of the material, as well as a characteristic magnitude of the cohesion forces, G_0.

Analysis shows that for a wide class of practically important cases of brittle and quasi-brittle fracture two basic hypotheses (Barenblatt, 1959) can be applied:

(1)   the *smallness* of the crack-head size d in comparison with the total crack length l ($d/l \ll 1$), and

(2)   the *autonomy* of the crack head, i.e., the property that in the mobile equilibrium state[†] the form of the crack head (and, consequently the distribution of the cohesion forces) is identical for all cracks in a given material under fixed external conditions (temperature, pressure and composition of the ambient medium, etc.).

[†] At the state of mobile equilibrium the cohesion forces reach a maximum value, so that the crack starts to extend even at slight increasing of the load.

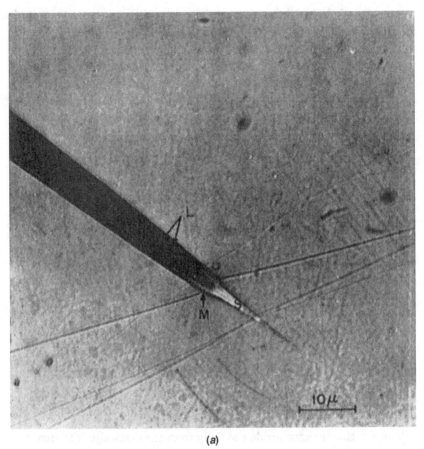

(a)

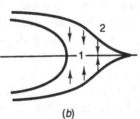

(b)

Figure 9.4. The crack head in a quasi-brittle fracture: (*a*) photograph of a crack head in organic glass (Van den Booghart, 1966), and (*b*) schematic of the cohesion forces in a quasi-brittle fracture.

The condition of boundedness of stress and strain at the crack tips proposed by S.A. Christianovich (Zheltov and Christianovich, 1955; Barenblatt and Christianovich, 1955) was also used in this approach.

The first hypothesis is in fact a definition of the class of cracks under consideration; the proposed approach is inappropriate for the consideration of crack formation and the early stages of its extension. The autonomy of the crack heads is explained by the very large magnitude of the cohesion forces, which are much larger than the characteristic magnitude of the applied loads σ_0, so that $\sigma_0/G_0 \ll 1$, and by the short distance-range action. Owing to these hypotheses the whole theory of brittle and quasi-brittle fracture is an intermediate-asymptotic one. It turns out that the quantities G_0 and d do not enter the theory separately. The only material property in this theory in addition to the ordinary elasticity constants (Young's modulus, Poisson's ratio) is the *fracture toughness*, or *cohesion modulus*:

$$K = \int\limits_0^d \frac{G(s)ds}{\sqrt{s}}, \quad K \sim G_0\sqrt{d}. \qquad (9.42)$$

The cohesion modulus is a material property and is independent of the shape of the crack as a whole, the form of the body, the loading, etc. namely because of the autonomy of the crack head, since the crack head size d and the distribution of cohesion forces $G(s)$ in the state of mobile equilibrium are identical for all cracks in a given material under given external conditions. The cohesion modulus (9.42) enters the problem statement in the following way. Consider a structure with *given* cracks and applied loads (without taking into account the cohesion forces). According to elasticity theory the tensile stress σ at the crack tip is singular:

$$\sigma \sim \frac{N}{\sqrt{s}} + O(1)$$

where s is the distance inside the body from the crack tip. The quantity N, called the *stress-intensity factor*, obtainable from the solution to the linear elasticity problem when the crack tip position is prescribed, depends linearly upon the applied loads. It depends also upon the shape of the structure and the cracks in it. A fundamental formula relating the stress-intensity factor N to the specific energy released at virtual crack extension (also obtainable from the solution to the linear elasticity problem) was obtained by Irwin (1957). In the simplest case of an isolated rectilinear crack N is a function of its length. According to the above hypotheses,

$$N = \frac{K}{\pi}$$

at the points of the crack contours where the mobile equilibrium state is reached. This relation in fact determines the position of the crack contours; in the simplest case it is an equation for the crack length. This very condition makes the problem of the elastic equilibrium of a body with cracks an essentially nonlinear free boundary problem in elasticity theory. The dimension of the cohesion modulus K is $FL^{-3/2}$, where F

is the force dimension. It characterizes the crack-extension toughness of a material and is a material-strength property that is independent of the elastic constants. It can be related (Barenblatt, 1959) to another material-strength property, the specific energy of crack formation, i.e., the energy expended per unit area of crack extension, introduced and used by Griffith in his fundamental paper (Griffith, 1920). The cohesion modulus K, introduced in the paper Barenblatt (1959) should be distinguished from another fracture toughness characteristic of the same dimensions, K_{IC}, introduced nearly simultaneously by Irwin (1960), and determined by the start of catastrophic crack extension. The latter requires instability of in the original mobile equilibrium state. The autonomy of the crack head is not achieved, generally speaking, if the original state at which the crack starts to extend is unstable. Therefore a large scatter in the experimental values of the property K_{IC} is observed.

If the cohesion forces are time independent (as well as the destruction of bonds in the crack head), the cohesion modulus is a constant material property. Fracture in such circumstances is called *static*. We give the values of the cohesion modulus under ordinary conditions for several materials: for organic glass $K \sim 10^2$ kgf/cm$^{3/2}$, for structural steel $K \sim 25 \times 10^4$ kgf/cm$^{3/2}$, for duraluminium $K \sim 10^4$ kgf/cm$^{3/2}$.

The cohesion modulus (fracture toughness) should be one of the governing parameters when the similarity laws of brittle and quasi-brittle fracture are formulated. The appearance of one governing parameter $K \sim G_0\sqrt{d}$ instead of two parameters d and G_0 is a typical manifestation of incomplete similarity.

9.2.2 Scaling in static fracture

Let us turn to several instructive examples.

1. The first, considered in the basic paper Griffith (1920), concerns an infinite body under plane strain with a single plane crack under the action of a uniaxial tensile stress σ, perpendicular to the crack surface (Figure 9.5(a)). The length l of the mobile-equilibrium crack is obviously governed by the dimensional quantities σ and K; dimensional elastic constants like Young's modulus do not enter the relations because the load, not the displacement, is prescribed. We obtain from dimensional analysis

$$l = \text{const} \frac{K^2}{\sigma^2}. \tag{9.43}$$

According to the complete solution given by Griffith (1920), the constant in (9.43) is $1/4\pi^2$. For the type of loading under consideration a catastrophic crack extension, leading to complete failure of the structure, starts at even the smallest increase in the load after the crack reaches the mobile-equilibrium state. Generally speaking, though, as the next example shows, this may not be the case.

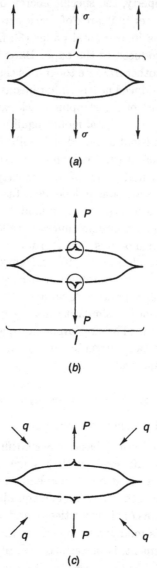

(a)

(b)

(c)

Figure 9.5. (a) The Griffith crack under the action of a uniform tensile stress. (b) The crack under the action of opposite concentrated forces. (c) The crack in a compressed body is supported by opposite concentrated forces.

2. Let us now consider, again for the conditions of a plane strain, an infinite body with a rectilinear crack supported by two concnetrated forces, equal in magnitude and opposite in direction, applied at the centres of the opposite crack sides (Figure 9.5b). The size l of the mobile-equilibrium crack depends on the cohesion modulus K and the force P per unit thickness ($[P] = FL^{-1}$). Dimensional analysis gives

$$l = \text{const}\, \frac{P^2}{K^2}. \tag{9.44}$$

According to the complete solution to the problem (Barenblatt, 1959a), the constant here equals 1. As may be seen, under this type of loading the reaching of mobile equilibrium is not related to the global failure of the structure: under further loading the crack length grows but the loading capacity of the structure is not exhausted. The formula (9.44) gives a good practical basis for experimental determination of the cohesion modulus (see Panasyuk, 1968).

3. If the resistance to a crack extension supported by the same concentrated forces is due to a compressive external pressure q (Figure 9.5(c)), and not to cohesion forces, as for instance in rock massifs, where the action of cohesive forces is negligible, then, according to dimensional analysis,

$$l = \text{const}\, \frac{P}{q}. \tag{9.45}$$

From the complete solution (Barenblatt, 1956) we obtain the constant in (9.45) as $2/\pi$. In this case unlike the previous one, the crack grows proportionally to the tearing force P, and not to its square.

4. In the remarkable experiments of Roesler (1956) and Benbow (1960), a punch with a small, flat point was pressed into the face of a sample of brittle material (glass or fused silica, see Figure 9.6). A conical crack was formed under the punch, and, as the load increased, the crack increased in size; the diameter of the base of the cone rapidly became much larger than the diameter of the punch.

It is natural to assume that the diameter D of the base of the conical crack depends on the load P and on the properties of the material, its fracture toughness (the cohesion modulus) K, which determines the resistance of the material to crack propagation, and Poisson's ratio ν. For sufficiently large loads, the diameter of the punch d is much smaller than the diameter of the base of the cone, and the size Δ of the fused

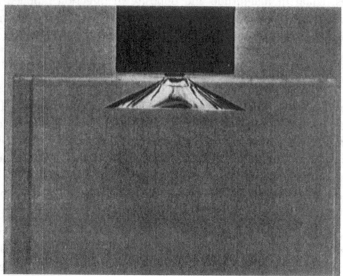

Figure 9.6. When a punch is pressed into a block of fused silica, a conical crack is formed. (From Benbow (1960)).

silica block is much larger than the diameter of the base of the cone, so that these two parameters may be *assumed* to be non-essential[†].

We therefore obtain the dimensions of the quantities involved:

$$[D] = L, \quad [P] = F, \quad [K] = FL^{-3/2}, \quad [\nu] = 1. \qquad (9.46)$$

Thus, the governing parameters P and K have independent dimensions, and dimensional analysis immediately yields

$$D = (P/K)^{2/3}\Phi(\nu). \qquad (9.47)$$

Benbow's analysis of the experimental data for punches of different sizes under various loads confirmed this relation (Figure 9.7). In fact, complete similarity in the parameters $d/(P/K)^{2/3} \ll 1$ and $\Delta/(P/K)^{2/3} \gg 1$ was tacitly assumed in this argument. As Figure 9.7 shows this assumption is apparently correct.

5. We consider now the general similarity rules for brittle (quasi-brittle) fracture. We repeat that when modelling fracture, one should keep in mind the fact that it is impossible to model a structure in the idealized

---

[†] As is known from the theory of elasticity, for a given load acting on a body the stress field in the body (which determines the size of the crack) does not depend on the second elastic constant, Young's modulus. This is the reason that Young's modulus here, as before, does not appear among the governing parameters.

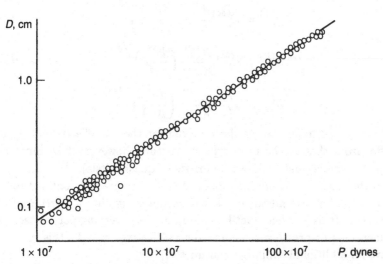

Figure 9.7. The experimental data on the propagation of a cone crack in a block of fused silica confirm law (9.47), which was derived using dimensional analysis (Benbow, 1960). The circles show the experimental points, while the solid line corresponds to the law $D \sim P^{2/3}$.

form in which it appears in the designer's drawing: there are always defects in any object – flaws and cracks. The surfaces of the cracks form an 'illegitimate' (although very important) part of the boundary of the body undergoing failure; the stress concentration on their contours is very high, so that these defects are precisely where failure begins. Therefore the first requirement for the model's structure is that it be geometrically similar to the prototype not only with respect to its 'legitimate' boundaries, but also with respect to the initial defects (cracks).

The loads acting on the structure are either *forces, tensions* (forces distributed along a line), or *stresses* (forces distributed over an area), depending on the way in which they are applied. Thus, the problem consists of determining the values of the force P_f, tension s_f, or stress σ_f corresponding to failure of the structure on a geometrically similar model. These quantities are determined by a characteristic length of the structure l (for instance, the specimen thickness), the fracture toughness (the cohesion modulus) K, and the dimensionless Poisson ratio ν:

$$P_f, s_f, \sigma_f = f(l, K, \nu). \tag{9.48}$$

The dimensions of the governing and determined parameters in the *LFT*

244   *9. Scaling in the deformation and fracture of solids*

class are $[P_f] = F$, $[s_f] = FL^{-1}$, and $[\sigma_f] = FL^{-2}$, $[l] = L$, $[K] = FL^{-3/2}$, $[\nu] = 1$, respectively. Thus

$$P_f^{(P)} = P_f^{(M)} \frac{K^{(P)}}{K^{(M)}} \left(\frac{l^{(P)}}{l^{(M)}} \right)^{3/2} , \qquad (9.49)$$

$$s_f^{(P)} = s_f^{(M)} \frac{K^{(P)}}{K^{(M)}} \left(\frac{l^{(P)}}{l^{(M)}} \right)^{1/2} , \qquad (9.50)$$

$$\sigma_f^{(P)} = \sigma_f^{(M)} \frac{K^{(P)}}{K^{(M)}} \left(\frac{l^{(P)}}{l^{(M)}} \right)^{-1/2} , \qquad (9.51)$$

(the fracture toughness of the material for the model's structure may differ from that of the prototype's structure; however, it is important that the model material also be brittle or quasi-brittle).

If the plastic deformations incident to failure are not confined to the narrow region around the crack but occupy a significant portion of the structure (this is called ductile failure), a new governing parameter with dimensions of pressure appears: the yield stress σ_Y. In this case, we have an additional similarity parameter,

$$\Pi_2 = \frac{\sigma_Y l^{1/2}}{K} = I , \qquad (9.52)$$

for which the name *Irwin parameter* or *Irwin number* has been proposed (in honour of the American scientist and engineer G.R. Irwin, who has made fundamental contributions to the study of fracture mechanics). In particular, Irwin (1960) was the first to establish the decisive influence of a parameter equivalent to I on the characteristics of ductile failure and on the transition from quasi-brittle to ductile failure.

The necessity of complying with the similarity conditions (equality of the Irwin similarity parameter) for the model and prototype structures creates particular difficulties in modelling. In fact, the Irwin parameter is physically the square root of the ratio of the length scale of the structure to the intrinsic structural parameter of the material K^2/σ_Y^2, which is, to an order of magnitude, equal to the length of the plastic zone near the crack tip. Thus, it is impossible to carry out tests in plastic materials on a model structure made of the same material under identical external conditions[†]. If the similarity conditions $\nu^{(M)} = \nu^{(P)}$ and $I^{(M)} = I^{(P)}$ are satisfied, the scaling of loads from model to prototype may be carried out using the same equations (9.49)–(9.51).

[†] Goldstein and Vainshelbaum (1978) have suggested that modelling should be carried out by testing models of the same material at different temperatures: the intrinsic scale of the material K^2/σ_Y^2, increases strongly with increasing temperature. This proposal rests on an essential assumption that all of the dimensionless characteristics of the material remain unchanged under temperature changes.

9.2.3 Scaling in fatigue

Fatigue – the gradual failure of a structure under a pulsating load, inevitable in contemporary structures, especially aircraft, is now the greatest challenge for engineers and mathematicians working in material science. In some sense it is an analogue of turbulence in solids (cracks and flaws forming a dynamic cascade corresponding to vortices in fluids (see the next chapter)).

In the absence of a theory, technological tests such as the traditional fatigue experiment are very important. In this experiment the specimen (a notched or slotted bar or plate) is loaded by a pulsating tensile load at constant frequency and amplitude. At the edge of the notch a fatigue crack is formed, and its propagation is recorded. In 'multicycle' fatigue experiments the number of cycles before the failure is of the order of millions. In the fundamental paper by Paris and Erdogan (1963) the following scaling law was discovered,

$$\frac{dl}{dn} = A(\Delta N)^m \, , \qquad (9.53)$$

where l is the crack length, n is the number of cycles, dl/dn is averaged over the cycle; $\Delta N = N_{max} - N_{min}$ is the stress intensity factor amplitude, and A, m are certain constants. The scaling relations (9.53) are well confirmed by experiment (Figure 9.8).

As we know, scaling laws never appear by chance but always reveal the self-similarity of a phenomenon; the scaling law (9.53) is no exception. Therefore, the question arises, what kind of self-similarity is this, and – most important – what are the constants A and m? More precisely, are they material properties or do they depend also on the specimen size? Usually designers estimate the lifetime of structures on the basis of the relations like (9.53); therefore, the question is of basic practical importance.

We consider here the phenomenon of fatigue crack propagation on the basis of the similarity approach (Barenblatt and Botvina 1981, 1983). The average velocity dl/dn can depend, in principle, upon the following quantities: $\Delta N = N_{max} - N_{min}$, $R = N_{max}/N_{min}$, the so called loading asymmetry (N_{max} and N_{min} are the maximum and minimum stress-intensity factors over a cycle), f, the frequency; t, the time; h, a characteristic specimen length scale (for instance, its thickness); the material properties, the yield stress σ_Y (detailed fatigue fracture surface analysis shows that local yield plays an important role), and a fracture toughness property. Here, for the latter, we take K_{IC} because plentiful experimental data is available in the literature.

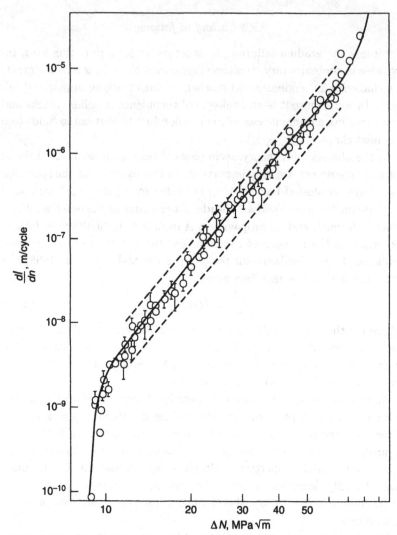

Figure 9.8. Kinetic diagram for fatigue crack growth in the aluminium alloy BT-3-1 (Botvina, 1989) confirms the scaling law (9.53) in the major part of the crack velocity range.

Dimensional analysis gives in a standard way

$$\frac{dl}{dn} = \left(\frac{\Delta N}{\sigma_Y}\right)^2 \Phi\left(\frac{\Delta N}{K_{IC}}, R, Z, ft\right) \qquad (9.54)$$

where

$$Z = \frac{\sigma_Y \sqrt{h}}{K_{IC}} \qquad (9.55)$$

is the basic similarity parameter. It is natural to consider an inter-mediate-asymptotic stage in the multicycle fatigue process where the influence of the argument ft disappears. The beginning of this stage is clearly marked by the appearance at the fracture surface of a regular system of striations.

The argument $\Delta N/K_{IC}$ is small, and it is expedient to consider the asymptotic relations for $\Delta N/K_{IC} \ll 1$. Two possibilities appear: complete similarity when the limit of the function Φ at $\Delta N/K_{IC} \to 0$ is finite and different from zero, and lack of complete similarity. In the case of complete similarity the following scaling relation is obtained for $\Delta N/K_{IC} \ll 1$:

$$\frac{dl}{dn} = \left(\frac{\Delta N}{\sigma_Y}\right)^2 \Phi_1(R, Z) \tag{9.56}$$

i.e., the scaling law (9.53) with $m = 2$; in practice this is hardly ever found, except for some aluminium alloys. According to the general procedure outlined in chapter 5 we now assume incomplete similarity in the parameter $\Delta N/K_{IC} \ll 1$ at the intermediate stage of the fatigue crack extension. We obtain

$$\Phi = \left(\frac{\Delta N}{K_{IC}}\right)^\alpha \Phi_1(R, Z), \quad \frac{dl}{dn} = \frac{(\Delta N)^{2+\alpha}}{\sigma_Y^2 K_{IC}^\alpha} \Phi_1(R, Z) \tag{9.57}$$

i.e., exactly the Paris–Erdogan scaling law (9.53) with

$$A = \frac{\Phi_1(R, Z)}{\sigma_Y^2 K_{IC}^\alpha}, \quad m = 2 + \alpha(R, Z). \tag{9.58}$$

Therefore the constants A and m depend not only upon the material properties and the asymmetry of loading, but, through the basic parameter Z, upon the specimen size h. The processed experimental data show that this dependence can be very strong (Figure 9.9) so that designers must exercise caution when using the results of the standard fatigue tests performed on small specimens for predicting the life-time of large structures.

9.2.4 Scaling in creep

The phenomenon of creep is characteristic for metals, in particular at elevated temperatures. In a common tensile test, under a constant stress σ, the strain

$$\epsilon = \frac{\Delta l}{l_0}, \tag{9.59}$$

where Δl is the specimen elongation and l_0 is the original specimen length is slowly growing with time; this is the phenomenon of creep. Usually three stages of creep are distinguished (Figure 9.10): *I*, unsteady creep, as a rule a short stage; *II*, steady creep; *III* terminal unsteady creep. Stage *II* is the longest and most interesting practically; at this

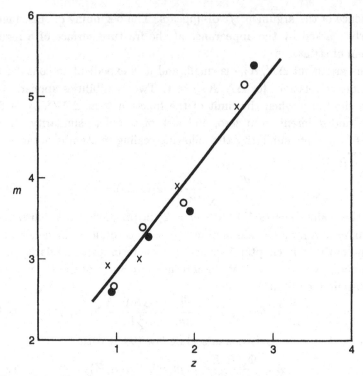

Figure 9.9. The dependence of the exponent in the Paris–Erdogan law (9.53) on the similarity parameter Z for 4340 steel for specimens of various orientations (A, L, T) with respect to rolling direction: o, A; •, L; ×, T. From Barenblatt and Botvina (1981) on the basis of data from Heiser and Mortimer (1972).

stage the strain rate $\dot{\epsilon}$ is constant in time. The relationship between this constant strain rate $\dot{\epsilon}$ and the stress σ, the creep constitutive equation, is an important material property. The creep phenomenon was discovered by Andrade (1910). In this paper (see also Bailey, 1929 and Norton, 1929) was proposed a scaling relation

$$\dot{\epsilon} = A\sigma^m \qquad (9.60)$$

where A and m are constants. Since that time many other constitutive equations have been proposed, of both scaling and non-scaling type. Sometimes scaling constitutive equations are considered simply as ordinary approximations among empiric formulae representing creep experimental data in different, non-scaling form. However, here again the

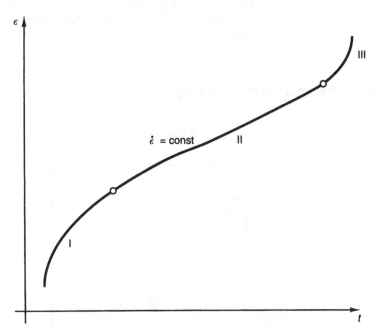

Figure 9.10. Various stages of creep. At the longest and most practically interesting stage, *II*, of steady creep, the strain rate is constant.

applicability of a scaling law demonstrates a deep physical property of steady creep: its self-similarity.

We shall derive the scaling law (9.60) from a basic assumption: *in the steady-creep process under consideration there exists no characteristic material time scale.*

Indeed, in this steady-creep process, in order to support a constant strain rate $\dot{\epsilon}_1$ a certain constant stress σ_1 has to be applied to the specimen. The stress σ_1 depends on $\dot{\epsilon}_1$ and perhaps on some additional parameters, among which, however, there exists no characteristic time scale. In fact, the stress in a steady-creep process is constant and, moreover, there is no characteristic time scale of the material governing the phenomenon, owing to the basic assumption. Let us now consider another steady-state process for the same material under the same external conditions (temperature, ambient atmosphere, etc.) but corresponding to a different strain rate $\dot{\epsilon}_2$; to support it a stress σ_2 is necessary. Owing to the lack of a characteristic time scale of the loading process or of the material, the ratio σ_2/σ_1 can only depend on the ratio $\dot{\epsilon}_2/\dot{\epsilon}_1$:

$$\frac{\sigma_2}{\sigma_1} = \varphi\left(\frac{\dot{\epsilon}_2}{\dot{\epsilon}_1}\right). \qquad (9.61)$$

If a certain characteristic time τ were available, the function φ would also depend on the argument $\tau\dot\epsilon_1$; the quantity $\dot\epsilon$ obviously has inverse time at its dimension. A similar equation is obtained for a third strain rate $\dot\epsilon_3$ and stress σ_3, so that

$$\frac{\sigma_3}{\sigma_1} = \varphi\left(\frac{\dot\epsilon_3}{\dot\epsilon_1}\right), \tag{9.62}$$

and we obtain from (9.61) and (9.62)

$$\frac{\sigma_3}{\sigma_2} = \frac{\varphi\left(\dot\epsilon_3/\dot\epsilon_1\right)}{\varphi\left(\dot\epsilon_2/\dot\epsilon_1\right)}. \tag{9.63}$$

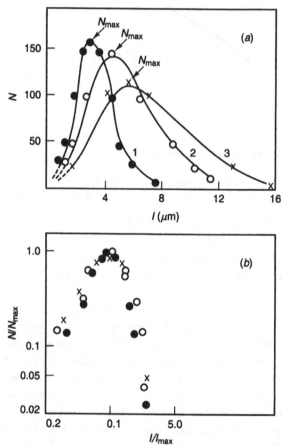

Figure 9.11. Pore size distributions in (a) natural coordinates, N being the number of pores of size l (Cane and Greenwood, 1975), and (b) universal coordinates, for crystalline iron specimens (0.006% carbon) tested in creep under 9.3 MPA tensile stress at 700 °C. (1) $\epsilon = 2.1\%$, $t = 23$ hours; (2) $\epsilon = 6.2\%$, $t = 142$ hours; (3) $\epsilon = 9.3\%$, $t = 262$ hours. N_{max}, maximum statistical frequency; l_{max}, corresponding pore size (Barenblatt and Botvina, 1986).

Note now that owing to the lack of characteristic time scale the process with strain rate $\dot{\epsilon}_1$ is not a distinguished one. Therefore the relation (9.62) will still be valid if σ_1 is replaced by σ_2 and $\dot{\epsilon}_1$ by $\dot{\epsilon}_2$:

$$\frac{\sigma_3}{\sigma_2} = \varphi\left(\frac{\dot{\epsilon}_3}{\dot{\epsilon}_2}\right), \qquad (9.64)$$

where φ is the same function as in (9.61), (9.62) and (9.63).

From (9.63) and (9.64) we obtain a functional equation for the function φ ($x = \dot{\epsilon}_2/\dot{\epsilon}_1$, $y = \dot{\epsilon}_3/\dot{\epsilon}_1$),

$$\frac{\varphi(y)}{\varphi(x)} = \varphi\left(\frac{y}{x}\right) \qquad (9.65)$$

(cf. chapter 1), whose solution is $\varphi(z) = z^\alpha$, whence, denoting σ_2 and $\dot{\epsilon}_2$ simply by σ and $\dot{\epsilon}$, and $n = 1/\alpha$, $A = \sigma_1^{-n}\dot{\epsilon}_1$, we obtain the relation (9.60).

An independent argument in favour of our basic assumption concerning the self-similarity of steady creep can be obtained from measuring pore size distributions at various strains within the steady-creep range (Figure 9.11). As it is seen, the pore size distribution transforms during the steady-creep process in a self-similar way.

An important point is related to the fracture process under conditions when the cohesion forces deteriorate in time (owing, for instance, to fluctuational breaking of bonds between the microstructural elements). Under wide assumptions, in this case the cohesion modulus K is not a constant, but a universal (for a given material under given external conditions) function of the velocity $u = dl/dt$ of the crack extension rate (l, crack length, t, the time): $K = K(u)$. The following scaling law was proposed for the latter relation:

$$K = Au^m$$

(Vavakin and Salganik, 1975; Parvin and Williams, 1975). This scaling law again demonstrates the self-similarity of the cohesion-force deterioration process and can be derived from the basic assumption of the lack of a characteristic time scale. This time, however, A and m are the material constants.

10

Scaling in turbulence

10.1 Homogeneous and isotropic turbulence

10.1.1 The problem of turbulence

This chapter differs from the previous ones in that scaling laws and
self-similar solutions of the first and second kind will be established
by making essential use of experimental data and without turning to
a mathematical formulation of the problem, which, for turbulence, is
lacking at the present time.

The problem of turbulence, to which this chapter is devoted, is considered with good reason to be the number-one problem of contemporary
classical physics. Discovered by Leonardo and baptized by Lord Kelvin[†],
it has attracted the greatest minds of the century, including such giants
as W. Heisenberg, A.N. Kolmogorov, G.I. Taylor, L. Prandtl, and Th.
von Kármán. Nevertheless, it remains an open problem: none of the
results available has been obtained from first principles. They are based
essentially on strong additional assumptions, which may or may not be
correct.

The phenomenon of turbulence, as is well known, consists in the following. As we have seen in chapter 1, the basic similarity parameter
that governs the global properties of the flow of an incompressible viscous fluid is the Reynolds number $\rho U l/\mu$ (ρ being the density and μ
the viscosity of the fluid, U a characteristic speed, and l a characteristic

---

[†] Professor U. Frisch showed me the place in the diaries of Leonardo where the word
'turbulent' was used in exactly the same sense as we use it now. An instructive
case of congeniality!

length scale of the flow). When the Reynolds number reaches a certain critical value $\mathrm{Re_{cr}}$, different for different flows (for example, for flow in a smooth cylindrical pipe of circular cross section, $\mathrm{Re_{cr}} \sim 10^3$, for flow in a boundary layer, $\mathrm{Re_{cr}} \sim 10^5$), the character of the flow changes suddenly and sharply. A stream that at subcritical values of the Reynolds number was regular and ordered – *laminar* – becomes essentially irregular both in time and in space. The flow properties for supercritical values of the Reynolds number undergo sharp and disorderly variations in space and time, and the fields of flow properties, – pressure, velocity, etc. – can to a good approximation be considered random. Such a regime of flow is called *turbulent*.

At the present time there exist only more or less plausible conjectures regarding the origin of turbulence, sometimes very interesting, but not having conclusive strength. Also, there exists no complete mathematical description of developed turbulent flows. Under these circumstances, in all attempts to create theoretical models valid for certain classes of turbulent flows though not pretending to be universal, similarity, scaling, and renormalization group considerations occupy a primary place.

Together with the majority of investigators, we shall start from the assumption that, for velocities small compared with the speed of sound, turbulent motion can actually be described by the equations of motion for a viscous incompressible fluid, i.e., by the Navier–Stokes equations for momentum balance and the continuity equation, which in rectangular Cartesian coordinates can be written in the form (Kochin, Kibel' and Roze, 1964; Batchelor, 1967; Germain, 1986a; Landau and Lifshitz, 1987)

$$\partial_t u_i + u_\alpha \partial_\alpha u_i = -\frac{1}{\rho}\partial_i p + \nu\Delta u_i\,,$$
$$\partial_\alpha u_\alpha = 0\,. \tag{10.1}$$

Here the u_i are the components of the velocity vector, $\nu = \mu/\rho$ is the kinematic viscosity, p is the pressure, and one sums over repeated Greek indices from one to three.

To construct, for instance numerically, a solution of these equations corresponding to some special realization of a developed turbulent flow is impossible in view of their extreme instability[†]. Hence, and also in view of the possibility noted above of considering the properties of a turbulent flow field as random, the description of turbulent flows is always given in

---

[†] All attempts at the direct numerical simulation of turbulent flows are related to values of the Reynolds number for which the flow cannot be considered as a developed one.

statistical terms. As is known (for details see, e.g., Monin and Yaglom, 1971, 1975), a sufficiently complete description of a developed turbulent flow is given by a set of mean quantities

$$\langle u_i(\mathbf{x}, t) \rangle, \quad \langle p(\mathbf{x}, t) \rangle \tag{10.2}$$

and moment tensors

$$
\begin{aligned}
B_{ijk} \ \ldots &= \ \langle u_i(\mathbf{x}, t) u_j(\mathbf{x}_1, t) u_k(\mathbf{x}_2, t) \ldots \ \rangle, \\
B_{pij} \ \ldots &= \ \langle p(\mathbf{x}, t) u_i(\mathbf{x}_1, t) u_j(\mathbf{x}_2, t) \ldots \ \rangle, \\
&\vdots \qquad\qquad\qquad\quad \vdots
\end{aligned} \tag{10.3}
$$

for all possible point systems: $\mathbf{x}, \mathbf{x}_1$; $\mathbf{x}, \mathbf{x}_1, \mathbf{x}_2$; $\mathbf{x}, \mathbf{x}_1, \mathbf{x}_2, \mathbf{x}_3, \ldots$.

Here the sign $\langle \ldots \rangle$ denotes the probability mean value. Taking probability mean values is used in theoretical work on turbulence as a natural method of averaging. In experimental practice, one uses volume or time means, the identification of these types of averaging with the taking of probability means being made on the basis of the so-called ergodic hypothesis.

A system of equations for the moments can be obtained by multiplying (10.1) by the velocity components at different points of the flow and subsequently averaging. This was done by Keller and Friedmann (1924). As a special case one obtains a non-closed set of equations for the mean quantities, first given in the fundamental paper of Reynolds (1895).

10.1.2 Homogeneous isotropic turbulence

Essential progress in the development of a statistical theory of turbulence occurred when Taylor (1935) introduced the idea of considering homogeneous isotropic turbulence. This idea gained additional fundamental significance after Kolmogorov (1941) and Obukhov (1941) predicted that at small scales all developed turbulent flows (i.e., flows at large Reynolds numbers) are *statistically identical* and therefore have the properties of homogeneity and isotropy. A flow is called *homogeneous* and *isotropic* if all its moment tensors remain unchanged upon translation, rotation, or mirror reflection with respect to some plane, of the system of points $\mathbf{x}, \mathbf{x}_1, \mathbf{x}_2, \ldots$. (To be unchanged means that in a coordinate system arranged relative to the transformed system of points in the same way as the original coordinate system was arranged relative to the original system of points, the values of the components of the tensor remain the same.) For a homogeneous isotropic flow the mean velocity vanishes, and the number of independent components of moment tensors is substantially reduced, as well as the number of quantities on which they depend. Thus in an arbitrary Cartesian coordinate system the components of the second-order moment tensor for the velocity field of a homogeneous isotropic flow are expressed in the following way:

$$B_{ij} = \langle u_i(\mathbf{x},t)u_j(\mathbf{x}+\mathbf{r},t) \rangle = (B_{LL}-B_{NN})\xi_i\xi_j/r^2 + B_{NN}\delta_{ij}. \quad (10.4)$$

Here $r = |\mathbf{r}|$ is the distance between points, the ξ_i are the components of the radius vector $\mathbf{r}$ joining the two points $\mathbf{x}$ and $\mathbf{x}_1 = \mathbf{x}+\mathbf{r}$, t is the time, and

$$\begin{aligned} B_{LL}(r,t) &= \langle u_L(\mathbf{x},t)u_L(\mathbf{x}+\mathbf{r},t) \rangle, \\ B_{NN}(r,t) &= \langle u_N(\mathbf{x},t)u_N(\mathbf{x}+\mathbf{r},t) \rangle, \end{aligned} \quad (10.5)$$

where u_L is the projection of the velocity vector in the direction of the radius vector $\mathbf{r}$ and u_N is its projection in the direction normal to the $\mathbf{r}$. Because of the incompressibility of the flow, the quantities B_{LL} and B_{NN} are connected by the relation

$$B_{NN} = B_{LL} + (r/2)\partial_r B_{LL}. \quad (10.6)$$

Thus, the second-order moment tensor for the velocity field is determined by a single scalar function of two scalar arguments, $B_{LL}(r,t)$. The situation is analogous for the two-point third-order moment tensor

$$B_{ij,k} = \langle u_i(\mathbf{x},t)u_j(\mathbf{x},t)u_k(\mathbf{x}+\mathbf{r},t) \rangle,$$

which, because of homogeneity, isotropy and incompressibility, can be expressed in terms of one component, a scalar function of the scalar arguments r and t; for example,

$$B_{LL,L}(r,t) = \langle u_L^2(\mathbf{x},t)u_L(\mathbf{x}+\mathbf{r},t) \rangle. \quad (10.7)$$

A similar reduction in the number of independent variables and independent components of moment tensors, due to homogeneity, isotropy and incompressibility, holds also for moments of higher order.

We turn now to the Navier–Stokes equations, multiply them by velocity components at successively increasing numbers of points, take the average, and use the symmetry relation following from the homogeneity and isotropy of the flow. Thus we obtain an infinite system of equations, which is, however, not closed at any finite stage because of the presence of quadratic-nonlinearity in the Navier–Stokes equations.

The first equation of this system, connecting the two-point second and third moments can be reduced to the form

$$\partial_t B_{LL}(r,t) = 2\nu \frac{1}{r^4}\partial_r r^4 \partial_r B_{LL} + \frac{1}{r^4}\partial_r r^4 B_{LL,L}(r,t). \quad (10.8)$$

This relation, connecting two unknown functions, is called the Kármán–Howarth equation. It should be noted that in the fundamental paper of von Kármán and Howarth (1938) this equation was presented in a different, less convenient form. It was first expressed in the form (10.8) by Loitsiansky (1939) and Millionshchikov (1939).

The problem in its complete form consists of solving the infinite system of equations for given initial conditions on the moments; this is the so-called problem of the decay of homogeneous isotropic turbulence. As a matter of fact, we have at best only very general information concerning the initial conditions, and so are unable to give the complete initial distribution of the moments. Therefore the asymptotics of the solution for large t, which 'remembers' only some basic properties of the initial conditions, is of particular interest. Under broad assumptions the asymptotics can be considered self-similar.

10.1.3 The decay of homogeneous isotropic turbulence

1. If at some stage of the motion the contribution of the third moments in the Kármán-Howarth relation (10.8) is small, then this relation becomes a closed equation for the second moment $B_{LL}(r,t)$,

$$\partial_t B_{LL} = 2\nu \frac{1}{r^4} \partial_r r^4 \partial_r B_{LL}, \qquad (10.9)$$

which coincides in form with the equation of heat conduction in five-dimensional space for the case of central symmetry. Self-similar solutions of this equation were obtained in the paper of von Kármán and Howarth (1938) (see also Sedov, 1944, 1959)); they have the form

$$B_{LL} = \frac{A}{(t-t_0)^n} f(\xi, n), \quad \xi = \frac{r}{\sqrt{\nu(t-t_0)}}, \qquad (10.10)$$

where A, n and t_0 are constants, and the function $f(\xi, n)$ satisfies the equation

$$\frac{d^2 f}{d\xi^2} + \left(\frac{4}{\xi} + \frac{\xi}{2} \right) \frac{df}{d\xi} + nf = 0 \qquad (10.11)$$

under the conditions

$$f(0, n) = 1, \quad f(\infty, n) = 0, \qquad (10.12)$$

the first of which is a normalization condition and the second of which is obtained from a natural assumption concerning the statistical independence of the velocities at infinitely distant points: $B_{LL}(\infty, t) = 0$. The function $f(\xi, n)$ so defined can be expressed, as is easily found (see Abramowitz and Stegun, 1970), in terms of a well-known special function, the confluent hypergeometric function $M(\alpha, \beta, z)$:

$$f = M(n, 5/2, -\xi^2/8). \qquad (10.13)$$

The spectrum of the eigenvalues n that determine the rate of decay of the second-order moments turns out, upon direct construction of the self-similar solution (10.10), to be continuous: a solution of (10.11) under the conditions (10.12) exists for any $n > 0$. The value of n that is actually realized must be determined by the initial conditions of the problem, for which (10.10) is a self-similar intermediate asymptotics.

If the initial distribution $B_{LL}(r, 0)$ is such that the quantity

$$\Lambda_0 = \int_0^\infty r^4 B_{LL}(r, 0) dr \qquad (10.14)$$

is finite and different from zero, i.e., $0 < \Lambda_0 < \infty$, then $n = 5/2$ and the asymptotics of the solution as $t \to \infty$ corresponding to such an initial distribution can be written in the form

$$B_{LL}(r, t) = \frac{\Lambda_0}{48\sqrt{2\pi\nu^5(t - t_0)^5}} \exp\left[-\frac{r^2}{8\nu(t - t_0)}\right]. \qquad (10.15)$$

Here the quantity

$$\Lambda = \int_0^\infty r^4 B_{LL}(r, t) dr \qquad (10.16)$$

is an integral of the motion, analogous to the total amount of heat in the theory of heat conduction; that is, it is independent of time, $\Lambda \equiv \Lambda_0$. Loitsiansky (1939) has shown that under certain assumptions this quantity remains independent of time even when third moments are taken into account.

One can prove, using properties of the confluent hypergeometric functions, that the solutions (10.10) with $n > 5/2$ have Λ equal to zero. These solutions are in a certain sense structurally unstable with respect to the initial conditions. In fact, if perturbations of such solutions have, say, small but finite Λ_0, then for sufficiently large t only the contribution of the perturbation will govern the decay law, since it corresponds to the smallest n:$n = 5/2$. For this reason self-similar solutions with $n > 5/2$ are of rather lesser interest. On the other hand there is considerable interest in solutions with $n < 5/2$, for which $\Lambda = \infty$. These solutions can be represented in the form

$$B_{LL}(r, t) = \frac{\Lambda_0}{\sqrt{\nu^5(t - t_0)^5}} \left(\frac{\sqrt{\nu(t - t_0)}}{l}\right)^{5-2n} f\left(\frac{r}{\sqrt{\nu(t - t_0)}}, n\right),$$

$$\qquad (10.17)$$

where Λ_0 and l are constants having dimensions $L^7 T^{-2}$ and L, respectively; they are chosen so that $A = \Lambda_0 l^{-(5-2n)} \nu^{5/2-n}$.

It is evident that all these solutions with $n \neq 5/2$ are self-similarities of the second kind, 'remembering' the characteristic length scale l of the initial distribution. (See chapter 3, where, for another problem, a completely analogous situation was analyzed). The situation is that the asymptotics of the dimensionless function

$$\Phi(\xi, \eta, \ldots), \quad \xi = \frac{r}{\sqrt{\nu(t - t_0)}}, \quad \eta = \frac{l}{\sqrt{\nu(t - t_0)}},$$

which appears upon applying dimensional analysis to the solution of the original non-self-similar problem, has, for small η, the form

$$\Phi(\xi, \eta, \ldots) \cong \eta^{2n-5}\Phi_1(\xi, \ldots).$$

Therefore the characteristic length scale l of the initial distribution appears in the constant A governing the solution, but only in combination with Λ_0; therefore it does not spoil the self-similarity.

We note that the stage of development of homogeneous isotropic turbulence at which the third-order moments are negligibly small is sometimes called the final stage of decay. It is sometimes argued that at this final stage the velocity is small and hence so are the third-order moments, which are of the order of the velocity cubed and therefore small compared with the second-order moments, which are of the order of the velocity squared. Such an argument is insufficient. Actually, a stage at which the third-order moments are negligibly small can occur only at the start of the motion with a special choice of initial conditions.

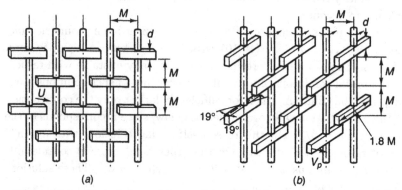

Figure 10.1. Turbulizing grids, used by S.C. Ling and co-workers: (*a*) passive; (*b*) active.

2. From the very first appearance in the papers of Taylor (1935) of the concept of homogeneous isotropic turbulent flow, attempts have been made to model this by the decay of turbulence in wind and water tunnels. There is a detailed summary of this work in the paper of Gad-el-Hak and Corrsin (1974). One should note especially the careful experiments performed by Ling and his associates (Ling and Huang, 1970; Ling and Wan, 1972) using a water tunnel – a long channel of square section into which water was introduced through a passive or active grid

of rods. Figure 10.1 shows the arrangements of grids used in these experiments: passive (Figure 10.1(a)) and active (Figure 10.1(b)). In the active grid the rods are equipped with agitating bars that perform oscillating motions at speeds and frequencies that can be varied. In the work of Gad-el-Hak and Corrsin (1974) a different type of active grid (a 'jet grid') was used; the rods of the grid were hollow and were provided with upwind or downwind controllable nozzles evenly distributed along each rod. Through the hollow rods and nozzles air was injected at varying rates into the flow. (In this work the experiments were performed in wind tunnels.) Thus in all these experiments turbulent fluctuations, introduced into the flow by a grid, then decayed as the fluid moved downstream. Here the fluctuations of velocity become close to isotropic even at small distances from the grid. Figure 10.2 shows the ratios of the mean square fluctuations of the measured longitudinal and transverse components of velocity (Ling and Huang, 1970); it is evident that they are close to unity. We see that if one takes as the time the quantity $t = x/U$ (U being the mean velocity of the flow and x the coordinate measured along the channel downstream from the grid), then the pattern of decay of turbulence along the channel corresponds sufficiently well to the scheme of decay of homogeneous isotropic turbulence in time. (The homogeneity was also specially checked by moving gauges, by means of which velocities were measured in the cross-flow planes $x = $ constant.) This idea for the implementation of homogeneous isotropic turbulent flow was proposed and realized for the first time by Taylor (1935).

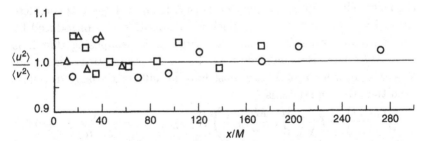

Figure 10.2. Velocity fluctuations in the turbulent flow behind a grid are nearly isotropic: o, $Re_M = 470$, $M = 1.78$ cm, $M/d = 2.8$, $U = 2.9$ cm/s; $\square$, $Re_M = 940$, $M = 3.56$ cm, $M/d = 2.8$, $U = 2.9$ cm/s; $\triangle$, $Re_M = 840$, $M = 3.18$ cm, $M/d = 5.0$, $U = 2.9$ cm/s. From Ling and Huang (1970).

The statistical properties – the moment tensors of the turbulent motion under consideration – are thus governed by the mean velocity U of the flow, the characteristic length scale M of the grid, the thickness d of the rods, the viscosity coefficient ν, and the quantities r and $t-t_0$, where t_0 is the effective origin of time, about whose determination something will be said below. Furthermore, for active grids of the type used by Ling and Wan (1972) the moments are governed also by the speed V_p and frequency of oscillation ω of the tips of the agitating bars; for the active grids used by Gad-el-Hak and Corrsin (1974) an additional governing parameter for the moment tensors is the injection ratio $J = Q_1/Q$ (Q_1 being the flux rate of gas supplied through the hollow rods of the grid and Q the flux rate of gas supplied to the grid).

Dimensional analysis gives for the two-point moments of second and third order,

$$B_{LL} = \frac{\nu}{t-t_0}\Phi_{LL}\left(\xi,\eta,\frac{M}{d},\frac{MU}{\nu},\ldots\right),\qquad(10.18)$$

$$B_{LL,L} = \left(\frac{\nu}{t-t_0}\right)^{3/2}\Phi_{LL,L}\left(\xi,\eta,\frac{M}{d},\frac{MU}{\nu},\ldots\right),\qquad(10.19)$$

where the Φ's are dimensionless functions of their dimensionless arguments

$$\xi = \frac{r}{\sqrt{\nu(t-t_0)}},\qquad \eta = \frac{M}{(x-x_0)} = \frac{M}{U(t-t_0)},$$

of the grid parameter M/d and the Reynolds number MU/ν of the grid, and also of the parameters characterizing the activity of the grid.

It is of interest to consider the motion at sufficiently large distances from the grid such that $M/U(t - t_0) \ll 1$ and one can assume that random details of the initial conditions at the grid no longer influence the flow. The simplest assumption is that for $\eta \ll 1$ there is complete similarity in the parameter η. Such an assumption was introduced by von Kármán (von Kármán and Howarth, 1938), supposing that it is satisfied for large Reynolds numbers. Under the assumption of complete similarity in η for $\eta \ll 1$, one must have at sufficiently large distances from the grid the relations

$$\frac{B_{LL}(r,t)}{B_{LL}(0,t)} = f\left(\xi,\frac{M}{d},\frac{MU}{\nu},\ldots\right),\qquad B_{LL}(0,t) = \frac{A}{(t-t_0)},\qquad(10.20)$$

$$\frac{B_{LL,L}(r,t)}{B_{LL}^{3/2}(0,t)} = g\left(\xi,\frac{M}{d},\frac{MU}{\nu},\ldots\right).\qquad(10.21)$$

Here A is a constant depending on the initial conditions at the grid. Equations (10.20) and (10.21) were proposed by Dryden (1943) and Sedov (1944).

Next in degree of complexity is the assumption of incomplete similarity in the variable η for $\eta \ll 1$. In this case one must have at large distances from the grid the relations

$$B_{LL}(r,t) = \frac{\nu M^\alpha}{U^\alpha (t - t_0)^{1+\alpha}} \quad F\left(\xi, \frac{M}{d}, \frac{MU}{\nu}, \dots\right), \tag{10.22}$$

$$\frac{B_{LL}(r,t)}{B_{LL}(0,t)} = f\left(\xi, \frac{M}{d}, \frac{MU}{\nu}, \dots\right), \tag{10.23}$$

$$B_{LL}(0,t) = \frac{A}{(t - t_0)^{1+\alpha}}, \tag{10.24}$$

$$B_{LL,L}(r,t) = \frac{\nu^{3/2} M^\alpha}{U^\alpha (t - t_0)^{3/2+\alpha}} g\left(\xi, \frac{M}{d}, \frac{MU}{\nu}, \dots\right), \tag{10.25}$$

$$\frac{B_{LL,L}(r,t)}{B_{LL}^{3/2}(0,t)} = B(t - t_0)^{\alpha/2} g\left(\xi, \frac{M}{d}, \frac{MU}{\nu}, \dots\right). \tag{10.26}$$

Here A, B and α are again constant quantities. The equality of the powers to which $\eta = M/U(t - t_0)$ appears in the expressions for $B_{LL}(r,t)$ and $B_{LL,L}(r,t)$ follows from the Kármán-Howarth equation (10.8), which relates these quantities.

We turn to the results of experiments. In Figures 10.3 and 10.4 are shown the results of measuring the correlation function

$$f = \frac{B_{LL}(r,t)}{B_{LL}(0,t)}$$

in the cases of a passive grid (Ling and Huang, 1970) and an active grid (Ling and Wan, 1972) as a function of $\xi = r/[\nu(t - t_0)]^{1/2}$ (the effective origin t_0 being appropriately defined, see below). It is evident that in each case the experimental points lie close to a single curve, different for each different case. This confirms the self-similarity of the correlation function f, but does not determine the character of the self-similarity of the moment tensors; it is evident from (10.20) and (10.23) that a corresponding result must hold in both cases, for complete as well as for incomplete similarity.

In Figure 10.5 and Figure 10.6 are shown the results of measuring the quantity $B_{LL}(0,t)$ for, respectively, passive grids of different types (Ling and Huang, 1970) and active grids (Ling and Wan, 1972). It is evident that in all cases the decay, even at small distances, follows the law

$$B_{LL}(0,t) = \frac{A}{(t - t_0)^n}, \quad n = 1 + \alpha. \tag{10.27}$$

The method for determining the effective origin of time t_0 is shown in Figure 10.7. The scaling law of decay $B_{LL}(0,t) = \langle u^2 \rangle \sim (t - t_0)^{-n}$ leads to the fact that for large t the quantity $[U^2/\langle u^2 \rangle]^{1/n}$ must be a

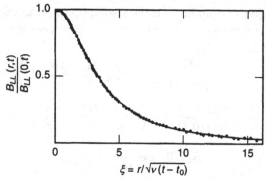

Figure 10.3. The correlation function f for flow behind a passive grid is self-similar: all experimental points lie on a single curve $f(\xi)$. From Ling and Huang (1970).

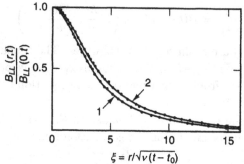

Figure 10.4. The velocity correlation functions for flow behind an active grid are self-similar: all experimental points lie on a single curve $f(\xi)$. Curve 1, $V_P/U = 3$; curve 2, $V_P/U = 17$. From Ling and Wan (1972).

linear function of time. Hence the intersections with the time axis of the straight lines drawn through the experimental points give the values of t_0.

We see that in all cases the exponent α turns out to be different from zero: it is equal to unity for passive grids, 0.73 for an active grid with $V_p/U = 3$, and 0.35 for an active grid with $V_p/U = 17$. This exponent thus depends on the initial conditions, i.e., the conditions at the grid.

The paper by Gad-el-Hak and Corrsin (1974) contains results of the data processing of other experiments of various researchers. In treating the variation of the quantity $B_{LL}(0,t)$ with t, this dependence was assumed to be a scaling law in accordance with (10.27). In some cases the turbulence was found to be weakly anisotropic, so the decay exponents are presented, in this paper, for all three components of velocity fluctuation. The experiments were performed on passive grids, as a rule, but the results of some experiments performed on active grids are also

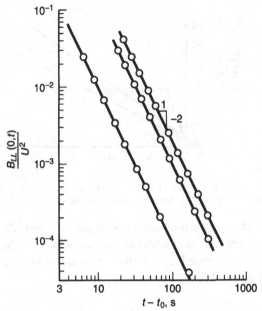

Figure 10.5. The moment $B_{LL}(0,t)$ behind a passive grid decays according to a scaling law, but the exponent is different from unity. Different curves correspond to different combinations of passive grids. From Ling and Huang (1970).

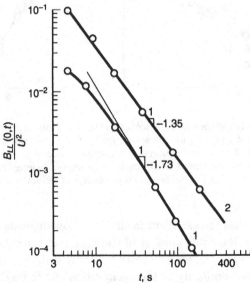

Figure 10.6. The moment $B_{LL}(0,t)$ behind an active grid decays according to a scaling law, but the exponent is different from unity. Curve 1, $V_P/U = 3$; curve 2, $V_P/U = 17$. From Ling and Wan, (1972).

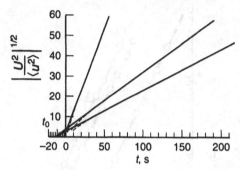

Figure 10.7. Determination of the effective origin of time, t_0, according to Ling and Huang (1970), for the case of a passive grid.

presented. In these experiments on active grids, as already mentioned, a grid of hollow rods was used with nozzles through which air was injected into the flow. The dependence of the exponent on the injection ratio J is shown in Figure 10.8.

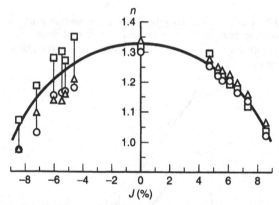

Figure 10.8. Dependence on injection rate J of the exponent n in the decay law for $B_{LL}(0,t)$, for an active grid. $J > 0$ corresponds to coflow injection and $J < 0$ to counterflow injection. For $J > 0$ the decay is isotropic and the exponent is different from unity. $\circ$, lengthwise component of velocity; $\triangle,\square$, transverse components of velocity. From Gad-el-Hak and Corrsin (1974).

It is evident that the exponent in the decay law depends on conditions at the grid (the Reynolds number of the grid, and the characteristics of its activity, J, V_p/U, ω, etc.). The exponent α turns out to be equal to zero, i.e., the self-similarity of the decay turns out to be complete, only in the case of enormously large Reynolds numbers of the grid, reached by Kistler and Vrebalovich (1966).

Unfortunately, third-order moments have been measured by almost no one: one of the few papers up to now in which measurements of third moments are given is that of Stewart (1951). In this paper the self-similarity of the correlation function is emphasized, and attention is especially given to the absence of a unique dependence of the quantity $B_{LL,L}(r,t)/B_{LL}^{3/2}(0,t)$ on the self-similar variable for different instants of time (Figure 10.9). This conforms to incomplete similarity of the decay (cf. (10.26)) and would not hold for complete similarity.

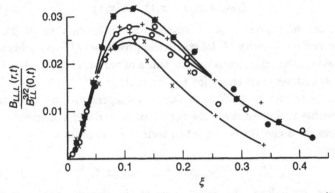

Figure 10.9. Unique dependence of the quantity $B_{LL,L}(r,t)/B_{LL}^{3/2}(0,t)$ on the self-similar variable is lacking – the curves for different moments of time do not coincide. $\times$, $t = 0.041$ s.; $\bullet$, $t = 0.0615$ s.; $\circ$, $t = 0.123$ s.; $+$, $t = 0.184$ s.; $\blacktriangle$ and $\not{\!\!\!}$, $t = 0.246$ s. From Stewart (1951).

Thus we have arrived at the conclusion that in experiments the decay of turbulenceis self-similar even at small distances from the grid, but this self-similarity is of the second kind, so that the influence of the initial scale – the length scale of the grid – never vanishes, but, because of the peculiarities of homogeneous isotropic turbulence, appears only in combination with various parameters. The exponent in the law of decay cannot be determined from considerations of dimensional analysis, but is selected from the continuous spectrum of possible values by the initial conditions (the conditions at the grid), the situation in principle being analogous to what we met above when considering the self-similar analogue of the Korteweg-de Vries equation.

The analysis of the self-similar decay of homogeneous isotropic turbulence presented above was given by Barenblatt and Gavrilov (1974).

10.1.4 Locally homogeneous and isotropic turbulence

The investigation of the local structure of turbulent flows of an incompressible viscous fluid at large Reynolds numbers in the papers of Kolmogorov (1941, 1962) and Obukhov (1941, 1962) also furnishes very instructive examples of self-similar intermediate asymptotics of various types. The outstanding work of their predecessor, the great British physicist L.F. Richardson (1922), who proposed a qualitative scheme of vortex cascade in turbulent flow, should also be acknowledged.

According to the basic hypothesis of Kolmogorov and Obukhov, at large Reynolds numbers hydrodynamic fields have the properties of local isotropy, homogeneity and stationarity. Local isotropy and homogeneity mean that the moment tensors, in which the relative velocities

$$\Delta_{\mathbf{r}}\mathbf{u} = \mathbf{u}(\mathbf{x}+\mathbf{r},t) - \mathbf{u}(\mathbf{x},t) \qquad (10.28)$$

appear, are homogeneous and isotropic at sufficiently small $|\mathbf{r}|$. The condition of stationarity of the statistical properties of local fields results from the fact that the characteristic times of variation for the local fields are much smaller than those for the basic flow.

Thus, as in the case of an ordinary homogeneous and isotropic incompressible turbulent flow, the tensor of second-order moments of the quantities $\Delta_{\mathbf{r}}\mathbf{u}$ can be expressed in terms of one of its components, for example,

$$D_{LL} = \langle (u_L(\mathbf{x}+\mathbf{r},t) - u_L(\mathbf{x},t))^2 \rangle \qquad (10.29)$$

(u_L being, as before, the component of the velocity vector $\mathbf{u}$ in the direction $\mathbf{r}$). The quantity D_{LL}, in principle, may depend on r, the modulus of the vector $\mathbf{r}$, and also on the kinematic viscosity of the fluid ν, the external scale Λ, and the energy transmitted per unit time from the large-scale motions to the fine-scale motions under consideration; this, by virtue of stationarity, is equal to the mean rate of turbulent energy dissipation per unit volume, $\langle \epsilon \rangle$. Introducing in place of the viscosity the linear scale λ of the motion in which viscous dissipation occurs,

$$\lambda = \nu^{3/4} \langle \epsilon \rangle^{-1/4}, \qquad (10.30)$$

the *internal Kolmogorov scale*, we have

$$D_{LL} = f(r, \langle \epsilon \rangle, \lambda, \Lambda). \qquad (10.31)$$

Dimensional analysis gives by the standard procedure

$$D_{LL} = \langle \epsilon \rangle^{2/3} r^{2/3} \Phi\left(\frac{r}{\lambda}, \frac{r}{\Lambda}\right). \qquad (10.32)$$

The relationships valid in the so-called inertial range of scales, i.e. for $\lambda \ll r \ll \Lambda$, are intermediate asymptotics of (10.32) as $r/\lambda \to \infty$ but $r/\Lambda \to 0$. (For large Reynolds numbers, $\lambda \lll \Lambda$.) In the classical version of the Kolmogorov-Obukhov theory an assumption is implicitly made that is equivalent to the assumption that there is a finite non-zero limit of $\Phi(r/\lambda, r/\Lambda)$ as $r/\lambda \to \infty$ and $r/\Lambda \to 0$, i.e., that there is complete self-similarity in both parameters r/λ and r/Λ. Therefore for $\lambda \ll r \ll \Lambda$ we obtain the famous 'two-thirds scaling law' of Kolmogorov,

$$D_{LL} = C\langle\epsilon\rangle^{2/3}r^{2/3}, \qquad (10.33)$$

where C is a universal constant that must be equal to $\Phi(\infty, 0)$.

In fact, the existence of complete similarity in the parameter r/Λ for small r/Λ is in some doubt owing to the so-called intermittency effect. Intermittency consists in the non-uniform spatial distribution of the energy transfer rate towards smaller vortices. Indeed this energy transfer rate is also a random quantity and the contribution of its fluctuations in larger scales than the scale of the 'equilibrium range' $r \ll \Lambda$, which is the only scale possessing local isotropy and homogeneity, can turn out to be essential. This point was attributed by A.N. Kolmogorov to L.D. Landau, although the relevant note in the book of Landau and Lifchitz (1987), (the first edition was published in 1944) in fact concerns a different matter.

We therefore assume that there is complete similarity in the parameter r/λ for $r/\lambda \gg 1$, and incomplete similarity in the parameter r/Λ for $r/\Lambda \ll 1$, so that, as $r/\lambda \to \infty$ and $r/\Lambda \to 0$,

$$\Phi\left(\frac{r}{\lambda}, \frac{r}{\Lambda}\right) \simeq C_1\left(\frac{r}{\Lambda}\right)^\alpha \qquad (10.34)$$

where C_1 and α are universal constants. Then (10.32) gives

$$D_{LL} = C_1\langle\epsilon\rangle^{2/3}r^{2/3+\alpha}\Lambda^{-\alpha}. \qquad (10.35)$$

But we find just such a relation in the refined Kolmogorov–Obukhov theory, which takes account of the influence of fluctuations in the energy dissipation (Kolmogorov, 1962; Obukhov, 1962). The constant α, according to this theory, is related to a coefficient in the relation for the variance of the energy transfer rate averaged on a scale r (see Monin and Yaglom, 1975),

$$\langle[(\ln\epsilon)']^2\rangle = A - \mu\ln r, \qquad (10.36)$$

so that $\alpha = \mu/9$. According to experimental data, $\mu = 0.4$, so that $\alpha \simeq 0.04$ and the dependence (10.35) actually differs from the two-thirds law only slightly.

In the paper by Castaing, Gagne and Hopfinger (1990) an alternative relation for the variance of the energy transfer rate averaged on a scale r was proposed,

$$\langle[(\ln\epsilon)']^2\rangle = (r/r_0)^{-\beta} \qquad (10.37)$$

(r_0 is a constant length parameter), which exactly corresponds to the incomplete similarity assumption. The experiments presented in this paper confirm the incomplete similarity relation (10.37) rather than the logarithmic relation (10.36). The authors also presented a model that leads to the relation $\beta = \beta_1 / \ln \text{Re}$, where β_1 is a constant, Re a global-flow Reynolds number based on the Taylor linear scale parameter. So, according to this work, the exponent α depends on the global-flow Reynolds number and therefore is not a universal constant.

Moreover, recently Barenblatt and Goldenfeld (1995) came, under certain additional assumptions, to the relation for D_{LL} in the inertial range

$$D_{LL} = \left[A_0 + \frac{A}{\ln \text{Re}} \right] (\langle \epsilon \rangle r)^{2/3} \left(\frac{r}{\Lambda} \right)^{\alpha / \ln \text{Re}} \qquad (10.38)$$

where A_0, A and α are universal constants. In this formula the correction $\alpha / \ln \text{Re}$ is not substantial in comparison with the exponent $2/3$ to which r enters (10.33) and can hardly be observed at present. The essential point is that according to (10.38) the correction to the Kolmogorov constant appears to be inversely proportional to $\ln \text{Re}$. This is consistent with the experimental data of Praskovsky and Oncley (1994).

10.2 Turbulent shear flows

10.2.1 Similarity laws for the velocity distribution in the wall region of a turbulent shear flow

A turbulent flow whose mean properties do not depend on the coordinate x in the direction of the mean velocity is called a *shear flow*. Thus, the mean velocity and all the other mean properties of a shear flow depend on only one coordinate, z, transverse to the mean flow and having its origin at the rigid wall that constrains the flow (Figure 10.10). This type of flow occurs in a channel or pipe far from the inlet, in the flow past a plate far from the leading edge, in the boundary layer of the atmosphere, etc.

In the vicinity of the wall bounding the flow, it can be assumed that the shear stress is constant, i.e., independent of the transverse coordinate z. The part of the shear flow in which this assumption is valid is called the *wall region.*

Thus, the properties of the motion at some point in the wall region of a turbulent shear flow are governed by the shear stress τ (which is, by assumption, constant), the properties of the fluid (its density ρ and kinematic viscosity ν), the distance z of the point under consideration from the wall, and some external flow scale Λ – the diameter of the pipe or depth of the channel, etc.

We shall adopt the gradient of the mean velocity u at a given point,

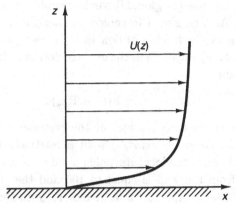

Figure 10.10. Shear flow.

$\partial_z u$, as the quantity being determined; it will become clear below why we do not choose the velocity itself. Thus, we have

$$\partial_z u = f(\tau, \rho, z, \nu, \Lambda).$$ (10.39)

The LMT class of systems of units is appropriate here. In this class, the governing parameters have the following dimensions:

$$[\tau] = \frac{M}{LT^2}, \quad [\rho] = \frac{M}{L^3}, \quad [\nu] = \frac{L^2}{T}, \quad [z] = [\Lambda] = L.$$ (10.40)

The dimensions of the first three governing parameters τ, ρ and z are obviously independent. The dimension of the determined parameter $\partial_z u$ and those of the last two governing parameters can be expressed in terms of the dimensions of the first three parameters in the following way:

$$[\partial_z u] = [\tau]^{1/2}[\rho]^{-1}[z]^{-1}, \quad [\Lambda] = [z], \quad [\nu] = [\tau]^{1/2}[\rho]^{-1/2}[z].$$ (10.41)

Thus, we obtain the following dimensionless form for the relation under study:

$$\Pi = \frac{z\partial_z u}{(\tau/\rho)^{1/2}} = \Phi(\Pi_1, \Pi_2),$$

$$\Pi_1 = \frac{\nu}{(\tau/\rho)^{1/2}z}, \quad \Pi_2 = \frac{\Lambda}{z}.$$ (10.42)

According to tradition, we shall now introduce the notation $u_* = (\tau\rho)^{-1/2}$; this quantity u_*, which has the dimension of velocity, is called the *dynamic*, or *friction*, *velocity*. It is both natural and convenient, for the analysis that follows, to transform from the parameter Π_1 to its reciprocal, $\Pi_1^{-1} = u_* z/\nu = \mathrm{Re}_l$, which serves as a local Reynolds number, and an analogous quantity $\Pi_2\Pi_1^{-1} = \mathrm{Re}_* = u_*\Lambda/\nu$, which serves as a global Reynolds number. (The local Reynolds number contains the local coordinate z, while the global Reynolds number contains the external length scale Λ. The global Reynolds number Re_* differs from that normally used in the hydraulics of flow in pipes in that it contains the dynamical velocity u_* rather than the average flow velocity in the pipe.)

Thus, we obtain

$$\Pi = \frac{z\partial_z u}{u_*} = \Phi_1(\text{Re}_l, \text{Re}_*).$$ (10.43)

We shall now estimate the values of the parameters Re_l and Re_*. For flowing water ($\nu = 10^{-2} \text{cm}^2/\text{s}$) with a relatively low dynamical velocity, $u_* = 10$ cm/s, the local Reynolds number at a distance of just one millimetre from the wall is equal to 100 and the global Reynolds number is 10 000 for a pipe 10 cm in diameter. Both these parameters are therefore large outside the immediate vicinity of the wall; it seems natural to investigate whether it is possible to use limiting similarity laws in this region. Note that it was precisely the desire to construct limiting similarity laws that led us to consider the gradient of the velocity rather than the velocity itself. The point is that the velocity at any distance from the wall, unlike the velocity gradient, obviously depends on the situation in the immediate vicinity of the wall, where the local Reynolds number cannot be assumed to be large.

Thus, under the assumption of complete similarity with respect to the local and global Reynolds numbers (this assumption dates back to von Kármán, 1930 and Prandtl, 1932b), we obtain the following result from (10.43):

$$\Pi = \frac{z\partial_z u}{u_*} = \Phi_1(\infty, \infty) = \text{const.}$$ (10.44)

The constant in (10.44) is traditionally denoted by $1/\kappa$; the constant κ is called the von Kármán constant. Clearly, under the assumption of complete similarity in Re_l and Re_* made above, the von Kármán constant must be universal, i.e., independent of the Reynolds number.

Integration of (10.44) yields a universal logarithmic law for the velocity distribution across the flow,

$$u = \frac{u_*}{\kappa} \ln z + \text{const},$$

which is usually written in the form

$$\varphi = \frac{1}{\kappa} \ln \eta + C_1,$$ (10.45)

where

$$\varphi = \frac{u}{u_*}, \quad \eta = \frac{u_* z}{\nu}, \quad C_1 = \text{const}.$$

At first glance, this universal logarithmic law seems to be fairly well confirmed by measured data on the mean velocity distributions in smooth pipes, boundary layers and channels (Figure 10.11), and a numerical value of approximately 0.4 is obtained for the von Kármán constant κ. However, more detailed analysis of the experimental data reveals a systematic dependence of the von Kármán constant on the Reynolds number for the flow. Thus, small but systematic deviations from the universal logarithmic law are observed in the velocity distribution.

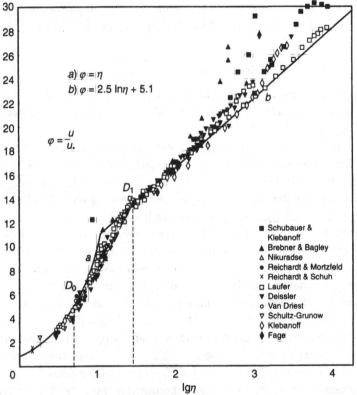

Figure 10.11. Universal dimensionless mean velocity profile $\varphi = u/u_*$ of turbulent flow close to a smooth wall according to the data of pipe-, channel-, and boundary-layer measurements (Kestin and Richardson, 1963). After Monin and Yaglom (1971, 1992): region a, $\varphi = \eta$; region b, $\varphi = 2.5 \ln \eta + 5.1$.

In accordance with the recipe outlined in chapter 5, we shall look (Barenblatt and Monin, 1979) at the possible assumption of incomplete similarity with respect to the local Reynolds number and lack of self-similarity with respect to the global Reynolds number. Under the assumption of incomplete similarity with respect to Re_l, (10.43) yields

$$\frac{z\partial_z u}{u_*} = \left(\frac{u_* z}{\nu}\right)^\alpha \Phi(\mathrm{Re}_*), \tag{10.46}$$

where the exponent α is also assumed to depend on the global Reynolds number Re_*. Integrating (10.46) and assuming, in agreement with experiment, that the constant of integration is equal to zero, we obtain the scaling velocity distribution

$$\varphi = C\eta^\alpha \qquad (10.47)$$

where we have defined $C = (1/\alpha)\Phi(\text{Re}_*)$. The global Reynolds number $\text{Re}_* = u_*\Lambda/\nu$ is a function of a commonly used Reynolds number for the flow, based on the mean flow velocity $\bar{u}$ rather than on the friction velocity:

$$\text{Re}_* = f(\text{Re}), \quad \text{Re} = \frac{\bar{u}\Lambda}{\nu};$$

this follows directly from dimensional analysis. Therefore the constants C and α in (10.47) can be considered also as functions of the flow Reynolds number Re.

Scaling laws for the velocity distribution in various turbulent shear flows have been suggested as empirical relationships for a long time. One recognises (cf. Schlichting, 1968; Hinze, 1959) that scaling (power) laws for velocity distributions, with exponents depending on global Reynolds number, are confirmed by experiment at least as well as is the universal logarithmic law. Nevertheless the latter is generally considered to have, in contrast to the scaling law, a theoretical basis whereas the scaling law is considered as simply an empirical relation. As a matter of fact, however, we have seen that the scaling law can be derived from the assumption of incomplete similarity of the flow in the local Reynolds number not less rigorously than the universal logarithmic law can be derived from the assumption of complete similarity, i.e., complete independence on the molecular viscosity. Therefore neither the logarithmic law (10.45) nor the power law (10.47) should be considered only as a convenient representation of the experimental data; both have rigorous theoretical foundations based, however, on different assumptions. Moreover, the assumption of the lack of a characteristic length in the flow leads in general to the scaling law, and as a special case to the logarithmic law[†].

[†] Indeed, according to this assumption the ratio of velocity gradients $\partial_y u$ at heights z_1 and z_2 is a function of z_1/z_2 only. This leads to the well-known functional equation, which we have used more than once,

$$\varphi\left(\frac{y}{x}\right) = \frac{\varphi(y)}{\varphi(x)}$$

whose solution is $\varphi(x) = x^n$, where the constant n remains indeterminate. If $n = -1$, the logarithmic law is obtained; if $n \neq -1$, the scaling law.

Therefore an important qualitative question arises: which of these assumptions is correct?

The results presented below (Barenblatt, 1991, 1993a; Barenblatt and Prostokishin, 1993) give some evidence in favour of the power-type law (10.47) with exponent α inversely proportional to the logarithm of the flow Reynolds number and constant C a linear function of this logarithm:

$$\alpha = \frac{3}{2 \ln \mathrm{Re}}, \qquad (10.48)$$

$$C = \frac{1}{\sqrt{3}} \ln \mathrm{Re} + \frac{5}{2} = \frac{\sqrt{3} + 5\alpha}{2\alpha}. \qquad (10.49)$$

As a matter of fact, the inverse proportionality of α to $\ln \mathrm{Re}$ was suggested by the idea that the local structure of developed turbulent flow should imprint the velocity distribution, and by the result of Castaing, Gagne and Hopfinger (1990) mentioned in subsection 10.1.4.

To check the dependence (10.48) all 16 sets of experimental data[†], available in the paper by Nikuradze (1932) of average velocity measurements in smooth cylindrical tubes at various distances from the wall and at various Reynolds numbers[§] were subjected to a stringent procedure for the verification of (10.47), (10.48). Namely, the functions $\varphi^{(2 \ln \mathrm{Re})/3}$ as functions of η were constructed and inspected. The question was whether the straight lines would be found for intermediate values of η. The processing of the experimental data clearly revealed (Figures 10.12(a)–(e)) such intermediate straight lines for all 16 sets. We note a good level of accuracy: the exponent $1/\alpha = (2 \ln \mathrm{Re})/3$ is large, of the order of 10 or so, therefore even small deviations in exponent from those that are needed could destroy the straight lines. The revealing of straight lines in the intermediate intervals can be considered as experimental verification of (10.48).

One point should be explained. There exists an obvious arbitrariness in the definition of Re – for instance, the maximum velocity can be taken instead of the mean velocity, or the radius instead of the diameter. For (10.48) it is immaterial, because in fact this relation should be considered as the first term in an asymptotic expansion, valid when $\ln \mathrm{Re}$, not only Re itself, is large, so that a different definition of Re will influence only higher-order terms in the expressions for C and α.

Further processing allowed one to obtain with rather good accuracy

---

[†] It is essential that they are presented in tabular form, contrary to the data of other experimentalists.

[§] Covering nearly three decimal orders of magnitude.

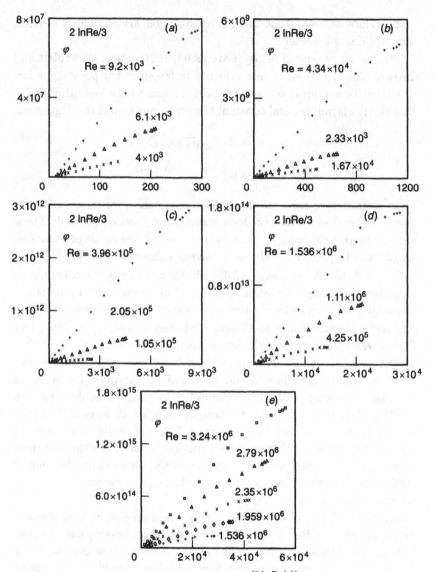

Figure 10.12(a)–(e). Each set of data for $\varphi^{(2\ln \mathrm{Re})/3}(\eta)$ reveals a straight line for an intermediate interval of values of η.

(Figure 10.13) the linear dependence (10.49) of the coefficient C in the power law (10.47) on $\ln \mathrm{Re}$. Therefore the power law (10.47) can be represented in the form

$$\varphi = \left(\frac{1}{\sqrt{3}}\ln \mathrm{Re} + \frac{5}{2}\right)\eta^{3/(2\ln \mathrm{Re})}. \qquad (10.50)$$

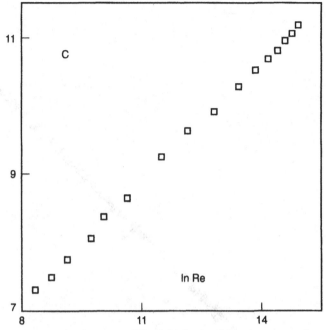

Figure 10.13. The function $C(\ln \mathrm{Re})$ obtained by the processing of experimental data.

Simple transformations allow to reduce the relation (10.50) to another form,

$$\Psi = \frac{1}{\alpha} \ln \frac{2\alpha\varphi}{\sqrt{3} + 5\alpha} = \ln \eta, \qquad (10.51)$$

which is quasi-universal, i.e. independent of Reynolds number. It means that if the relations (10.50), (10.51) are correct, all experimental points for sufficiently large η and various Reynolds numbers (large enough to correspond to developed turbulence) should lie on a single curve in the Ψ, $\ln \eta$-plane, in fact, a straight line, the bisectrix of the first quadrant. Figure 10.14 shows that the overwhelming majority of the 256 experimental points available in the tables of the paper by Nikuradze (1932) occur, for large η, close to the bisectrix in accordance with (10.51). Certain points that occur rather far from the bisectrix correspond to measurements either in close proximity to the wall, where η is not sufficiently large for turbulence to have developed, or at comparatively low Reynolds numbers for which the turbulence is not fully developed. (Some errors

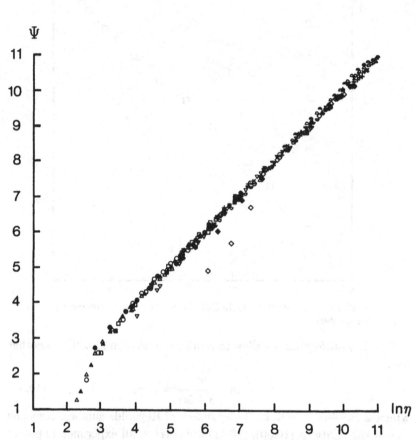

Figure 10.14. The experimental points in the coordinates Ψ, $\ln\eta$ settle down for large η close to the bisectrix of the first quadrant, confirming the quasi-universal form of the scaling law. The values of Re are as follows: $\triangle$, 4×10^3; $\blacktriangle$, 6.1×10^4; $\circ$, 9.2×10^3; $\bullet$, 1.67×10^4; $\square$, 2.33×10^4; $\blacksquare$, 4.34×10^4; $\blacksquare$, 1.05×10^5; $\blacktriangledown$, 2.05×10^5; $\triangledown$, 3.96×10^5; $\blacktriangledown$, 7.25×10^5; $\lozenge$, 1.11×10^6; $\blacklozenge$, 1.536×10^6; $+$, 1.959×10^6; $\times$, 2.35×10^6; $\cap$, 2.79×10^6; $\blacktriangle$, 3.24×10^6.

in the measurements close to the wall, where the distance is comparable with the gauge size are also possible.)

As we have seen, in the φ, $\ln\eta$-plane the curves that represent the power laws for varying Reynolds numbers

$$\varphi = \left(\frac{1}{\sqrt{3}} \ln \text{Re} + \frac{5}{2}\right) e^{(3\ln\eta)/(2\ln\text{Re})} = F(\ln\eta, \text{Re}), \qquad (10.52)$$

form a family of which the Reynolds number is the parameter. The

family possesses an envelope that satisfies both equation (10.52) and
the equation $\partial_{\mathrm{Re}}F = 0$. The latter equation can easily be reduced to
the form

$$\frac{3\ln\eta}{2\ln\mathrm{Re}} = \frac{\sqrt{3}}{10}\ln\eta\left[\left(1 + \frac{20}{\sqrt{3}\ln\eta}\right)^{1/2} - 1\right]. \qquad (10.53)$$

Eliminating $\ln\mathrm{Re}$ from (10.52) and (10.53) we obtain the equation of
the envelope; this is represented in Figure 10.15. As is seen it is close
to the straight line representing the universal logarithmic law with an
empirically fitted constant $C_1 = 5.5$, even for rather moderate $\ln\eta$[†].
This seems to be natural because the envelope is the locus of the points
where the derivative with respect to Reynolds number vanishes. It was,
however, the first of the assumptions on which the derivation of the
logarithmic law (10.45) was based.

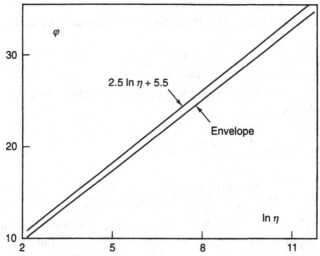

Figure 10.15. The envelope of the scaling law curves for fixed Re in the
φ, $\ln\eta$-plane is very close to the generally accepted universal logarithmic
law even at moderate $\ln\eta$.

Moreover, if we let $\ln\eta$ tend to infinity while remaining on one of the
curves (10.52) at Re = const, the function $\Phi(\eta, \mathrm{Re})$ in the general similarity relation (10.43) obviously tends to infinity. However, if we let $\ln\eta$
tend to infinity while remaining on the envelope, i.e., we let $\ln\mathrm{Re}$ tend
simultaneously to infinity, we will obtain a finite limit for $\Phi(\eta, \ln\mathrm{Re})$.

[†] The envelope is even closer to the straight line with the constant $C_1 = 5.1$, used
for instance in the book by Monin and Yaglom (1971) (see Figure 10.11).

Indeed, for large $\ln \mathrm{Re}$ the relation (10.53) gives $(3\ln\eta)/(2\ln\mathrm{Re}) \to 1$, whence and from (10.52) we obtain

$$\varphi = \frac{\sqrt{3}e}{2}\ln\eta + \mathrm{const}, \quad \Phi(\eta,\mathrm{Re}) \to \frac{\sqrt{3}e}{2}. \qquad (10.54)$$

Therefore at large $\ln\eta$ all assumptions leading to the universal logarithmic law are fulfilled on the envelope and this envelope can be identified with this law. This allows us to obtain the value of the von Kármán constant:

$$\kappa = \frac{2}{\sqrt{3}e} \simeq 0.424. \qquad (10.55)$$

Nikuradze (1932) himself, on the basis of processing his experimental data, arrived at the value $\kappa = 0.417$.

On the basis of the proposed scaling law (10.50) for the average velocity distribution a corresponding skin friction law can be proposed and compared with experimental data.

We define the skin friction dimensionless coefficient λ in the traditional way:

$$\lambda = \frac{\tau}{\rho\bar{u}^2/8} = 8\left(\frac{u_*}{\bar{u}}\right)^2. \qquad (10.56)$$

According to (10.50) the following relation for the bulk average velocity $\bar{u}$ is obtained:

$$\begin{aligned}
\bar{u} &= \frac{8}{d^2}\int_0^{d/2} u(z)\left(\frac{d}{2}-z\right)dz \\
&= u_*\frac{\sqrt{3}+5\alpha}{\alpha}\left(\frac{u_*d}{\nu}\right)^\alpha\frac{1}{2^\alpha(1+\alpha)(2+\alpha)}.
\end{aligned} \qquad (10.57)$$

In deriving (10.57) the fact was used that for developed turbulent flows in pipes we can neglect the contribution to the bulk flow rate of the viscous layer near the wall as well as the contribution of the region near the tube axis where the law (10.50) is not valid.

From (10.48),

$$\mathrm{Re} = \bar{u}d/\nu = e^{3/2\alpha} \qquad (10.58)$$

whence and from (10.57) we obtain

$$\frac{u_*d}{\nu} = \left[e^{3/2\alpha}2^\alpha\frac{\alpha(1+\alpha)(2+\alpha)}{\sqrt{3}+5\alpha}\right]^{1/(1+\alpha)}, \qquad (10.59)$$

and the final relation for the dimensionless skin friction coefficient λ corresponding to the scaling law (10.50) takes the form

$$\lambda = \frac{\tau}{\rho\bar{u}^2/8} = 8\left(\frac{u_*}{\bar{u}}\right)^2 = \frac{8}{\Psi^{2/(1+\alpha)}} \qquad (10.60)$$

where

$$\Psi(\alpha) = \frac{e^{3/2}(\sqrt{3} + 5\alpha)}{2^{\alpha}\alpha(1 + \alpha)(2 + \alpha)}, \quad \alpha = \frac{3}{2\ln \mathrm{Re}}. \tag{10.61}$$

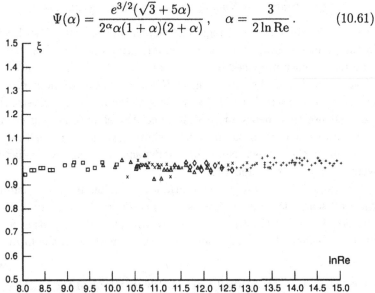

Figure 10.16. The experimental data for various pipes and various Reynolds numbers confirm the skin friction law (10.60), (10.61) which follows from scaling law (10.50) with rather good accuracy. The values of d are as follows: $\square$, 1 cm; $\triangle$, 2 cm, $\Diamond$, 3 cm; $\times$, 5 cm; $+$, 10 cm.

A comparison of values of λ predicted by (10.60) with experimental values (Nikuradze (1932), table 9) is presented in Figure 10.16. It was found to be less instructive to compare predictions (10.60), (10.61) with experimental data directly, since they coincide with very high accuracy. Therefore a different, more objective, form of comparison has been chosen: the quantity

$$\xi = \frac{\lambda_e}{\lambda} = \frac{\lambda_e}{8}\Psi^{2/(1+\alpha)}, \tag{10.62}$$

where λ_e is the experimentally determined value of the skin friction coefficient plotted as a function of $\ln \mathrm{Re}$. In table 9 of the paper, Nikuradze (1932) a total of 125 points is available, corresponding to various Reynolds numbers in a rather wide range from slightly supercritical ($\mathrm{Re} = 3.07 \times 10^3$) to very large ($\mathrm{Re} = 3.23 \times 10^6$). Ideally the quantity ξ would be equal to unity. As Figure 10.16 shows, nearly all deviations from this value lie within the experimental scatter range.

The coincidence of the experimental data concerning skin friction with predictions based on the scaling law (10.50) also contributes to the verification of this law.

*10.2.2 The wall region of a turbulent shear flow – semi-empirical
theory and Kolmogorov's similarity hypothesis*

We now consider the wall region of the turbulent shear flow described
in the preceding section in more a quantitative way, based on the semi-
empirical theory proposed by Kolmogorov (1942). Kolmogorov's theory
is based on closing the equations for the conservation of momentum and
turbulent energy with the help of certain similarity hypotheses. Later
we shall use this theory more than once; therefore it is appropriate to
demonstrate here its basic ideas for a simple example, the wall region of
a turbulent shear flow. Note that, independently, closely similar ideas
were developed by Prandtl (1945).

We exclude from our consideration the region in close vicinity to
the wall where the viscous stresses are comparable with the turbulent
stresses due to momentum transfer by vortices. The equation for momentum balance for a shear flow can then be written in the form

$$-\rho\langle u'w'\rangle = \tau\,, \tag{10.63}$$

where u' and w', respectively, are the fluctuations in the velocity components longitudinal and transverse to the wall, ρ is the density of the
fluid, and τ is the shear stress; in (10.63) we neglect the contribution of
the viscous stresses in comparison with the turbulent Reynolds stresses.
The equation for turbulent energy balance for a steady shear flow can
be written in the form (Monin and Yaglom, 1971)

$$\langle u'w'\rangle\partial_z u + \partial_z\left\langle\left(\frac{p'}{\rho} + \frac{u'^2 + v'^2 + w'^2}{2}\right)w'\right\rangle + \epsilon = 0\,. \tag{10.64}$$

Here v' is the fluctuation in the transverse velocity component, p' the
fluctuation in pressure, and ϵ the mean rate of dissipation of turbulent
energy per unit mass of fluid. Equation (10.64) reflects the simple fact
that the local balance of turbulent energy for a steady shear flow consists
of the generation of turbulent energy by the mean motion (the first
term), the diffusive influx of turbulent energy (the second term), and
the dissipation of turbulent energy into heat (the third term). For the
problem of interest the transfer of turbulent energy by diffusion is small
and we can neglect it.

We introduce the coefficient k of momentum exchange by the relation

$$\langle u'w'\rangle = -k\partial_z u\,. \tag{10.65}$$

We stress that for a shear flow the relation (10.65) is simply a redesig-
nation and does not involve any additional hypothesis.

As was said earlier, a developed turbulent flow contains a great number of vortices, creating irregular rapidly varying motion. The basic

idea of Kolmogorov was that it is possible to assume as a first (perhaps, 'naive', as he used to say) approximation that the local structure of this set of vortices is statistically the same for all developed turbulent flows as far as dimensionless quantities are concerned. Therefore according to Kolmogorov's (1942) hypothesis, the momentum exchange coefficient and the rate of energy dissipation ϵ at a given point in the flow are determined only by the local values of two kinematic quantities of different dimensions, e.g. the mean turbulent energy per unit mass,

$$b = \tfrac{1}{2}\langle u'^2 + v'^2 + w'^2 \rangle,$$

and the turbulence external length scale l. (Another possibility will be considered later). Dimensional analysis leads in the standard way to the relations

$$k = l\sqrt{b}, \quad \epsilon = \gamma^4 b^{3/2}/l, \tag{10.66}$$

where, by virtue of identification of the length scale to within a constant factor, the constant in the first relation can be taken to be equal to unity, and the constant γ is close to 0.5 by estimates from experimental data (see Monin and Yaglom, 1971).

Substituting (10.65) and (10.66) into (10.63) and (10.64), and neglecting in the latter equation the contribution of the diffusion of turbulent energy, we get a system of equations in the form

$$l\sqrt{b}\partial_z u = u_*^2, \quad l\sqrt{b}(\partial_z u)^2 - \frac{\gamma^4 b^{3/2}}{l} = 0, \tag{10.67}$$

where $u_* = (\tau/\rho)^{1/2}$ is the friction velocity.

This system is still not closed, since the turbulence length scale l is not yet defined. In accordance with subsection 10.2.1, in the wall region of a turbulent shear flow, where the friction velocity u_* is constant, the turbulence length scale l depends on the friction velocity u_*, the kinematic viscosity ν, the vertical coordinate z, and the external length scale Λ. Dimensional analysis gives

$$l = z\Phi_l(\mathrm{Re}_l, \mathrm{Re}_*), \tag{10.68}$$

where, as before, $\mathrm{Re}_l = u_* z/\nu$ and $\mathrm{Re}_* = u_*\Lambda/\nu$ are the local and global Reynolds numbers. Under the assumption of complete similarity in both Reynolds numbers, at large Re_l and Re_* the function Φ_1 is identically equal to a constant that it is convenient to denote by $\kappa\gamma$, γ being the constant introduced earlier and κ a new constant (the von Kármán constant), so that

$$l = \kappa\gamma z. \tag{10.69}$$

Substituting (10.69) into (10.67), we get

$$u = \frac{u_*}{\kappa} \ln z + \text{const}, \quad b = \frac{u_*^2}{\gamma^2}, \qquad (10.70)$$

i.e., the logarithmic law for the velocity distribution (10.45), obtained earlier from more general considerations. If we assume that for this flow similarity of the flow in the local Reynolds number is incomplete, then the relation for the turbulent length scale assumes the form

$$l = z \left(\frac{u_* z}{\nu} \right)^{-\alpha} \Phi_l(\text{Re}_*), \qquad (10.71)$$

and from this and (10.67) we find

$$\frac{z \partial_z u}{u_*} = \left(\frac{u_* z}{\nu} \right)^{\alpha} \Phi(\text{Re}_*), \quad \Phi(\text{Re}_*) = \frac{\gamma}{\Phi_l(\text{Re}_*)}, \qquad \alpha = \alpha(\text{Re}_*),$$
$$(10.72)$$

i.e. the scaling law (10.47), obtained earlier from more general considerations.

10.2.3 Unsteady phenomena in the viscous sublayer of a turbulent shear flow

In the last few decades fundamental investigations of turbulent shear flows in the immediate vicinity of a wall have been published. (See Kline, Reynolds, Schraub and Runstadler, 1967; Corino and Brodkey, 1969; Kim, Kline and Reynolds, 1971; Offen and Kline, 1975.) By a skilful combination of visualization methods (hydrogen bubbles and tracing pigments) and thermoanemometric methods it was shown in these papers that turbulent flow close to the wall has a complicated, essentially unsteady, and spatially inhomogeneous structure.

The question concerns the phenomena in a viscous sublayer where the global characteristics of the flow are governed by the shear stress τ, the density ρ, and the kinematic viscosity ν of the fluid, and also by some external length scale Λ, for example the momentum thickness of the boundary layer. Thus, the kinematic properties must be governed only by the friction velocity $u_* = (\tau/\rho)^{1/2}$, the external length scale Λ, and the kinematic viscosity ν. It turns out that, in the range of thickness of the order of some tens of the characteristic linear scale ν/u_* of the viscous sublayer, there arise with a statistically determined frequency local separations of the flow, as a result of which horseshoe-shaped vortices are generated, move deep into the flow, and in their own right stimulate the occurrence of new local separations. This generates a chequered pattern of longitudinal strips of the retarded flow. Interactions between the horseshoe vortices that arise, lead to the local loss of stability and bursting. As is convincingly shown in the papers of Kline and his associates, basically it is just these bursts that determine the generation of turbulence close to a rigid boundary in a turbulent shear flow. An illustration of the character of the local flows that arise is given by the photograph in Figure 10.17, which shows the twisting by these flows of originally vertical lines of hydrogen bubbles.

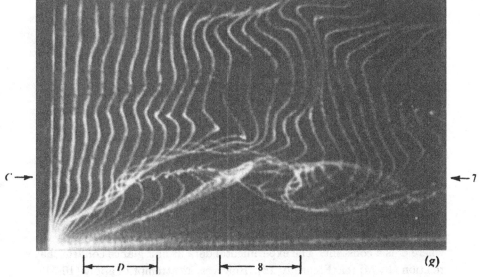

Figure 10.17. In the wall region of a turbulent shear flow there exists a complicated unsteady and spatially inhomogeneous flow. Shown in the photograph is the twisting by a vortex of initially vertical lines of hydrogen bubbles. From Kim, Kline and Reynolds (1971).

Despite the complicated character of the local flows in the viscous sublayer of a turbulent shear flow, some of their statistical characteristics are well described by scaling laws. We demonstrate this here for the mean time T_B between bursts, i.e., the mean period of the cyclic process occurring close to the wall. This quantity can depend, according to the above, on the friction velocity u_*, the kinematic viscosity ν, and the external length scale Λ, whence, applying dimensional analysis, we obtain

$$T_B = \frac{\nu}{u_*^2} \Phi\left(\frac{u_*\Lambda}{\nu}\right). \tag{10.73}$$

The parameter $u_*\Lambda/\nu$ is very large, of order 100 or more; therefore it was natural to take, as a first assumption, one equivalent to the hypothesis of complete similarity in this parameter. This gives

$$T_B = C\frac{\nu}{u_*^2}, \tag{10.74}$$

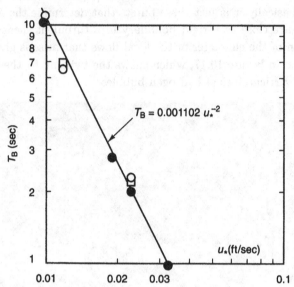

Figure 10.18. Experimental data at first glance confirm complete similarity of the dependence of the time T_B between bursts on the parameter
$u_* \Lambda/\nu$, for large values of this parameter. From Kim, Kline and Reynolds
(1971).

where C is a constant. The experimental data at first glance confirm the
relation (10.74) (cf. Figure 10.18). However, an attempt to apply (10.74)
to the experiments of B.J. Tu and W.W. Willmarth (see Rao, Narasimha
and Badri Narayanan, 1971), in which significantly higher friction velocities were achieved, led to errors of more than an order of magnitude.
Actually, as was shown by Rao, Narasimha and Badri Narayanan (1971),
there is no complete similarity in the parameter $u_* \Lambda/\nu$. Consideration
of more complete experimental data in the latter paper and that by Kim,
Kline and Reynolds (1971) led to the relation

$$\Phi = 0.65 \left(\frac{U\Lambda}{\nu} \right)^{0.73} \tag{10.75}$$

(see Figure 10.19). Here U is the free-stream velocity and Λ the momentum thickness of the boundary layer. As is well known the ratio
u_*/U is close to a power-law function of the global Reynolds number in
a restricted interval of this parameter. This relation therefore reveals
what appears to be incomplete similarity with respect to the parameter
$u_* \Lambda/\nu$.

10.2.4 Decay of a turbulent burst in a fluid at rest

1. The problem of the decay of instantaneously formed turbulent bursts

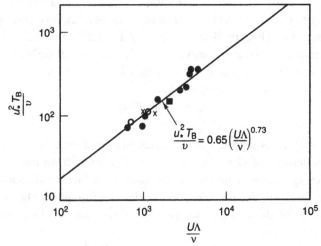

Figure 10.19. Experiments over a wider range of values of the parameter $U\Lambda/\nu$ show incomplete similarity in this parameter. From Kim, Kline and Reynolds (1971).

is of special interest for turbulence studies. Indeed, the formation of turbulent bursts by intersecting or self-intersecting vortices is (see the preceding subsection) one of the leading mechanisms of turbulence generation. From the theoretical viewpoint this problem deserves consideration as a fundamental local disturbance, like the heat source or the very intense explosion at the early stages of a nuclear blast considered in chapter 2. We will present here the solution for a symmetric initial form, based on several assumptions concerning the turbulence behaviour. Initially the burst has, generally speaking a non-symmetric form (Figure 10.20(a)). However, it then takes a more or less symmetric form and only thereafter begins to extend.

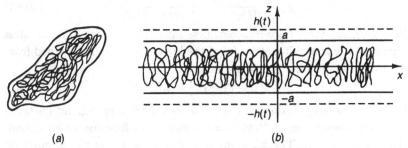

Figure 10.20. Two stages of a turbulent burst in a fluid at rest. (a) The burst initially has an arbitrary form; (b) the burst now has the form of a statistically horizontally uniform layer.

Thus, we shall consider the evolution (extension and decay) in a homogeneous quiescent fluid of a statistically horizontally-homogeneous turbulent burst (Figure 10.20(b)) enclosed initially in a layer between the horizontal planes $z = a$ and $z = -a$. Let the initial turbulent energy per unit area of the plane layer boundary be

$$Q = \int_{-a}^{a} b(z,0)dz \qquad (10.76)$$

where $b(z,t)$ is, as before, the turbulent energy of unit mass of fluid.

The dimension of Q is clearly L^3T^{-2}. Evidently from two kinematic quantities Q and a a kinematic quantity of arbitrary dimension can be composed,; so, for instance, the initial conditions for the turbulent energy and the dissipation rate per unit mass can be presented in the form

$$b(z,0) = \frac{Q}{a}u_0(\zeta), \quad \epsilon(z,0) = \frac{Q^{3/2}}{a^{5/2}}v_0(\zeta). \qquad (10.77)$$

Here $\zeta = z/a$, $u_0(\zeta)$ and $v_0(\zeta)$ are dimensionless even functions, identically equal to zero at $|\zeta| \geq 1$.

Thus, all kinematic statistical properties of the motion for arbitrary time t at an arbitrary point z are determined by the parameters

$$Q, t, z, a. \qquad (10.78)$$

The first two of these governing parameters have independent dimensions. Dimensional analysis gives, accordingly, the following relations for the turbulent energy of unit mass, $b(z,t)$, and the dissipation rate of turbulent energy per unit mass, $\epsilon(z,t)$:

$$b = \frac{Q^{2/3}}{t^{2/3}}B(\xi,\eta), \quad \epsilon = \frac{Q^{2/3}}{t^{5/3}}E(\xi,\eta) \qquad (10.79)$$

where

$$\xi = \frac{z}{Q^{1/3}t^{2/3}}, \quad \eta = \frac{a}{Q^{1/3}t^{2/3}}. \qquad (10.80)$$

Here $B(\xi,\eta)$, $E(\xi,\eta)$ are dimensionless functions of their dimensionless arguments. For a half-width $h(t)$ of the layer a relation is obtained from the same dimensional considerations,

$$h = Q^{1/3}t^{2/3}H(\eta), \qquad (10.81)$$

because, evidently h does not depend on z and does depend on three other arguments. Here $H(\eta)$ is a dimensionless function of its dimensionless argument. The basic interest is in considering the asymptotic solution at large times when the layer thickness is much larger than the initial one, $h(t) \gg a$, i.e., the asymptotics of the functions (10.79), (10.81) at $\eta \ll 1$.

2. The simplest assumption is that for $\eta \ll 1$ there is complete similarity in the parameter η. This means that finite limits different from zero exist of the functions $B(\xi, \eta)$, $E(\xi, \eta)$, and $H(\eta)$, as $\eta \to 0$,

$$B(\xi) = \lim B(\xi, \eta), \quad E(\xi) = \lim E(\xi, \eta), \quad \xi_0 = \lim H(\eta), \quad (10.82)$$

so that the self-similar asymptotic solution takes the form

$$b = \frac{Q^{2/3}}{t^{2/3}} B(\xi), \quad \epsilon = \frac{Q^{2/3}}{t^{5/3}} E(\xi), \quad h = \xi_0 Q^{1/3} t^{2/3}. \quad (10.83)$$

This assumption is, however, incorrect. Indeed, the equation of turbulent energy balance for the case of a horizontally-homogeneous shearless flow assumes the form

$$\partial_t b + \partial_z q_b = -\epsilon \quad (10.84)$$

where q_b is the turbulent flux of turbulent energy. Let us integrate this equation within the limits $z = -h(t)$ to $z = h(t)$ taking into account that at the boundaries of the layer $z = \pm h(t)$ both turbulent energy and its flux vanish:

$$b(\pm h(t), t) = 0, \quad q_b(\pm h(t), t) = 0. \quad (10.85)$$

We obtain that the time derivative of the total turbulent energy per unit area of the layer is negative,

$$\frac{d}{dt} \int_{-h(t)}^{h(t)} b(z, t) dz = - \int_{-h(t)}^{h(t)} \epsilon(z, t) dz < 0 \quad (10.86)$$

because ϵ, the dissipation rate of turbulent energy, is a positive quantity within the turbulent layer. At the same time it follows from the asymptotic solution (10.83) that the turbulent energy per unit area

$$\int_{-h}^{h} b(z, t) dz = \frac{Q^{2/3}}{t^{2/3}} \int_{-h}^{h} B(\xi) dz = Q \int_{-\xi_0}^{\xi_0} B(\xi) d\xi = \text{const } Q \quad (10.87)$$

is time independent. The contradiction obtained proves that the assumption of complete similarity is incorrect.

3. We assume, therefore, incomplete similarity with respect to the parameter η, i.e., we assume that the functions $B(\xi, \eta)$, $E(\xi, \eta)$ and $H(\eta)$ at $\eta \to 0$ have power-type asymptotics,

$$B = \eta^{\lambda_1} B_0 \left(\frac{\xi}{\eta^{\nu_1}} \right), \quad E = \eta^{\lambda_2} E_0 \left(\frac{\xi}{\eta^{\nu_1}} \right), \quad H = \text{const } \eta^{\nu_2}, \quad (10.88)$$

where λ_1, λ_2, ν_1, ν_2 are constants.

We can use the fact that the quantities b, ϵ and q should satisfy the turbulent energy balance equation (10.84) and therefore that this equation should be reduced to an ordinary differential equation involving the functions of one variable ξ/η^{ν_1}: $B_0(\xi/\eta^{\nu_1})$ and $E_0(\xi/\eta^{\nu_1})$. This allows us to express three of the constants λ_1, λ_2, ν_1, ν_2 via one of them, say ν_1. The asymptotic solution is reduced to the form

$$b = At^{-2\mu}F(\zeta), \quad \epsilon = A^2 t^{-2\mu-1}G(\zeta), \quad h = At^{1-\mu}. \qquad (10.89)$$

Here

$$F(\zeta) = \text{const}_1 B_0(\zeta), \quad G(\zeta) = \text{const}_2 E_0(\zeta),$$

$$\zeta = \frac{z}{h} = \frac{z}{At^{(1-\mu)}}, \quad A = \text{const}\, Q^{(1-\mu)/2} a^{(3\mu-1)/2}, \quad \mu = \frac{1+2\nu_1}{3}.$$
$$(10.90)$$

We must emphasize two essential points: the parameter μ, which determines the layer extension and turbulent energy decay rate cannot be obtained from dimensional considerations, and both the initial bulk energy of the turbulent layer per unit area, Q, and the initial layer thickness a enter the constant A in some powers.

To determine the parameter μ an eigenvalue problem must be stated and solved.

First we shall use the (b, ϵ) semi-empirical turbulence model (the traditional notation is the k, ϵ model). This model, based on the idea of A.N. Kolmogorov (1942), was elaborated by D.B. Spalding, B.E. Launder and their associates (Launder, Morse, Rodi and Spalding, 1972; Launder and Spalding, 1974; Hanjalic and Launder, 1972). For a comprehensive review see Reynolds (1976). According to this model the dissipation rate ϵ is taken in addition to the specific turbulent energy as the second kinematic quantity necessary to determine all the kinematic properties of turbulent flow. In addition to the equation for turbulent energy balance (10.84) the equation for balance of the dissipation rate is used. This equation is obtained basically in the same way as the turbulent energy balance equation and, for the case under consideration of the horizontally-homogeneous layer and shearless flow, assumes the form (see, e.g., Reynolds, 1976)

$$\partial_t \epsilon + \partial_z q_\epsilon = -U. \qquad (10.91)$$

Here q_ϵ is the turbulent flux of the dissipation rate and U is the rate of homogeneization of turbulent energy; they are the one-point moments of the velocity and the velocity-gradient fluctuations.

Let us introduce turbulent exchange coefficients, k_b for the turbulent energy and k_ϵ for the dissipation rate, according to the relations

$$k_b = -\frac{q_b}{\partial_z b}, \quad k_\epsilon = -\frac{q_\epsilon}{\partial_z \epsilon}. \qquad (10.92)$$

We emphasize that for the case of a horizontally-homogeneous layer these relations do not contain any additional assumptions. According to the Kolmogorov similarity hypothesis, which is assumed to be applicable in the turbulent part of the flow, the turbulent vortex field in the developed flow is statistically self-similar. Therefore all kinematic flow properties are determined, each to within a constant, by any two of them having different kinematic dimensions. We take as such the governing properties b and ϵ.

Dimensional analysis gives

$$k_b = \alpha \frac{b^2}{\epsilon}, \quad k_\epsilon = \beta \frac{b^2}{\epsilon}, \quad U = \gamma \frac{\epsilon^2}{b} \tag{10.93}$$

where the coefficients of α, β, γ, according to the assumed hypothesis, should be universal constants. Roughly speaking it is so, and these constants have been determined by comparison with experiment for some special cases (see, e.g., Reynolds, 1976).

Thus, equations (10.84) and (10.91) take the form

$$\partial_t b = \alpha \partial_z \left(\frac{b^2}{\epsilon} \partial_z b \right) - \epsilon \tag{10.94}$$

$$\partial_t \epsilon = \beta \partial_z \left(\frac{b^2}{\epsilon} \partial_z \epsilon \right) - \gamma \frac{\epsilon^2}{b} \tag{10.95}$$

and (10.94), (10.95) form a closed set of equations for the turbulent energy and turbulent energy dissipation rate.

4. After substitution of the solution in the form (10.89) into the system (10.94), (10.95), a set of ordinary differential equations is obtained for the functions $f = \beta F$ and $g = \beta G$:

$$\frac{\alpha}{\beta} \frac{d}{d\zeta} \left(\frac{f^2}{g} \frac{df}{d\zeta} \right) + (1 - \mu)\zeta \frac{df}{d\zeta} + 2\mu f - g = 0 \tag{10.96}$$

$$\frac{d}{d\zeta} \left(\frac{f^2}{g} \frac{dg}{d\zeta} \right) + (1 - \mu)\zeta \frac{dg}{d\zeta} + (1 + 2\mu)g - \gamma \frac{g^2}{f} = 0 \tag{10.97}$$

By symmetry, only one half of the layer, $0 \leq z \leq h(t)$, need be considered, with boundary conditions at $z = 0$ and $z = h(t)$. At the boundary $z = h(t)$ the turbulent energy b, the dissipation rate ϵ, and their fluxes q_b and q_ϵ must be continuous. However, outside the layer there is fluid at rest. Therefore at $z = h(t)$ the following conditions must be fulfilled.

$$b = 0, \quad \epsilon = 0, \quad q_b = 0, \quad q_\epsilon = 0. \tag{10.98}$$

From (10.98) and from (10.92) and (10.93) we find boundary conditions for the set (10.94), (10.95):

$$b = 0, \quad \epsilon = 0, \quad \frac{b^2}{\epsilon} \partial_z b = 0, \quad \frac{b^2}{\epsilon} \partial_z \epsilon = 0 \quad \text{at } z = h(t). \tag{10.99}$$

Using the self-similar representation of the solution (10.89) we obtain a first group of boundary conditions for the system of ordinary equations (10.96), (10.97):

$$f = g = 0, \quad \frac{f^2}{g}\frac{df}{d\zeta} = 0, \quad \frac{f^2}{g}\frac{dg}{d\zeta} = 0 \quad \text{at } \zeta = 1. \tag{10.100}$$

Furthermore, owing to symmetry with respect to the middle plane of the layer, $z = 0$, the fluxes are equal to zero:

$$q_b = q_\epsilon = 0 \quad \text{at } z = 0. \tag{10.101}$$

From (10.101) and from (10.92) and (10.93) we obtain the second group of boundary conditions, this time at the boundary $\zeta = 0$, for the system (10.96), (10.97):

$$\frac{f^2}{g}\frac{df}{d\zeta} = 0, \quad \frac{f^2}{g}\frac{dg}{d\zeta} = 0 \quad \text{at } \zeta = 0. \tag{10.102}$$

We have obtained, therefore, a boundary-value problem (10.100), (10. 102) for the system of second-order ordinary differential equations (10.96), (10.97), which contains a parameter μ. We shall restrict ourselves further to the case $\alpha = \beta$: the recommended values of the parameters α and β give for the ratio α/β an interval between 0.7 and 1.2.

A more detailed investigation shows that there exists a class of solutions f, g, that are positive at $0 \leq \zeta \leq 1$, identically equal to zero at $\zeta \geq 1$, continuous, having continuous quantities:

$$\frac{f^2}{g}\frac{df}{d\zeta}, \quad \frac{f^2}{g}\frac{dg}{d\zeta} \tag{10.103}$$

(the fluxes must also be continuous).

In the vicinity of the point $\zeta = 1$ at $\zeta < 1$ the solution satisfying the conditions (10.100) is expanded in series:

$$f = c(1 - \mu)(1 - \zeta) + \dots, \quad g = c^2(1 - \mu)(1 - \zeta) + \dots. \tag{10.104}$$

Here the positive quantity c, along with μ, is a parameter of the problem. Each pair of values c, μ determines uniquely non-trivial solutions to the system (10.96), (10.97) satisfying the conditions (10.100). It is necessary to find the values of the parameters c, μ for which the solutions also satisfy the conditions (10.102). Thus, as it is customary for self-similar solutions of the second kind we have obtained a nonlinear eigenvalue problem for determining the time degrees in the self-similar variables.

This problem is easily solved analytically. Indeed, let us assume that $g = cf$ over the whole interval $0 \leq \zeta \leq 1$. Substituting this relation into (10.96), (10.97) we obtain for the function $f(\zeta)$ two second-order ordinary differential equations that coincide if $c = 1/(\gamma - 1)$. For this value of c the equation for f takes the form

$$\frac{1}{c}\frac{d}{d\zeta}\left(f\frac{df}{d\zeta}\right) + (1 - \mu)\zeta\frac{df}{d\zeta} + 2\mu f - cf = 0. \tag{10.105}$$

Integrating from $\zeta = 0$ to $\zeta = 1$ and using boundary conditions (10.100) and (10.102), we obtain

$$(1 - 3\mu + c)\int_0^1 f\,d\zeta = 0 \tag{10.106}$$

whence

$$\mu = \frac{1 + c}{3} = \frac{\gamma}{3(\gamma - 1)} \tag{10.107}$$

because the function f is positive and so is its integral in (10.106). For the values $c = 1/(\gamma - 1)$ and $\mu = \gamma/3(\gamma - 1)$ the equation (10.105) can be simply integrated:

$$f = D(1 - \zeta^2), \quad g = D(1 - \zeta^2)(\gamma - 1)^{-1} \quad (0 \le \zeta \le 1).$$
$$f = g \equiv 0 \quad (\zeta \ge 1) \tag{10.108}$$

where $D = (2\gamma/3 - 1)/2(\gamma - 1)^2$. Thus, the solution (10.89) assumes the final form

$$b = \frac{A^2}{\alpha t^{2\mu}}D\left(1 - \frac{z^2}{h^2}\right), \quad \epsilon = \frac{A^2}{\alpha t^{2\mu+1}}\frac{D}{(\gamma - 1)}\left(1 - \frac{z^2}{h^2}\right) \quad (0 \le z \le h(t))$$
$$h(t) = At^{(2\gamma-3)/3(\gamma-1)}, \quad b = \epsilon = 0 \quad (z \ge h(t))$$
$$A = \text{const } Q^{(2\gamma-3)/6(\gamma-1)}a^{1/2(\gamma-1)} \tag{10.109}$$

In particular, for the recommended value $\gamma = 2$ the form of the solution is especially simple:

$$b = \frac{A^2}{\alpha t^{4/3}}D\left(1 - \frac{z^2}{h^2}\right), \quad \epsilon = \frac{A^2}{\alpha t^{7/3}}D\left(1 - \frac{z^2}{h^2}\right) \quad (0 \le z \le h(t))$$
$$h(t) = At^{1/3}, \quad A = \text{const } Q^{1/6}a^{1/2}. \tag{10.110}$$

So, the solution is essentially different from (10.83), which was obtained from 'naive' dimensional considerations.

5. Another closing of the system can be obtained within the frames of a different turbulence model, the so-called b, l model based on the same similarity hypothesis but a different choice of second governing kinematic quantity: the external length scale l instead of ϵ.

According to the dimensional analysis the relations

$$k_b = l\sqrt{b}, \quad \epsilon = c_1 b^{3/2}/l \tag{10.111}$$

hold, where c_1 is a constant (the constant in the relation for k_b can be assumed to be equal to unity by corresponding renormalization of the length scale). The equation for turbulent energy balance (10.84) takes the form

$$\partial_t b = \partial_z l \sqrt{b} \partial_z b - c_1 \frac{b^{3/2}}{l}. \tag{10.112}$$

To close this equation the simplest assumption, $l = \alpha_1 h$, $\alpha_1 = \text{const}$ is made: we assume that the length scale is constant over the turbulent layer and is proportional to its thickness. So, we assume that the characteristic size of the vortices is a fixed fraction of the layer thickness and so we neglect the time of adjustment of the vortices to the layer thickness. The equation (10.112) then assumes a closed but non-local form:

$$\partial_t b = \frac{2}{3} \alpha_1 h(t) \partial_{zz}^2 b^{3/2} - \frac{c_1}{\alpha_1} \frac{b^{3/2}}{h(t)}. \tag{10.113}$$

The non-locality of (10.113) is related to the fact that the width $h(t)$ entering the right-hand side is a global functional of the solution $b(z,t)$.

The asymptotic solution to the equation (10.113) is represented, as before, in the form (10.89):

$$b = A^2 t^{-2\mu} f(\zeta), \quad \zeta = \frac{z}{h}, \quad h = At^{1-\mu}.$$

For determining the function f and the parameter μ the following relations are obtained from equation (10.113), the boundary conditions (10.85) and the symmetry conditions:

$$\frac{d}{d\zeta}\left(\alpha_1 \sqrt{f} \frac{df}{d\zeta}\right) + (1-\mu)\zeta \frac{df}{d\zeta} + 2\mu f - \frac{c_1}{\alpha_1} f^{3/2} = 0;$$

$$\frac{df^{3/2}}{d\zeta} = 0 \quad \text{at } \zeta = 0; \quad f = 0, \quad \frac{df^{3/2}}{d\zeta} = 0 \quad \text{at } \zeta = 1. \tag{10.114}$$

These relations again form a nonlinear eigenvalue problem. Due to the group invariance property the eigenvalue μ depends only on the combination c_1/α_1^2. The eigenvalue problem is easily solved numerically; its solution is represented in Figure 10.21.

The solutions based on the two closure hypotheses, i.e., the b,l and b,ϵ models, are basically identical. The most essential differences are the sharper decrease of turbulent energy near the boundary and the vanishing of the length scale $l = c_1 b^{3/2}/\epsilon$ at the boundary in the b,ϵ model.

6. Batchelor and Linden (1992) noted that in the problem of turbulent burst evolution there exists an invariant of instructive form.

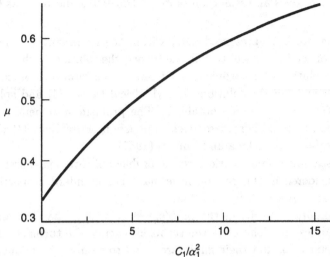

Figure 10.21. The eigenvalue μ in the b, l model versus the parameter c_1/α_1^2.

Consider the turbulent energy per unit area at time t:

$$Q(t) = \int\limits_{-h}^{h} b(z, t)dz. \qquad (10.115)$$

At the asymptotic stage of the motion, we have, according to the solution (10.89),

$$Q(t) = \text{const}\, A^3 t^{1-3\mu}, \quad h = At^{1-\mu}, \qquad (10.116)$$

where the constant equals $\int\limits_{-1}^{1} f(\zeta)d\zeta$. Therefore the quantity

$$Q(t)[h(t)]^{(3\mu-1)/(1-\mu)} = \text{const}\, A^{2/(1-\mu)} = \text{const}_1 Q a^{(3\mu-1)/(1-\mu)} \qquad (10.117)$$

is preserved in time at the asymptotic stage of turbulent-burst evolution. At the initial, non-self-similar stage of turbulent-burst evolution the relations (10.116) and (10.117) do not hold, but it is instructuve that the value of the integral (10.117) is preserved if instead of the initial distribution of turbulent energy its distribution at an arbitrary moment of time is taken.

Let us consider the relevant constants. Numerical calculation of the eigenvalue in the second variant (the b, l model) shows that the value $\mu = 2/3$, obtained in the b, ϵ model for the recommended value $\gamma = 2$ corresponds to $c_1/\alpha_1^2 \simeq 17$. According to (10.111), in the b, l model $l = c_1 b^{3/2}/\epsilon$, $k_b = c_1 b^2/\epsilon$. Comparing with (10.93), we obtain $c_1 = \alpha$. Therefore $\alpha_1 = (\alpha/17)^{1/2}$. The recommended (Reynolds, 1976) range of values of α gives an estimation of c_1 as 0.06–0.085 and of α, as 0.063–0.071.

7. Thus, we have presented here, within the framework of the basic models of semi-empirical turbulence theory, the solution to the problem of the evolution of a turbulent burst having the form of a plane layer. In the same way the solutions for cylindrical ($\mu = 3/4$) and spherical ($\mu = 4/5$) bursts can be obtained. The problem statement and the solutions presented here were given in papers by Barenblatt (1983) and by Barenblatt, Galerkina and Luneva (1987).

The statement and solutions to the problem of turbulent-burst evolution considered in this section have many non-standard properties deserving special attention, as follows.

An important point in the problem statement is that the ambient fluid is quiescent. Therefore the turbulent energy, the turbulent energy dissipation rate, and their fluxes are equal to zero at the boundary of the burst. Zero is also a number, and this means that no additional constants enter the problem statement except the constants appearing in the equations. The values of these constants are more or less established.

Equation (10.113) for the turbulent energy in the b, l-model is non-traditional also from the mathematical viewpoint. We emphasize again that it is non-local because on its right-hand side we have a global functional of the solution. Therefore the work of Kamin and Vázquez (1992) was very important. They proved the existence and uniqueness of the solution to the initial-value problem for this equation. An especially important result of this work is that a self-similar solution of the second kind presented in point (5) above is indeed the asymptotics to the solution of the initial-value problem at $t \to \infty$. The coefficient A is the integral of this solution: taking the solution at any time instead of the initial condition we obtain the same value of A as for the asymptotics. The problems of the existence and uniqueness of the solution of the initial-value problem for (10.113) were considered also by Grebenev (1992).

In the paper by Hastings and Peletier (1992) some rigorous estimates for self-similar solutions of (10.113) were obtained. Chen and Goldenfeld (1992) applied to the initial-value problem for the non-local equation (10.113) the renormalization group technique and obtained by the ϵ-expansion method a very accurate result for the eigenvalue.

The system (10.94), (10.95) of quasi-linear parabolic equations, describing the evolution of the burst within the framework of the b, ϵ model has also attracted the attention of mathematicians. Bertsch, dal Passo and Kersner (1994) proved the existence and uniqueness of the solution to the initial-value problem for this system in the case $\alpha = \beta$. They also proved that the self-similar solution of the second kind (10.109) is the asymptotics of the solution to the initial-value problem at $t \to \infty$, A being the integral of the solution to the initial-value problem. Hulshof (1993) considered self-similar solutions for the system (10.94), (10.95) at $\alpha/\beta \neq 1$, and proved the existence of solutions with 'compact support', i.e., $h(t) < \infty$ for $\alpha/\beta < 2$. M. Bertsch suggested that for $\alpha/\beta > 2$ the asymptotics is self-similar, but the thickness of the burst $h(t)$ instantaneously becomes infinite as for the linear heat conduction equation. This hypothesis has found confirmation in the numerical computations of V.M. Prostokishin (1994).

Scaling in geophysical
fluid dynamics

11.1 Scaling laws for the atmospheric surface layer

Geophysical fluid dynamics has become in the last few decades a broad subject (see Pedlosky, 1979) with many applications in earth sciences and in engineering practice. In all branches of geophysical fluid dynamics using similarity considerations, scaling laws and self-similar solutions play an important, often decisive role. We have chosen in this chapter for demonstration's sake some topics from geophysical fluid mechanics related mainly to geophysical turbulence.

The *surface layer* of the atmosphere is usually modelled (see, e.g., Monin and Yaglom, 1971) by a turbulent flow that is statistically horizontally-homogeneous and stationary, and is bounded below by a horizontal plane. The shear stress τ in the surface layer is also assumed to be constant. The essential difference from the flow in the wall region considered in section 10.2 consists in the presence in the surface layer of thermal stratification – temperature inhomogeneity over the height of the layer[†]. The stratification is stable if the temperature increases with height and unstable in the opposite case. Owing to the thermal inhomogeneity, a vertical displacement of fluid particles, produced by a vertical velocity fluctuation, is accompanied by work done against the force of gravity (or extracted, depending on whether the stratification

---

[†] We do not consider here such supplementary factors as moisture, dust etc.

is stable or not). This work is either taken from the turbulent energy or added to it, thus influencing the turbulence level, i.e. the transfer of heat, mass and momentum, and consequently also influencing the vertical distribution of the mean longitudinal velocity across the flow. The effectiveness of the influence of thermal stratification on the balance of turbulent energy is governed by the product of the coefficient of thermal expansion of the air and the acceleration of gravity, the so-called *buoyancy parameter*. The air in the atmospheric surface layer is usually considered to be a thermodynamically ideal gas, for which the coefficient of thermal expansion is equal to $1/T$, where T is the absolute temperature. The atmospheric surface layer is not thick, so the variation in mean pressure and the corresponding variation in the density can be neglected. In general, the variations of density and absolute temperature in the surface layer are considered to be small, and their influence on the dynamics of the flow is taken into account only through the buoyancy, which governs the contribution of thermal stratification to the turbulent energy balance. Thus the state of motion at some point of the flow in the atmospheric surface layer is governed by the following quantities: (1) the friction velocity $u_* = \sqrt{\tau/\rho_0}$; (2) the reference density ρ_0; (3) the dynamic temperature T_*, introduced by analogy with the friction velocity through the relation

$$T_* = -\frac{\langle w'T' \rangle}{u_*} \tag{11.1}$$

(where w' is the vertical velocity fluctuation, T' the temperature fluctuation, and the quantity $\langle w'T' \rangle$ coincides to within a constant factor with the vertical heat flux, which is also assumed constant over the surface layer), the dynamic temperature T_* being positive in the case of stable stratification $(\partial_x T > 0)$ and negative for unstable stratification; (4) the buoyancy parameter $\beta = g/T_0$ (where g is the acceleration due to gravity and T_0 is the reference temperature, which does not appear separately anywhere, since a change in T_0 turns out to influence the flow dynamics only through the buoyancy parameter, i.e., in combination with the force of gravity); (5) the vertical coordinate z; (6) the molecular kinematic viscosity of air, ν; (7) the molecular thermal diffusivity of air, χ; and (8) the external geometric length scale Λ (e.g. the height of the atmospheric surface layer).

The standard procedures of dimensional analysis give

$$\Pi_u = \frac{z\partial_z u}{u_*} = \Phi_u \left(\frac{z}{L_0}, \frac{u_* z}{\nu}, \frac{u_* \Lambda}{\nu}, \mathrm{Pr} \right), \tag{11.2}$$

$$\Pi_T = \frac{z\partial_z T}{T_*} = \Phi_T\left(\frac{z}{L_0}, \frac{u_* z}{\nu}, \frac{u_* \Lambda}{\nu}, \mathrm{Pr}\right), \qquad (11.3)$$

where $\mathrm{Pr} = \nu/\chi$ is the Prandtl number and L_0 is the thermal length scale[†],

$$L_0 = u_*^2/\beta T_* . \qquad (11.4)$$

The existing similarity theory for flows in the surface layer of the atmosphere, which owes its origin to the pioneering work of Prandtl (1932a) and the works of A.S. Monin and A.M. Obukhov (see Obukhov, 1946; Monin, 1950; Monin and Obukhov, 1953, 1954), is based on the assumption of complete similarity of the flow in both Reynolds numbers, the local one, $\mathrm{Re}_l = u_* z/\nu$, and the global one, $\mathrm{Re}_* = u_* \Lambda/\nu$. The plausibility of such an assumption and, consequently, of neglecting the dependence on Re_l and Re_* in (11.2) and (11.3) is usually argued on the basis of the very large values of both Reynolds numbers (for the local one, very large values outside a small region close to the surface itself whose height does not exceed a few millimeters). Here the assumption of the existence of finite limits of the functions Φ_u and Φ_T as $\mathrm{Re}_l \to \infty$ and $\mathrm{Re}_* \to \infty$ is accepted implicitly. If the functions Φ_u and Φ_T tend to finite limits as $\mathrm{Re}_l \to \infty$ and $\mathrm{Re}_* \to \infty$ in accordance with the assumption of complete similarity, then for sufficiently large Re_l and Re_* a universal similarity law, independent of the Reynolds numbers, must hold:

$$z\partial_z u/u_* = \Psi_u(z/L_0, \mathrm{Pr}), \qquad (11.5)$$

$$z\partial_z T/T_* = \Psi_T(z/L_0, \mathrm{Pr}). \qquad (11.6)$$

This is called in the literature the Monin–Obukhov similarity law. In the special case when thermal stratification of the flow disappears, we again arrive at the universal logarithmic law, considered in subsection 10.2.1.

The considerations presented in subsection 10.2.1 show that even in the case of a thermally neutral flow one detects a weak dependence of the universal function on both Reynolds numbers. This weak dependence allowed us to introduce the assumption of incomplete similarity of the flow in the local Reynolds number, which is apparently not contradicted by the experimental data on flows in smooth pipes, etc. It is natural to make a similar assumption for thermally stratified flows in the surface layer of the atmosphere (Barenblatt and Monin, 1976, 1979a).

[†] This definition of the thermal length scale follows Yaglom (1974) and is somewhat different from the conventional one.

11.2 Flows with strongly stable stratification

As discussed earlier, in spite of many years of continuous effort by many researchers, including top-level people in physics, mathematics, and mechanics, hydrodynamic turbulence remains a challenge for even the simplest case, a homogeneous incompressible fluid. Fluid lamination by density under the gravity field – *stratification* – additionally complicates the pattern of turbulent flows and leads to the appearance of substantially new effects.

Flow stratification introduces into our considerations a characteristic vertical length scale – the vertical length at which the density variation reaches a magnitude at which it will influence the flow dynamics. Stratification is considered to be strong if the characteristic vertical flow length scale is substantially larger than this scale. In this and the following sections scaling laws are considered for phenomena related to turbulent flows with strongly stable stratification. Some of these phenomena are of basic interest for geophysical fluid mechanics.

First, one of the simplest stratified flows is considered, turbulent flow in the wall region, where the strongly stable stratification is due to suspended small heavy particles. Examples of such flows are sediment transporting rivers, high-energy benthic bottom layers in the ocean, and dust storms. Considering flow stratification by suspended particles is attractive owing to the simple formulae at which we arrive, which allow one to explain a seemingly paradoxical phenomenon: under certain conditions the heavy particles can accelerate the flow!

In the case of stable stratification created by temperature and/or salinity an essentially new factor appears, internal waves. An instructive problem will be considered preliminarily: heat transfer in the oceanic *upper active layer* where the temperature (and salinity) distribution is subject to seasonal variations. As is known the density of sea water differs from the density of fresh water by three to four per cent, whereas its density fluctuations have the order of tenths of one per cent. Nevertheless these tiny density variations can influence the flow dynamics in an essential way. The temperature distribution over depth in the upper active ocean layer has the typical form represented schematically in Figure 11.1. The *upper homogeneous layer* where temperature and salinity[†], and, consequently, density are nearly uniformly distributed, is

---

[†] For simplification's sake we will speak further only about temperature stratification.

the result of turbulent mixing. This mixing is carried out by the simultaneous actions of shear and of convection, which causes heavier fluid particles to sink. These particles thus come to the flow depth from the surface layer where the fluid is heavier, because this layer is cooled, and its salinity increased by evaporation from the oceanic surface and also by the breaking of surface waves. The depth of the homogeneous layer is time dependent: in moderate latitudes it grows in the fall-winter period and decreases in spring. The upper homogeneous layer is supported by a region where the temperature variation is sharp, the *upper thermocline*, which terminates at the depth where seasonal temperature variations vanish. This depth, i.e., the depth of the oceanic upper active layer, has an order of magnitude of about 200–250 m.

Analysis of the mean temperature distribution in the strongly stably stratified upper thermocline shows that an adequate model of this distribution is the Hertz travelling thermal wave, whereas the magnitude of the effective thermal diffusivity coefficient appears to be constant, to rather high accuracy. This coefficient is of order $10^{-1} - 1\mathrm{cm}^2/\mathrm{s}$, surprisingly, at first sight, being intermediate in value between the turbulent thermal diffusivity coefficient in the upper homogeneous layer, estimated as 10^3 cm^2/s, and the molecular thermal diffusivity coefficient, of order 10^{-3} cm^2/s. At the same time high-precision measurements show (this is discussed in detail in the monograph Fedorov (1976)) that the instantaneous temperature distribution over the depth is never smooth (as appeared to be the case earlier when these measurements were performed by highly inertial gauges), but has a rather step-wise character: intervals where the temperature is nearly constant alternate with intervals where there are large temperature, salinity, and, consequently, density gradients.

This is explained by a peculiar intermittence of the turbulence in flows with strongly stable stratification. The turbulence in such flows is not uniformly distributed everywhere, but is concentrated within patches. It is generated and then decays rather rapidly and is closely connected with internal waves. The interaction between internal waves and turbulence is illustrated by an instructive experiment performed by O.M. Phillips, which is described and analyzed below.

Internal waves also determine the very structure of turbulence in a strongly stable stratified flow. For various reasons the internal waves break, forming patches of mixed fluid that collapse through being squeezed at the level of their density. This collapsing of mixed fluid patches is also considered below. It is shown that at all its basic stages the collapse is characterized by various scaling laws. The most durable stage of collapse is the final, viscous one, where the drag to the patch-extension driving force is due mainly to viscous force. It may occur that the patch extension at this stage is so slow that the patch can seem to the unprepared observer as being non-extending.

It is essential to note that the fluid inside the patch is uniform or close to being uniform, whereas outside the patch it is strongly and stably stratified. Therefore outside the patch the turbulence must spend part of its energy in working against the buoyancy force, whereas inside the patch it does not. Hence outside the patches the turbulence decays rapidly, and inside it is supported at a higher level. This very effect gives rise to the peculiar, intermittent, archipelago-like character of turbulence in a flow with strongly stable stratification.

The inhomogeneous and strongly anisotropic character of turbulence under strongly stable stratification was predicted by A.N. Kolmogorov in the late nineteen-forties. The existence of pancake-form turbulent patches (*blini*, from a similar Russian word) in the atmosphere and ocean under strong stratification conditions was established by Phillips (1967).

Turbulence in a shear flow, if it is sufficiently strong, can support sediment which would fall down in a non-turbulent or weakly turbulent flow. The capturing of falling sediment by turbulent patches can make the latter visible by contrast in an ambient fluid with weaker turbulence. It is plausible that this explains at least partially discoidal formations in the atmosphere, which have attracted rather wide attention, as well as 'turbidity clouds' in the ocean.

11.3 The regime of limiting saturation of a turbulent shear flow laden with sediment

We turn now to the consideration of a flow laden with small suspended particles. The volume and mass concentrations of particles are assumed to be very small (e.g. in rivers carrying a large amount of sediment, their volume and mass concentrations rarely exceed several ten-thousandths). Nevertheless, the dynamic action of the particles on the flow can turn out to be crucial, owing to the vast influence of the force of gravity. Furthermore, the particles are assumed to be much smaller than the internal turbulence scale; therefore the viscous relaxation time of the particles is negligible and one can assume that the horizontal components of the instantaneous velocities of the particles and the fluid coincide, and that the vertical ones differ by a constant quantity a, the velocity of free fall of the particles in the unbounded fluid.

We consider again the wall region of the flow (for example, the surface layer of the atmosphere or the bottom layer of a channel). The momentum equation in this region remains just the same as for the pure fluid, since the influence of the particles on the density of the mixture is negligibly small.

The equation of conservation of mass for the load is obtained by setting equal to zero the total flux of particles through unit horizontal area. This flux is the sum of the flux of turbulent transport of particles, $\langle s'w' \rangle$, and the flux of settling particles, $-as$, so that

$$\langle s'w' \rangle - as = 0. \tag{11.7}$$

Here s and s' are the mean volume concentration of particles and its fluctuation, respectively.

Finally, the steady equation of turbulent energy balance assumes, if one neglects the contribution of the diffusion of turbulent energy, the form

$$\langle u'w' \rangle \partial_z u + \epsilon + \sigma \langle s'w' \rangle g = 0. \tag{11.8}$$

Here $\sigma = (\rho_p - \rho)/\rho$ is the relative excess of the density of particles ρ_p over the density of the fluid ρ and ϵ is the dissipation rate of turbulent energy. The last term expresses the rate at which turbulent energy is consumed in the suspending of particles by the flow; the other terms were explained in subsection 10.2.2. Despite the smallness of the concentration of particles in the flow, this term can have a significant value, since the force of gravity is very large.

Equation (11.8) can be put into the form

$$\langle u'w' \rangle \partial_z u (1 - \mathrm{Ko}) + \epsilon = 0, \tag{11.9}$$

where the dimensionless parameter

$$\mathrm{Ko} = -\frac{\sigma g \langle s'w' \rangle}{\langle u'w' \rangle \partial_z u}, \tag{11.10}$$

called the Kolmogorov number, expresses the relative consumption of turbulent energy on the suspending of particles by the flow. This parameter gives a natural criterion for the dynamic activity of the load, i.e., the influence of the suspended particles on the dynamics of the flow. A similar parameter for stratification due to temperature and/or salinity (see below) is called the Richardson number. We introduce, in analogy with the coefficient of momentum exchange, the coefficient of load exchange, according to the relation

$$\langle s'w' \rangle = -k_s \partial_z s. \tag{11.11}$$

We now assume, in the spirit of the Kolmogorov similarity hypothesis of semi-empirical turbulence theory, that this coefficient too, like the coefficient of momentum exchange and the mean rate of dissipation, depends only on the local turbulent energy of a unit mass and on the length scale of the turbulence. By dimensional analysis we then obtain

$$k_s = \alpha_s l \sqrt{b}, \tag{11.12}$$

where α_s is a constant.

In the problem being considered of a laden flow, in contrast to the flow of a pure fluid an additional parameter has appeared, the Kolmogorov number Ko, so that for the external scale of turbulence we have according to dimensional analysis,

$$l = z \Psi(\text{Re}_l, \text{Re}_*, \text{Ko}).$$

Under the assumption of complete similarity in the local and global Reynolds numbers[†] the turbulent length scale can be represented through a universal function of the Kolmogorov number,

$$l = \kappa \gamma z \Phi_l(\text{Ko}), \tag{11.13}$$

where $\Phi_l(0)$ is obviously equal to one. The turbulence scale decreases under the influence of the load, so the function Φ_l must decrease when its argument increases. Thus, under the assumption made, the basic system of equations for the wall region of a laden turbulent shear flow assumes the form

$$l\sqrt{b}\partial_z u = u_*^2, \quad \alpha_s l \sqrt{b}\partial_z s + as = 0, \quad b = \frac{u_*^2}{\gamma^2}(1 - \text{Ko})^{1/2},$$

$$l = \kappa \gamma z \Phi_l(\text{Ko}), \quad \text{Ko} = \frac{\sigma gas}{u_*^2 \partial_z u}. \tag{11.14}$$

The system of equations (11.14) has some characteristic properties. First of all, it contains only the gradient of the velocity $\partial_z u$, and not the velocity itself. Furthermore, for the case of an unrestricted supply of particles on the underlying surface, in view of the back influence of the particles on the dynamics of the flow we can anticipate the existence of flow regime in which the flow absorbs the maximum possible amount of the sediment load for given friction velocity and other parameters.

[†] As was shown in chapter 10, this assumption is apparently invalid for neutral flow. However, we neglect here further discussion of this matter, for two reasons. First, the quantitative difference is not essential here and does not play any significant role: we do not have at our disposal sufficiently precise experimental data to distinguish complete and incomplete similarity in sediment-laden flow. Second, what we need here is a qualitative explanation, and this one is more transparent for the case of complete similarity.

This regime, which we shall call the *regime of limiting saturation*, must be described by a singular solution of (11.14), which in turn must be determined by the parameters appearing in the differential equations themselves. Thus the determination of the regime of limiting saturation does not require the prescribing of any boundary condition for the sediment concentration.

An essential point is that the system (11.14) is invariant with respect to the transformation group

$$s = S/\alpha, \quad z = \alpha Z, \quad u = U + \beta \tag{11.15}$$

($\alpha > 0$ and β being the group parameters), so that substituting (11.15) into (11.14) we obtain the same system (11.14) but in the variables S, U, Z. Let the singular solution corresponding to the regime of limiting saturation determine the velocity gradient and load concentration by the relations

$$\partial_z u = f(z), \quad s = g(z). \tag{11.16}$$

But the singular solution is determined only by the system itself and therefore it also must be invariant with respect to the group (11.15), i.e., it can be expressed in the form

$$\partial_z U = f(Z), \quad S = g(Z),$$

where f and g are the same functions as in the relations (11.16). Expressing U, S and Z in terms of u, s, z, and α, we get for the functions f and g the functional equations

$$f(z) = \alpha f(\alpha z), \quad g(z) = \alpha g(\alpha z). \tag{11.17}$$

The solution of these functional equations is found in an elementary way,

$$f = \frac{C_1}{z}, \quad g = \frac{C_2}{z}, \tag{11.18}$$

where C_1 and C_2 are constants subject to determination. Substituting into (11.15) the relations

$$\partial_z u = \frac{C_1}{z}, \quad s = \frac{C_2}{z}, \quad \text{Ko} = \frac{\sigma g C_2}{C_1^2} \equiv \text{const}, \tag{11.19}$$

we obtain

$$C_1 = \frac{u_*}{\kappa(1 - \text{Ko})^{1/4}\Phi_l(\text{Ko})} = \frac{u_*}{\kappa\omega}, \quad \omega = \frac{a}{\alpha_s \kappa u_*}, \tag{11.20}$$

whence we find a finite equation for determining the Kolmogorov number Ko, which is constant in the regime of limiting saturation:

$$\omega = (1 - \text{Ko})^{1/4}\Phi_l(\text{Ko}). \tag{11.21}$$

But Φ_l is a non-increasing function of its argument, $\Phi_l(0) = 1$, and Ko by its physical meaning lies between zero and unity. Hence it follows that for $\omega > 1$ there exists no root of (11.21), and for $\omega < 1$ a unique root exists. Therefore a necessary condition for the existence of a regime of limiting saturation is

$$\omega = \frac{a}{\alpha_s \kappa u_*} < 1. \tag{11.22}$$

The physical meaning of the condition (11.22) is transparent. In fact, the value of the friction velocity u_* is proportional to the mean square velocity fluctuation. Thus if the fluctuation is large, so that during the time in which a fluid mass is lifted up by the turbulent fluctuation the heavy particles inside it have no time to fall (the velocity a of free fall being relatively small), the particles come into the main core of the flow and become suspended in it. In the opposite case, the particles are transported by the flow in the bottom layer, do not reach the main core of the flow, and do not influence the flow dynamics in the main part of the stream.

From the first equation of (11.19), taking into account (11.20), we get for $\omega < 1$

$$u = \frac{u_*}{\kappa \omega} \ln z + \text{const}. \tag{11.23}$$

This means that for the flow in the regime of limiting saturation, which can be realized for $\omega < 1$, the velocity distribution remains logarithmic, just as in a pure fluid, but a reduction of the von Kármán constant has occurred: instead of κ it is now equal to $\kappa\omega$. Therefore *under the same external conditions (the same friction velocity) the flow accelerates under the action of particles in comparison with the flow of pure fluid.*

Since the capture of particles by the flow is realized by turbulent fluctuations, the turbulent energy must decrease. Actually, the turbulent energy per unit mass for the regime of limiting saturation is equal to

$$b = b_0(1 - \text{Ko})^{1/2}, \tag{11.24}$$

where $b_0 = u_*^2/\gamma^2$ is the turbulent energy for the flow of pure fluid with the same friction velocity. But the turbulent flow drag depends on the intensity of the fluctuations, so it turns out that the *suspended particles decrease the turbulent drag.* It is clear that this conclusion is valid only under the conditions indicated above of horizontal or nearly horizontal flow and of small volume and mass concentrations of particles, etc. Under such conditions a drag reduction in the flow and an apparent decrease in the von Kármán constant under the action of suspended particles have been observed by experimentalists (Vanoni, 1946; Einstein and Ning Chen, 1955).

The theory presented here of the transport of particles by a turbulent flow was developed by Kolmogorov (1954), and Barenblatt (1953, 1955);

the derivation of the equations for the regime of limiting saturation on the basis of group considerations was given by Barenblatt and Golitsyn (1974).

Dust storms in the Earth's atmosphere, as well as in the atmosphere of Mars, can be explained by the effect discussed above (Barenblatt and Golitsyn, 1974). It also explains the acceleration of flows in rivers carrying a relatively large amount of sediment, which had been repeatedly noticed by hydrologists. Apparently flow acceleration due to strongly stable stratification by particles, temperature, and salinity can explain (Barenblatt, Galerkina and Lebedev, 1992, 1993) the formation of high-energy benthic boundary layers in the ocean. The discovery of these layers (Nowell and Hollister (1985), see also the rest of the issue of *Marine Geology* in which this paper is published; Weatherly and Kelly (1982)) was one of the most important events in modern oceanology.

11.4 Upper thermocline in the ocean –
the travelling thermal wave model

Stratification in the ocean is established due to the non-uniform distribution over depth of temperature and/or salinity. In contrast with stratification due to suspended particles, considered in the previous section, here we have the mixing of stratifying agent and fluid on the molecular level. Hence the turbulent exchange intensity and consequently the heat and mass exchange intensities are closely related to internal waves. We shall see this as we consider an instructive problem regarding the temperature distribution in the oceanic upper thermocline in moderate latitudes. For the fall-winter period the upper thermocline is most clearly manifested and its lowering occurs.

Heat and mass exchange on the oceanic surface, including the falling of heavy cold fluid particles formed from breaking waves, leads to the formation of a peculiar oceanic boundary layer where temperature (and salinity) are influenced by the water-air interface. This layer – the upper active layer of the ocean – consists of the upper homogeneous layer, where the temperature is nearly constant, and the upper thermocline supporting it where, on the contrary, temperature variation is sharp (Figure 11.1). The depth of the upper active layer in the open ocean is at least an order of magnitude less than the total oceanic depth. Therefore the upper thermocline can be considered as a semi-infinite region $h < z < \infty$ (z is the depth measured from the oceanic surface, h

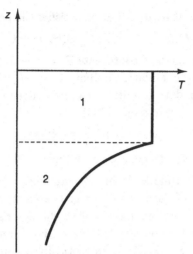

Figure 11.1. Schematic representation of the upper active layer of the ocean: (1) upper homogeneous layer; (2) upper thermocline.

is the depth of the upper homogeneous layer). The excess temperature – the difference between the current temperature and the mean annual temperature at a given point – therefore vanishes at infinity. If the natural observation data are averaged over a period of a month, the influence of short-time processes (daily variations, short-time random temperature anomalies, etc.) will disappear. The averaged parameters of the oceanic upper active layer, such as the speed of lowering of the upper homogeneous layer, $u = dh/dt$, and its excess temperature Θ_0 will be functions of some dimensionless 'slow' time T:

$$u = u(T), \quad \Theta_0 = \Theta_0(T). \tag{11.25}$$

Let us neglect the influence of horizontal inhomogeneity, and let us assume furthermore that the motions governing the turbulence, and in consequence the heat and mass exchange mechanisms in the upper thermocline, are statistically steady, spatially homogeneous and small scale. Then, under these assumptions the heat conduction equation for the averaged excess temperature $\Theta(z, t)$ is obtained:

$$\partial_t \Theta = \kappa \partial_{zz}^2 \Theta. \tag{11.26}$$

Here t is the ordinary dimensional time, and κ is the effective thermal diffusivity coefficient, which is constant according to our assumptions. Note, that in further considerations the variability of this coefficient, and, in particular, its dependence on temperature and/or temperature gradient could be taken into account without additional mathematical difficulty. Further analysis will show, however, that there is no need for this complication.

The excess temperature distribution satisfies the boundary conditions

$$\Theta(h, t) = \Theta_0(T), \quad \Theta(\infty, t) = 0 \tag{11.27}$$

Let us introduce a moving coordinate $\xi = z - h$ reckoned from the thermocline's upper boundary, so that $\Theta = \Theta(\xi, t)$. Then equation (11.26) and the boundary conditions can be written in the form (cf. the analogous consideration in section 7.5)

$$\partial_t \Theta - u(T) \partial_\xi \Theta = \kappa \, \partial_{\xi\xi}^2 \Theta \tag{11.28}$$

$$\Theta(0, t) = \Theta_0(T), \quad \Theta(\infty, t) = 0. \tag{11.29}$$

The characteristic time scale in this problem $\tau = \kappa/u^2$, as will be seen, is of the order of days. The averaged excess temperature Θ is not influenced by short-time processes, so the characteristic time scale of t should be considered to be larger than τ. For large t/τ the solution to the problem (11.28), (11.29) asymptotically becomes steady, so that the time derivative in (11.28) disappears. The equation obtained can be integrated simply, and under boundary conditions (11.29) the solution achieves the form

$$\Theta = \Theta_0 \exp(-u\,\xi/\kappa). \tag{11.30}$$

It is convenient to introduce a universal variable,- the relative excess temperature $\theta = (\Theta_0 - \Theta)/\Theta_0$. In terms of θ the solution (11.30) can be written in the form

$$\theta = 1 - \exp(-u\,\xi/\kappa), \tag{11.31}$$

i.e., the relative excess temperature distribution appears to be self-similar. Self-similarity of the relative excess temperature distribution in the upper thermocline was found empirically by Kitaigorodsky and Miropolsky (1970), by processing natural data. Linden (1975) performed a successful attempt to process, on the basis of the self-similarity hypothesis, data obtained in his laboratory experiments concerning the salinity profile in the laboratory model of the upper thermocline. However, the self-similarity was not associated in these papers with a definite physical mechanism. In the preceding chapters we have seen repeatedly that self-similarity never occurs by chance, it always means that there exists a certain stabilization of the process. The settling down of the average temperature field to a steady travelling wave is another instructive example confirming this general rule. The model of a travelling wave for the upper thermocline was proposed and shown to agree with experimental data – both laboratory and natural data – in Barenblatt (1978b); independently and simultaneously an analogous model was proposed by Turner (1978). Note that the steady-state solution (11.30) for the heat conduction equation with a moving heat source was found by the famous German physicist H. Hertz in the last century and since then has been used in many branches of mathematical physics. In particular, it plays a fundamental role in the theory of combustion, where it describes the temperature distribution ahead of the combustion region where the chemical reaction has not as yet started (see chapter 7). The travelling thermal wave model has also been much used in physical oceanology. Munk (1966) applied a similar model in a different physical situation; following his figurative expression, we might say that this model comes from oceanographic antiquity.

Integrating (11.31) we obtain

$$\frac{\kappa}{u} = \int_0^\infty (1 - \theta)d\xi \,, \tag{11.32}$$

so that the relation (11.31) can be represented in universal form:

$$\ln\frac{1}{1-\theta} = \frac{\xi}{\int_0^\infty (1-\theta)d\xi} = \zeta \,. \tag{11.33}$$

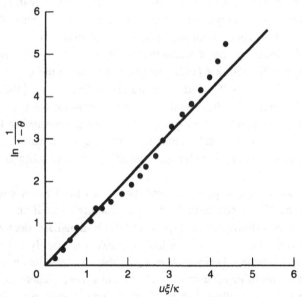

Figure 11.2. Results of the processing of natural data confirm the travelling wave model for the upper thermocline.

Therefore, if the proposed model is an adequate one then in the reduced coordinates

$$\ln \frac{1}{1-\theta}, \ \zeta$$

the experimental points should settle down along the bisectrix of the first quadrant. Efimov and Tsarenko (1980) processed numerous natural data using these coordinates and were able to confirm the travelling-wave model for the upper thermocline. In Figure 11.2 are presented Efimov's results of the processing of natural data by B.N. Filyushkin of the averaged October 1968–72 temperature profile at the 'Echo' weather station (34°0′ N and 48°0′ W). As is seen these data confirm the thermal travelling wave model for the upper thermocline and the constancy of the effective vertical temperature diffusivity coefficient[†]. The data presented allow one to give some instructive estimates. For natural measurements at the 'Echo' station the κ/u value, determined according to formula (11.32) is about 2×10^3 cm. The estimates for the averaged velocity of lowering of the thermoclines upper boundary give $u \sim 10^{-4} - 10^{-3}$ cm/s. Hence, using the previous estimate we find $\kappa \sim 10^{-1} - 1$ cm^2/s. It is important to note that this value is intermediate between the magnitude of the molecular thermal diffusivity coefficient of water, of order 10^{-3} cm^2/s, and the magnitude of the turbulent temperature diffusivity coefficient in the upper homogeneous layer, of order 10^3 cm^2/s. This estimate obtained for the effective vertical temperature diffusivity coefficient agrees with Munk's (1966) estimate for the value at an intermediate depth of the Pacific and corresponds to Stommel's (1958) global estimate. An analogous intermediate value is obtained for the data from laboratory experiments. This gives some basis for assuming (Phillips, 1977) that heat (and mass) exchange in the upper thermocline is governed by intermittent turbulence, related to the breaking of internal waves.

Internal waves are a phenomenon specific to fluid flows with stable stratification. When the stratification is 'discrete', i.e., if the fluid consists of several homogeneous layers and their densities decrease from bottom to top, the energy of the internal waves is mainly concentrated close to the interfaces between layers. In the case of continuous stable stratification the internal waves fill the whole space, if there exist in the fluid some sources of disturbance. Such sources always exist in the atmosphere and in the ocean (let us note even the tides, which provide in

---

[†] The systematic deviations in the upper part of this graph, corresponding to the lowest part of the thermocline, are apparently due to the processing procedure.

the ocean internal waves of huge amplitude and length), so that internal waves are a phenomenon occurring everywhere in the ocean and also in the atmosphere (see, e.g., Gossard and Hooke, 1975). The interaction between internal waves and turbulence is decisive for the whole pattern of turbulence in a stratified fluid.

11.5 Strong interaction of turbulence with internal waves. Deepening of the turbulent region

The strong nonlinear interaction of internal waves with turbulence leads to essentially new effects compared with those due to turbulence in neutral fluid flows. One such effect was demonstrated by Phillips (1976). The principal scheme of his remarkable experiment is as follows.

At the interface of two layers of liquids of different density internal waves are produced (Figure 11.3). After the establishing of steady waves turbulence was initiated by an oscillating grid at the upper boundary. Gradually the turbulent region propagated downwards from the upper boundary of the upper layer. Its lower boundary appears to be very sharp. The basic effect demonstrated was that when the turbulent front approached the interface of the two layers, the waves became smoothed out and disappeared practically instantaneously.

Note that in fact the tank in Phillips' experiment had the form of a circular cylinder, and turbulence was stirred up by a rotating disc bearing metallic ripples which covered the upper boundary of the upper layer, so that a rotating shear mean flow appeared. For theoretical consideration (Barenblatt, 1977) the problem statement was simplified: the fluid layers were assumed to be horizontally homogeneous, and turbulence was assumed to be stirred by an oscillating grid without shear. In fact, the simplified stirring scheme corresponds to a real experiment (Turner, 1968, 1973; Thompson and Turner, 1975).

Consider first the propagation of turbulence, stirred without the formation of shear flow, at the boundary $z = 0$ of an infinitely deep statistically horizontally-uniform layer of constant density. Since the flow is shearless, turbulence generation by a mean flow does not occur and the turbulent energy balance equation in the turbulent region can be written in the form

$$\partial_t b + \partial_z \left\langle \left(\frac{p'}{\rho} + \frac{u'^2 + u'^2 + w'^2}{2} \right) w' \right\rangle + \epsilon = 0 \qquad (11.34)$$

(the notation is the same as in subsection 10.2.2). We cannot neglect

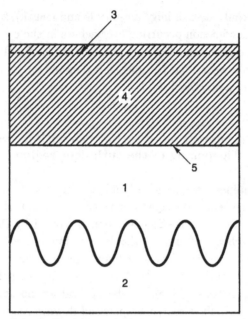

Figure 11.3. Principal scheme of Phillips' experiment: (1) light fluid
layer; (2) heavy fluid layer; (3) oscillating grid producing turbulence; (4)
turbulent region; (5) turbulent front.

the effect of turbulent energy diffusion here because this effect governs
the 'entrainment' – the extension of the turbulent region.

The relation for the diffusion flux of turbulent energy can be written
in the form

$$\left\langle \left(\frac{p'}{\rho} + \frac{u'^2 + u'^2 + w'^2}{2}\right) w' \right\rangle = -k_b \partial_z b \qquad (11.35)$$

where k_b is the turbulent energy exchange coefficient. We emphasize that
the relation (11.35) simply introduces a new quantity k_b not related to
any additional assumption.

Let us assume as before (subsection 10.2.2) the Kolmogorov similarity
hypothesis. This gives (within the frame of the b, l model)

$$k_b = l\sqrt{b}, \quad \epsilon = \frac{c_1 b^{3/2}}{l} \qquad (11.36)$$

where c_1 is again a constant; according to subsection 10.2.4 the estimated
range of its value is 0.06–0.085. Equation (11.34) then assumes the form

$$\partial_t b - l\sqrt{b} \partial_z b + c_1 \frac{b^{3/2}}{l} = 0. \qquad (11.37)$$

Equation (11.37) is relevant to the nonlinear heat conduction equations considered in chapter 2. Initial turbulence in the flow field is

absent, according to the experimental conditions, so the initial value of turbulent energy can be assumed equal to zero. The energy flux (per unit mass) q at the boundary is assumed to be constant (in fact, special organization of the experiment is required to maintain the energy flux at the boundary as time independent). Thus, the initial and boundary conditions appropriate to the problem under consideration take the form

$$b(z,0) = 0, \quad \left(l\sqrt{b}\partial_z b\right)_{z=0} = -q \tag{11.38}$$

so that the turbulent energy depends on the quantities t, q, z whereas the turbulence length scale l, which we assume constant over the depth of turbulent region, depends on t and q. This assumption of a length scale that is constant over depth is related to an important idea of Townsend (1976) concerning the governing role of large vortices in turbulent exchange processes.

As in the problems of a strong thermal wave, considered in chapter 2, and of turbulent-burst evolution, considered in chapter 10, the solution is represented by a finite function, different from zero in a finite region $0 \le z \le h(t)$ only.

Dimensional analysis gives

$$b = q^{2/3} f\left(\frac{z}{h(t)}\right), \quad l = \alpha_1 h(t), \quad h(t) = \xi_0 q^{1/3} t. \tag{11.39}$$

Here the constant α_1 can be considered as a known universal constant; the other constant ξ_0 is determined in the course of solution. The estimates performed in subsection 10.2.4 gave α_1 the range $\alpha_1 = 0.063 - 0.071$. Substituting (11.39) into (11.37) and (11.38) we obtain for the function f an ordinary differential equation

$$\frac{\alpha_1}{\xi_0}\frac{d}{d\xi}\left(\sqrt{f}\frac{df}{d\xi}\right) + \xi\frac{df}{d\xi} - \frac{c_1}{\alpha_1\xi_0}f^{3/2} = 0, \quad \xi = \frac{z}{h(t)}, \tag{11.40}$$

and boundary conditions

$$\left(\alpha_1\sqrt{f}\frac{df}{d\xi}\right)_{\xi=0} = -1, \quad f(1) = 0, \quad \frac{df^{3/2}(1)}{d\xi} = 0. \tag{11.41}$$

We recall that, like the solution to similar problems in chapters 2 and 10, the function f is different from zero for $0 \le \xi \le 1$, identically equal to zero for $\xi \ge 1$, and continuous, as is the derivative $df^{3/2}/d\xi$; the last two conditions (11.41) follow from these continuity conditions and the first condition (11.38). The constant ξ_0 is obtained as an eigenvalue of the problem (11.40), (11.41).

Therefore under the assumptions we have accepted the turbulent energy at a certain time instant t is different from zero in a region of finite depth $h(t)$ growing linearly with time, so that the entrainment speed appears to be constant:

$$h = \xi_0 q^{1/3} t, \quad \frac{dh}{dt} = \xi_0 q^{1/3}. \tag{11.42}$$

The bulk turbulent energy in the turbulent region per unit boundary area grows linearly with time,

$$\int_0^h b \, dz = \sigma q t, \quad \sigma = \xi_0 \int_0^1 f(\xi) d\xi, \tag{11.43}$$

whereas the integral dissipation rate per unit area is time independent:

$$\int_0^h \epsilon \, dz = \tau q, \quad \tau = \frac{c_1}{\alpha_1} \int_0^1 f^{3/2} d\xi. \tag{11.44}$$

Integrating (11.40) from $\xi = 0$ to $\xi = 1$ and using the boundary conditions (11.41) we find an obvious relation between τ and σ: $\sigma = 1 - \tau$.

We note that the problem under consideration could be solved within the framework of the b, ϵ turbulence model. However, in the latter model the dissipation-rate flux must somehow be prescribed at the boundary, and up to now we have no physical argument adequate to determine this quantity. This point is in general a difficult aspect of using the b, ϵ model. The case of turbulent-burst evolution considered in subsection 10.2.4 was a lucky exception because this quantity turned out to be equal to zero due to symmetry.

We return to Phillips' experiment. The dynamics of the internal waves at the interface between the heavy and light fluids is described by simple potential theory and, as is known, the energy density of the waves rapidly (exponentially) decays with distance from the interface. Therefore, the pattern of steady waves at the interface is not influenced by turbulence until the turbulent front, propagating with finite velocity, reaches the vicinity of the interface. It is plausible to assume that turbulence creates a turbulent flux $\mathbf{j}$ of wave motion energy E_w directed opposite to its gradient (Benilov, 1973):

$$\mathbf{j} = -k_w \operatorname{grad} E_w. \tag{11.45}$$

Here k_w is the coefficient of wave energy exchange, which is introduced in the same way as the exchange coefficients of mass, temperature, momentum, etc. were introduced earlier. Therefore, applying here for qualitative reasoning the Kolmogorov similarity hypothesis we obtain the estimate

$$k_w \sim l\sqrt{b}. \tag{11.46}$$

Thus, until the turbulent front approaches the region of large wave energy gradient, i.e., the close vicinity of the interface, the coefficient k_w remains equal to zero because the turbulent energy is zero; also, when the turbulent front is far from the interface, the gradient of the wave energy in the turbulent region is exponentially small. Therefore the wave energy flux is small everywhere. However, when the turbulent front approaches the close vicinity of the interface the wave energy exchange coefficient there becomes different from zero, and, because close to the interface the gradient of wave energy is not small, the turbulent flux of wave energy increases sharply (Figure 11.4). Due to this flux the wave energy becomes more or less uniformly distributed over the whole upper layer. Thus, when the turbulent front arrives at the interface, the wave producer becomes ineffective: the wave energy due to turbulence becomes uniformly distributed over the whole upper part of the tank and not concentrated close to the interface as it was before the turbulent region approached the interface. This is the reason for the smoothing of the waves at the interface.

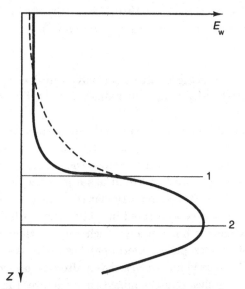

Figure 11.4. Wave energy redistribution over the whole volume of the turbulized fluid: (1) turbulent front; (2) interface.

It is clear that the special circumstances of the Phillips experiment are immaterial here: we have met, when analyzing this experiment, a phenomenon of a quite general nature. Indeed, we have seen how the turbulent region, extending in the ambient fluid, and sharply bounded by a front when the turbulence intensity in the ambient fluid is small, 'sucks' and redistributes the energy of waves approaching the region involved in wave motion. This leads to rapid smoothing of the waves. The phenomenon which we have just discussed is often observed when looking at the sea surface from the rear of a ship: the smooth mirror-like surface over the turbulent wake of the ship is in a sharp contrast with the ambient rippled surface. Also, those who practise water skiing observe a smooth track behind the speed-boat. This is due to the sucking of wave energy by the turbulent wake of the speed-boat. However, the skier feels also a frequent weak tremor due to turbulence.

Note that the remarkable fact of the presence of a sharp boundary between the turbulent and non-turbulent regions, which was basic in our previous considerations and has been obtained here from a mathematical model, has repeatedly been noted by experimentalists (see, especially, the paper by Kovasznay *et al.* (1970)[†], and Turner's (1973) monograph where a review of earlier work can also be found).

11.6 The breaking of internal waves and extension of mixed-fluid patches in a stably stratified fluid

The interaction of internal waves and turbulence in a fluid having strongly stable stratification is not restricted to the redistribution of wave energy by turbulence considered in the previous section. In fact, turbulence in a stably stratified fluid has a peculiar, intermittent, spatial structure. Observations, in particular those in the upper layer of the ocean, show that it is concentrated in 'patches' of turbulence, pancake-form layers extending horizontally much further than their thickness (Fedorov, 1976). These pancake-form patches occur as sharply bounded and relatively long-lived formations. Even after the decay of turbulence, the fluid in the patches remains mixed (homogenized) for a rather long time. Therefore the origin and evolution of mixed-fluid patches in stably stratified fluids is of considerable interest, in particular in connection with the oceanic fine structure and microstructure.

[†] In this paper some proposals are also discussed concerning the mathematical modelling of the sharp boundary.

The initial formation of mixed-fluid patches is related to the destruction of internal waves which can be due to various reasons: simple breaking (Woods, 1968), shear or convective instability (Belyaev and Gesenzwei, 1978; Korotaev and Panteleev, 1977), resonance interaction, etc. After the destruction of internal waves patches of mixed fluid appear (Figure 11.5) and extend, being gradually squeezed by the ambient stratified fluid and intruding into it. It is clear that the intrusions will be squeezed at the vertical level $z = z_1$ where the density of the stratified fluid is equal to the mixed-fluid density. The mixing of fluid within the patch establishes an excess pressure there, which forms the driving force of intrusion: owing to this force the extension of the patch goes on.

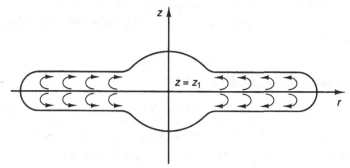

Figure 11.5. Mixed-fluid-patch intrusion into the ambient stratified fluid.

It is natural to distinguish three stages in the collapse-spreading of mixed fluid patches in a stably stratified fluid: (1) the initial stage, essentially a non-steady one, when the driving force of the intrusion substantially exceeds the drag forces; (2) an intermediate steady stage, when the driving force is in equilibrium with the form drag and the wave drag due to radiation of internal waves by spreading patches; and (3) a final, viscous stage, when internal waves are not radiated and the driving force is in equilibrium with the viscous drag. As we will see, each of these basic stages is governed by its own scaling laws. Between the basic stages there are some non-self-similar intermediate periods. After the viscous stage (3) the patch becomes mixed due to diffusion with the ambient fluid and disappears. Classification of these collapse stages of mixed-fluid patches goes back to the paper of Wu (1969). In this paper an important fact was observed: the volume of a patch remains constant from the early stages until the last observed extension period.

Let us consider these stages in sequence. In the first stage a particle in the mixed fluid will be falling, or being raised, to the plane of its density level $z = z_1$, with subsequent spreading along this plane. Therefore the variation rate of the patch's planform area S at the first stage is proportional to the current patch area times the speed of fluid inflow to the plane $z = z_1$. Under conditions of free fall the speed of fluid inflow is proportional to the product of time and free fall acceleration. In turn the free fall acceleration is proportional to N^2, where N is a basic stratification parameter, the so-called Brunt–Väisälä frequency, defined by the relation[†]

$$N^2 = \frac{g}{\rho}\left|\frac{d\rho}{dz}\right|. \tag{11.47}$$

N is equal to the frequency of the linear internal wave that arises when the stably stratified fluid at rest is weakly disturbed[‡]. The patch's vertical size is not large, therefore the variation of N in the vertical direction can be neglected. Therefore for the first stage the following relation is obtained:

$$\partial_t S \sim SN^2 t. \tag{11.48}$$

Integrating, we obtain for small Nt

$$\frac{(S - S_0)}{S_0} \sim N^2 t^2. \tag{11.49}$$

Here S_0 is the initial patch planform area. Thus, at the first stage the variation of the patch's characteristic length L is proportional to time squared:

$$\frac{L - L_0}{L_0} \sim N^2 t^2, \quad \frac{dL}{dt} \sim L_0 N^2 t. \tag{11.50}$$

Indeed, for the case of a patch having the form of an elongated cylinder with horizontal axis, typical for a wake, S is proportional to L and the relation (11.50) is obtained from (11.49) directly. However, if the length sizes of the patch planform are about equal, for instance if the patch planform is circular, $S \sim L^2$. When $L - L_0 \ll L_0$, $S - S_0 \sim L^2 - L_0^2 \sim 2(L - L_0)L_0$, whence and from (11.49) the relation (11.50) is obtained again. Relations of the form (11.50) were obtained by Wu (1969) by processing the data of laboratory experiments with wakes. They were confirmed by the semi-quantitative theoretical analysis of Kao (1976), and by numerical calculations. They appear to be valid up to $t \sim 2.5/N$.

At the intermediate stage the intrusion-driving force is in equilibrium

---

[†] In the relation (11.47) and in what follows g is the acceleration due to gravity, and ρ is the potential density, i.e., the density obtained when the fluid pressure is reduced adiabatically to a certain standard value.

[‡] A typical value of N is for the atmosphere is 10^{-2} s^{-1} and for the ocean is 10^{-3} s^{-1}.

with the form drag and the wave drag, so the governing parameters will be the basic stratification parameter N and the current mean patch thickness h. Dimensional analysis gives

$$\frac{dL}{dt} \sim Nh\,. \tag{11.51}$$

We now note that at this stage the patch extension rate is different for different planforms. In fact, as was said previously, the patch volume V is constant. Therefore for a wake-like patch having cross-sectional area S, $h \sim S/L$, and for a patch of circular planform $h \sim V/L^2$, and we obtain for these cases, correspondingly,

$$\frac{dL^2}{dt} \sim SN, \quad L \sim [SN(t - t_0)]^{1/2}\,, \tag{11.52}$$

$$\frac{dL^3}{dt} \sim VN, \quad L \sim [VN(t - t_0)]^{1/3}\,. \tag{11.53}$$

Here t_0 is the time of the origin of the second stage. Again, relations of the type (11.52) were obtained by Wu (1969) by processing laboratory experimental data, and confirmed by the semi-quantitative theoretical analysis of Kao (1976). They appear to be valid in the interval $3/N < t < 25/N$.

Let us consider the final, viscous stage of mixed-fluid patch collapse (Barenblatt, 1978a).

Under the assumptions made earlier in the section, the equation of mass conservation in the mixed fluid patch has the form

$$\partial_t h + \mathrm{div}\, h\mathbf{u} = 0\,, \tag{11.54}$$

where $h(x, y, t)$ is the local patch thickness, x, y are the rectilinear coordinates in the $z = z_1$ plane, t is the time, and $\mathbf{u}$ is the mixed-fluid velocity, averaged over a given vertical line inside the patch. For the derivation of this equation it is enough to consider the mass balance in an elemental volume, or particle, of the patch, shown in Figure 11.6, and to take into account that no ambient fluid entrainment and no viscous erosion of the patch is occuring at this stage.

To determine the average velocity $\mathbf{u}$ consider the system of forces acting on the elemental particle (see Figure 11.6). It is bounded by the patch's upper and lower surfaces and by a cylindrical surface around an elemental area δS in the plane $z = z_1$. The intrusion-driving force for this particle is due to the gradient of the quantity ph, where p is the excess pressure over that of the ambient stratified fluid, averaged over the vertical within the patch:

$$\mathbf{F}_m = -(\mathrm{grad}\, ph)\delta S\,. \tag{11.55}$$

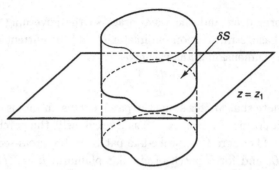

Figure 11.6. Elemental particle of a mixed-fluid patch.

Indeed, on each vertical line on the cylindrical boundary of the elemental particle a force $ph\mathbf{n}$ per unit contour length is acting here; $\mathbf{n}$ is a unit vector normal to the contour δC of the elemental area δS. The driving forces at the upper and lower patch boundaries are zero because the pressure there coincides with the pressure in the ambient stratified fluid. Applying Gauss's formula we obtain

$$\mathbf{F}_m = -\int_{\delta C} ph\,\mathbf{n}\,ds \simeq -(\mathrm{grad}\,ph)\delta S\,.$$

The drag force $\mathbf{F}_r$ acting on the particle can be calculated in the following way. At the stage under consideration the patch collapse is proceeding slowly, so that it is possible to neglect the accelerations within the patch and simplify the equations of motion. Integrating these simplified equations over the patch thickness we obtain that the drag force is proportional to the derivative over z of the actual fluid velocity at the upper or lower patch boundary. The velocity distribution can easily be found given boundary conditions at the upper and lower patch boundaries. Experiments and numerical calculations show that the fluid velocity at the upper and lower patch boundaries is much lower than the average velocity within the patch. This is due to the fact that above and below the patch the fluid motions are, during the collapse, in the reverse direction to the fluid motions inside the patch. Therefore, it can be assumed with adequate accuracy that the fluid velocity vanishes at the upper and lower patch boundaries. Integrating the simplified equations of motion under this condition it is possible to obtain the velocity distribution over the patch thickness, and, consequently, the relation for the drag force.

We will, however, derive this relation directly, using dimensional analysis. In fact, the drag force per unit planform area is governed by the local mean fluid velocity $\mathbf{u}$, the fluid's dynamic viscosity μ inside the patch, and the local patch thickness h. Dimensional analysis gives us that this force is proportional to $\mu\mathbf{u}/h$. Therefore the viscous drag force $\mathbf{F}_r$ acting on the elemental particle of the patch is equal to

$$\mathbf{F}_r = c\frac{\mu\mathbf{u}}{h}\delta S \tag{11.56}$$

where c is a constant, which can be found by comparison with the well-known solution to the problem of viscous fluid motion between plane walls, according to which $\mathbf{F}_r = 12\mu\mathbf{u}\delta S/R$. Therefore we obtain that $c = 12$.

From the equality of the drag force (11.56) and the driving force (11.55) on the particle, we obtain

$$\mathbf{u} = -\frac{h}{c\mu}\,\mathrm{grad}\,ph. \tag{11.57}$$

Only the mean excess pressure in the mixed fluid remains to be determined. The density distribution in the stratified fluid near the $z = z_1$ plane, along which the patch is extending, can be considered as linear one owing to the small patch thickness. Evidently the extending patch is symmetric with respect to the plane $z = z_1$, so this divides the patch into two symmetric parts. Let us denote by p_1 and ρ_1 the pressure and density of the stratified fluid at the level $z = z_1$. Then, integrating the hydrostatic equation we obtain a relation for the pressure variation in the stratified fluid:

$$p = p_1 - \rho_1 g(z - z_1) + \rho_1\frac{N^2(z - z_1)^2}{2}. \tag{11.58}$$

Here, as before, N is the Brunt–Väisälä frequency, so $N^2 = ag$ and $a = |d\rho/dz|/\rho_1$. Furthermore, the pressure at a patch boundary coincides with the pressure of the ambient stratified fluid at the same height. Thus, the pressures at the upper point, $z = z_1 + h/2$, and the lower point, $z = z_1 - h/2$ on a vertical line within the patch are equal to

$$p_u = p_1 - \rho_1 g\frac{h}{2} + \rho_1\frac{N^2 h^2}{8}, \quad p_l = p_1 + \rho_1 g\frac{h}{2} + \rho_1\frac{N^2 h^2}{8}. \tag{11.59}$$

The pressure inside the patch is distributed hydrostatically because the fluid density in the patch is constant and equal to ρ_1:

$$p = p_1 - \rho_1 g(z - z_1) + \frac{\rho_1 N^2 h^2}{8}. \tag{11.60}$$

The mean pressures averaged over a vertical line inside the patch and over a similar depth in the ambient fluid are correspondingly equal to

$$p_{ai} = p_1 + \rho_1\frac{N^2 h^2}{8}, \quad p_{as} = p_1 + \rho_1\frac{N^2 h^2}{24}. \tag{11.61}$$

The difference between these quantities gives the mean excess pressure over a given vertical line inside the patch:

$$p = p_{ai} - p_{as} = \frac{\rho N^2 h^2}{12} . \tag{11.62}$$

From relations (11.57) and (11.62) we obtain

$$\mathbf{u} = -\frac{\rho_1 N^2}{12c\mu} h \, \mathrm{grad} h^3 = -\frac{\rho_1 N^2}{4c\mu} h^3 \, \mathrm{grad} h . \tag{11.63}$$

After substituting this relation into the equation for mass conservation in the patch, we obtain for the patch thickness a nonlinear equation of heat conduction type:

$$\partial_t h = \kappa \Delta h^5 , \quad \kappa = \frac{N^2}{20c\nu} , \tag{11.64}$$

where Δ is the two-dimensional Laplace operator and ν is the kinematic viscosity of the fluid inside the patch. In particular, for symmetric rectinlinear and axisymmetric one-dimensional motions, equation (11.64) assumes the respective forms

$$\partial_t h = \kappa \, \partial_{xx}^2 h^5 \tag{11.65}$$

$$\partial_t h = \frac{\kappa}{r} \partial_r (r \partial_r h^5) \tag{11.66}$$

where x is a horizontal Cartesian coordinate and r is the horizontal polar radius. We met an analogous equation earlier, in chapter 2, when considering the initial stage of a nuclear explosion.

If the initial length sizes of the patch planform are about the same, it is natural to expect that the patch will become axisymmetric, presumably during the end of the second stage and the final viscous stage. The results of a numerical computation performed by E.I. Tikhomirova are presented in Figure 11.7: equation (11.64) was solved numerically[†] for a non-symmetric initial patch-thickness distribution $h(x, y, 0)$. As is seen, after even slight spreading the patch planform becomes indistinguishable from circular; therefore the patch collapse can be considered as axisymmetric and equation (11.66) can be used for its description. The equation for patch-volume conservation assumes the form

$$2\pi \int_0^\infty h(r, t) r \, dr = V = \mathrm{const} . \tag{11.67}$$

We are interested, first of all, in the intermediate-asymptotic stage of

---

[†] Numerical computation methods for the degenerate case – the vanishing of h at the boundary of the disturbed region – were used in this computation, developed by Samarsky and Sobol' (1963).

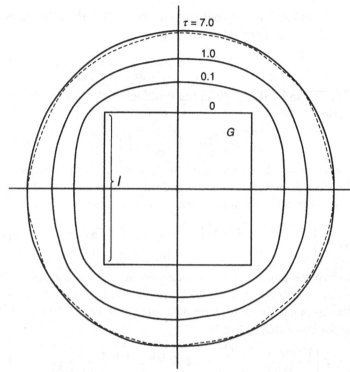

Figure 11.7. Numerical solution to equation (11.64) for a non-symmetric initial condition: $h(x,y,0) = h_0$ for points x,y inside the square G; $h(x,y,0) \equiv 0$ for points outside G; $\tau = \kappa h_0^4 t / l^2$. (The calculation was performed by E.I. Tikhomirova.)

patch extension when the patch planform diameter substantially exceeds its initial diameter. At this stage the random details of the initial patch-thickness distribution cease to be important. Therefore, as in Chapter 2, for an asymptotic description of the viscous stage of collapse we can assume an initial distribution in the form of a concentrated instantaneous source:

$$h(r,t_1) \equiv 0 \text{ at } r \neq 0, \quad 2\pi \int_0^\infty h(r,t_1) r\, dr = V \qquad (11.68)$$

where t_1 is the moment taken as the beginning of the self-similar viscous stage.

The problem solution here is completely analogous to that for the instantaneous heat source problem in Chapter 2. The solution to equation (11.66) under the initial condition (11.68) depends on the quantities

$t - t_1$, κ, V and r. Dimensional analysis shows that the solution is self-similar and can be represented in the form

$$h = \left[\frac{V}{2\pi\kappa(t - t_1)}\right]^{1/5} f(\zeta), \quad \zeta = \frac{r}{[V^4\kappa(t - t_1)/16\pi^4]^{1/10}}. \quad (11.69)$$

From (11.69) and (11.66) an ordinary differential equation follows for the function $f(\zeta)$:

$$\frac{d^2 f^5}{d\zeta^2} + \frac{1}{\zeta}\frac{df^5}{d\zeta} + \frac{\zeta}{10}\frac{df}{d\zeta} + \frac{1}{5}f = 0. \quad (11.70)$$

This equation, when multiplied by ζ, is reduced to an equation in total derivatives. After integrating and using the conditions (11.67) and (11.68) we obtain a very simple relation for the function f:

$$f(\zeta) = \begin{cases} (\frac{10^{1/5}}{6})^{1/4}\left(1 - \frac{\zeta^2}{\zeta_0^2}\right)^{1/4} & \text{for } 0 \leq \zeta \leq \zeta_0, \\ 0 & \text{for } \zeta \geq \zeta_0 = \frac{10^{3/5}}{2} \approx 2. \end{cases} \quad (11.71)$$

Therefore the patch has at each moment a finite radius $r_0(t)$; this is the peculiar property of the nonlinear equation (11.64) that distinguishes the latter from the linear heat conduction equation (cf. chapter 2). The relation for the patch radius is

$$r_0(t) = 2\left[\frac{V^4\kappa(t - t_1)}{16\pi^4}\right]^{1/10} \simeq 0.55 V^{4/10}\nu^{-1/10}\left[\frac{N^2}{(t - t_1)}\right]^{1/10} \quad (11.72)$$

which reveals that it grows with time very slowly. For the maximum patch thickness at time t we obtain the expression

$$h(0, t) = h_0(t) = \left(\frac{10^{1/5}}{6}\right)^{1/4}\left[\frac{V}{2\pi\kappa(t - t_1)}\right]^{1/5} \simeq 1.5\left[\frac{V\nu}{N^2(t - t_1)}\right]^{1/5}, \quad (11.73)$$

so that its decay in time also proceeds very slowly. It is of interest to consider the form of the patch's cross-section as represented in the reduced self-similar variables $r/r_0(t)$, $h/h_0(t)$ on Figure 11.8. The patch thickness is nearly constant and only at the very edge does it shrink abruptly, so that the patch has in fact a discoidal form.

Similarly, if the patch has a wake-like form, as do the patches of mixed fluid in the wakes of aircraft, often seen in the sky, equation (11.65) can be used, x being the horizontal coordinate along the wake's symmetry axis. The conservation condition for the volume of the mixed-fluid patch assumes the form

$$\int_{-\infty}^{\infty} h(x, t)dx = S = \int_{-\infty}^{\infty} h(x, 0)dx = \text{const}; \quad (11.74)$$

here S is the patch's cross-sectional area. The initial conditions for

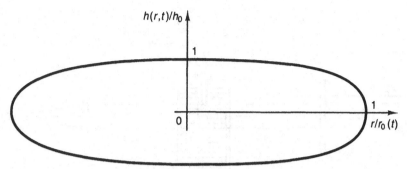

Figure 11.8. The form of a mixed-fluid patch at the viscous stage is nearly a discoidal.

an intermediate-asymptotic solution of instantaneous source type are represented in the form

$$h(x, t_1) \equiv 0 \text{ at } x \neq 0, \qquad \int_{-\infty}^{\infty} h(x, t_1)dx = S, \qquad (11.75)$$

and the solution, which may be obtained analogously to the previous one, is

$$h = \left[\frac{S^2}{4\kappa(t - t_1)} \right]^{1/6} f_1(\zeta), \quad \zeta = \frac{x}{[S^4\kappa(t - t_1)/16]^{1/6}} \qquad (11.76)$$

$$f_1(\zeta) = \begin{cases} A\left(1 - \frac{\zeta^2}{\zeta_0^2}\right)^{1/4} & \text{for } (0 \leq \zeta \leq \zeta_0) \\ 0 & \text{for } (\zeta \geq \zeta_0) \end{cases} \qquad (11.77)$$

$$\zeta_0 = 15^{1/6} \left[\frac{2\Gamma(7/4)}{\Gamma(1/2)\Gamma(5/4)} \right]^{2/3} \approx 1.72, \quad A = \left(\frac{\zeta_0^2}{15} \right)^{1/4} \approx 0.67.$$

so that the relation for the edge of the patch $x = \pm x_0(t)$ is

$$x_0(t) = \zeta_0 \left[S^4\kappa(t - t_1)/16 \right]^{1/6}, \qquad (11.78)$$

and maximum patch thickness decays with time as follows:

$$h(0, t) = h_0(t) = A \left[\frac{S^2}{4\kappa(t - t_1)} \right]^{1/6}. \qquad (11.79)$$

Comparison of the intermediate asymptotics (11.72) and (11.78) with the intermediate asymptotics for the previous stage, (11.52), (11.53), leads to an important conclusion: after transition to the viscous stage, spreading of the patch slows down sharply, so that if observed at time intervals that are not too large, the patch can seem invarying. The scaling law of 1/6 (11.76) was obtained in a semi-quantitative way by

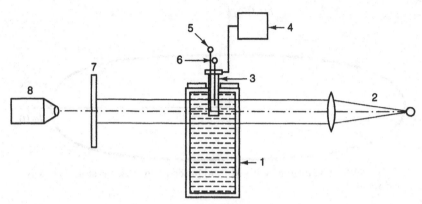

Figure 11.9. The scheme of the laboratory set-up for investigating the spreading of a mixed-fluid patch of circular planform (Zatsepin, Fedorov *et al.*, 1978). 1, the tank with stratified fluid; 2, shadowgraph; 3, glass tube; 4, DC electric motor; 5, mixer; 6, piston; 7, screen; 8, photographic or movie camera.

Maxworthy (1973) and, using the more detailed analysis presented here, by Barenblatt (1978a).

The relation (11.72) for the spreading of a patch of circular planform looks remarkably simple, so its experimental checking was important. This was performed in an elegant laboratory experiment by Zatsepin, Fedorov, *et al.* (1978). The following experimental scheme was used (Figure 11.9). Into an open Plexiglass tank filled by linearly density-stratified fluid a vertical hollow cylindrical tube was slowly introduced, under the fluid level. The fluid in the lower part of the tube, separated by a piston at rest, was then stirred by a special mixer. After allowing sufficient time for the fluid motion to decay the tube was slowly raised, releasing a patch of mixed fluid which started to spread into the ambient stratified fluid. The experimental set-up allowed the observation and recording of the two last stages of the patch's collapse. Photographs similar to those presented in Figure 11.10 demonstrate that, as expected, soon after the collapse starts, spreading of the patch slows sharply and the patch at this stage has the form of a disc with blunted edges, similar to that presented in Figure 11.8. The patch volume, kinematic fluid viscosity, fluid, and tube diameter were the same in all experiments. Therefore, if the relation (11.72) is valid then experimental points plotted in the coordinates

$$\log[2r_0(t)/D], \ \log[N^2(t - t_1)] \tag{11.80}$$

should lie on a single straight line with slope 1/10. This is confirmed by the graph of Figure 11.11. The value $t_1 = 10$ s was obtained by the

Figure 11.10. The shadow image of a spreading mixed-fluid patch (after Zatsepin, Fedorov *et al.*, 1978).

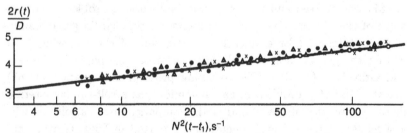

Figure 11.11. Results of experiment confirm the one-tenth law. •, $N = 0.63\ \text{s}^{-1}$, ×, $N = 1.00\ \text{s}^{-1}$, △, $N = 0.58\ \text{s}^{-1}$ (after Zatsepin, Fedorov *et al.*, 1978).

same method of extrapolation as in Figure 10.7. Thus, the one-tenth law (11.72) received experimental confirmation by Zatsepin, Fedorov *et al.* (1978). It turned out also that the duration of the final viscous stage is at least one order of magnitude longer than the duration of the two first stages; it reached in these experiments the value of about $300/N$.

It may be of interest for some applications to consider the case where the stratification is not continuous but stepwise: a two-layer fluid of density ρ_1 in the upper layer and density ρ_2 in the lower layer. Then for a viscous intrusion of intermediate density ρ_3 along the interface between the two fluid layers an equation similar to (11.64) is obtained but with a different exponent, 4 instead of 5 (Barenblatt, 1978a; Huppert, 1982):

$$\partial_t h = \kappa \Delta h^4, \quad \kappa = \frac{(\rho_2 - \rho_3)(\rho_3 - \rho_1)}{8c\mu(\rho_2 - \rho_1)} g. \tag{11.81}$$

In the special case $\rho_2 = \infty$ (viscous flow along the solid interface),

$$\kappa = \frac{(\rho_3 - \rho_1)g}{8c\mu}. \tag{11.82}$$

In the above-mentioned papers an axisymmetric solution of (11.81) was also presented, which corresponds to the spread of an initially concentrated viscous intrusion:

$$h = \left[\frac{V}{2\pi\kappa(t - t_1)}\right]^{1/4} f(\zeta), \quad \zeta = \frac{r}{[V^3\kappa(t - t_1)/8\pi^3]^{1/8}},$$

$$f(\zeta) = \begin{cases} (\frac{8^{1/4}}{5})^{1/3} \left(1 - \frac{\zeta^2}{\zeta_f^2}\right)^{1/3} & (0 \le \zeta \le \zeta_f), \\ 0 & (\zeta \ge \zeta_f = \frac{8^{3/4}}{\sqrt{3}}). \end{cases} \tag{11.83}$$

Thus the intrusion spreads along the interface of the layers according to

$$r_f(t) = 8^{3/4}3^{-1/2}[V^3\kappa(t - t_1)/8\pi^3]^{1/8}. \tag{11.84}$$

The solution (11.83) was compared with experiment in Huppert (1982); a good agreement was found.

In the paper Diez, Gratton and Gratton (1992) the same equation as (11.81) was derived and the axisymmetric focussing problem – an analogue of the Guderley very intense implosion problem in gas dynamics – was obtained. (For a rigorous investigation of the focussing problem for the general porous-medium equation see the paper by Aronson and Graveleau (1993).) The focussing problem for equation (11.81), like the Guderley problem in gas dynamics, has a self-similar intermediate asymptotics of the second kind. Comparison of this solution with a specially performed experiment, also reported in Diez, Gratton and Gratton (1992), leads to good agreement. It is an instructive example of the comparison of a self-similar solution of the second kind with an experiment that is physical, not numerical.

11.7 Several phenomena related to turbulence in a stably stratified fluid

Under strong stable stratification turbulence is suppressed owing to a large consumption of turbulent energy in work done against the buoyancy force. In fact, the equation for the turbulent energy balance in a stratified fluid has the form

$$\partial_t b = -\langle u'w'\rangle \partial_z u - \epsilon - \partial_z[\langle b'w'\rangle + \langle p'w'\rangle/\rho] - \overline{\rho'w'}g/\rho. \qquad (11.85)$$

The last term is new in comparison with the same equation for a homogeneous fluid and determines the energy consumption rate for the work against the buoyancy force. Owing to the large value of the acceleration due to gravity, g, the contribution of this term is substantial in spite of the fact that the magnitude of the density fluctuations ρ' is small in comparison with the mean density ρ, and the contribution of ρ' to other terms in the equation is small and can be neglected. Therefore the turbulence cannot exist under natural conditions for long times in the whole fluid volume (Monin and Ozmidov, 1981). In fact, turbulence is concentrated in pancake-form layers, vertically uniform owing to turbulent mixing and separated by relatively thin layers (laminar sheets, see Woods, 1968), where a sharp variation in temperature, salinity, density, electroconductivity, and sound speed, as well as other thermodynamic properties, and sometimes also in flow velocity, is concentrated. This laminated vertical structure of flow fields, which is revealed by steps on vertical profiles of temperature, density, and other thermodynamic properties, is called, depending on the vertical scale, the *microstructure* or *fine structure* of flow fields. Numerous field experiments specially performed from research vessels by the method of continuous vertical sounding have revealed that this phenomenon exists practically everywhere in the world's oceans.

Smoothing the distributions of density or temperature over depth one obtains a curve that characterises large-scale stratification (a similar smooth curve is obtained when averaging data over time). The stability parameter of the shear flow for a stably stratified fluid is the Richardson number

$$\text{Ri} = \frac{|d\rho/dz|\,g}{\rho(\partial_z u)^2} = \frac{N^2}{(\partial_z u)^2}. \qquad (11.86)$$

Instability in a shear flow is commonly related to reaching a critical value $\text{Ri}_{cr} = 0.25$ in the Richardson number: this was a theoretical result of Miles (1961, 1963). In stable flows $\text{Ri} > \text{Ri}_{cr}$. As a rule, large scale stratification is stable from the viewpoint of this criterion, so Ri > 0.25. However, if the microstructure is taken into account, intervals with Ri < 0.25 are observed on the graph Ri(z), i.e., instability regions. Apparently, in some of these intervals turbulence generation at the moment of sounding was observed[†]. This generation of turbulence is related to internal waves: the Richardson number is obviously minimal at the crests and hollows of internal waves. Moreover, other such mechanisms of turbulence generation are plausible, such as the breaking of internal waves, their resonance interaction, etc. For mixed-fluid patches in the atmosphere and ocean that appear after the breaking of internal waves, the rapid generation of a continuous spectrum, i.e., the formation of developed turbulence, is characteristic even for short times after breaking (Belyaev, Losovatsky and Ozmidov, 1975; SethuRaman, 1980). This is understandable, because after fluid stirring following the breaking of internal waves the fluid in the stirred region becomes density homogeneous, so the energy consumption rate due to work against the buoyancy flow (the last term of equation (11.85)) vanishes there. Therefore in the patch of mixed fluid formed in a shear flow conditions arise for the generation and rather long-term existence of turbulence at higher levels than in the ambient stratified fluid. In fact, the turbulent energy inflow due to the work done by the Reynolds stresses against the velocity gradient (the first term on the right-hand side of equation (11.85)) is consumed in the patch of mixed fluid by viscous dissipation into heat only, which is a relatively small effect; the third term on the right-hand side of equation (11.85), which is related to turbulent energy transfer, vanishes after integration over the patch thickness, so it does not influence the mean value of the turbulent energy. Thus, mainly owing to internal waves, the turbulence in fluids with strongly stable stratification has a specifically intermittent, 'archipelago' character. Furthermore, the patches of mixed fluid collapse basically in the same way as described in the previous section, where the turbulence inside the patch was not accounted for explicitly. The difference consists only in the time variation in the turbulent viscosity. Note its effect should not be substantial. This is seen already from the fact that the fluid viscosity within the patch enters the patch extension law (11.72) to the degree 1/10, so its variation over three orders of magnitude changes the radius only by a factor of

---

[†] Loss of stability in a steady homogeneous shear flow considered in Miles' (1961) paper, does not mean the origin of turbulence. Therefore the reaching of the critical value $\text{Ri}_{cr} = 0.25$ as the condition of turbulence generation should be considered with a certain care. Analysis of experiments in the ocean shows (Monin and Ozmidov, 1981) that the critical Richardson number related to the origin of turbulence is close to 0.1.

two. It is plausible therefore that the turbulent patches in a fluid with strong stratification also have the form of circular discs and expand very slowly at the final stage. It is also plausible that the microstructure and fine structure in the ocean are related to pancake-form patches of fluid having constant temperature and density. These patches are in fact patches of mixed fluid, of various length scales, which arise mainly due to the breaking of internal waves, and are at the longest, final, viscous, stage of their evolution. The above considerations also give us a plausible explanation (Barenblatt and Monin, 1979b) of discoidal formations in the atmosphere, which have repeatedly attracted attention in recent years ('flying saucers').

Indeed, under strong stable stratification in the atmosphere (Sethu-Raman, 1980) and in the ocean (Woods, 1968) internal wave breaking is going on and so patches of mixed air or fluid appear. The patches of mixed fluid collapse, reach a discoidal form, and under sufficiently strong shear become turbulized. In this respect the paper of SethuRaman (1980) is of special interest. SethuRaman observed and registered the breaking of internal waves, as well as the creation and long-term existence of localized turbulence patches, under conditions of strong temperature inversion (strongly stable stratification) and strong shear of wind velocity (Figure 11.12). It is essential to note that, as SethuRaman's observations showed, sharply localized turbulence arising as a result of internal wave breaking (Figure 11.12(c)) is not a burst at the very moment of wave breaking, which then rapidly decays (cf. Figure 11.12(a)). On the contrary, it develops rather slowly and is maintained for rather a long time (more than an hour). The creation and long-term maintenance of localized turbulence can be explained, according to what was said above, by the work done by the Reynolds stresses on the strong wind-velocity shear at small viscous dissipation, when turbulent-energy consumption by the work against the buoyancy force has disappeared. Indeed, after wave breaking and subsequent stirring the air density in the patch should be homogeneous or close to homogeneous.

Discoidal patches of stirred turbulized air in the atmosphere sometimes, although rather rarely, become visible. Such visibility can be explained in the following way.

Let us assume that for some reason a certain amount of suspended particles, e.g. an aerosol, appears in the atmosphere. Under ordinary conditions these particles fall more or less uniformly over the area (Figure 11.13). To support the particles in a suspended state it is necessary (see section 11.3) for the ratio of the velocity of free fall to the mean

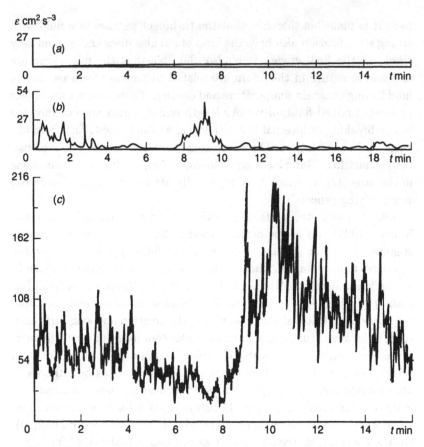

Figure 11.12. A turbulence-intensity measure – the dissipation rate ϵ – at different stages of internal wave breaking. (a) $z = 183$ m, no internal waves; (b) $z = 40$ m, the beginning of internal wave breaking, $z = 40$ m, (c) after internal wave breaking. (SethuRaman, 1980).

square velocity fluctuation to be less than a certain critical value. In the stratified air between the patches, the turbulence, according to what was said above, is low and cannot support the particles. In contrast, the turbulence in the patches can be sufficiently high to suspend the particles. The suspended particles increase the optical thickness of the patch and it becomes visible when the ambient air is illuminated by the Sun or Moon. Indeed, for the patch to be visible it is enough that its optical thickness τ is around one hundredth to one tenth. The necessary concentration of particles can be estimated by the relation $n = \tau/2\pi r^2 H$, where H is the effective patch thickness and r is the particle radius; we obtain for $H \sim 10^3$ cm and particles of size 3 μm (having a velocity of free fall

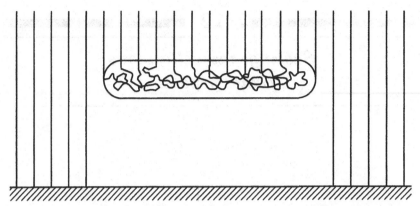

Figure 11.13. In a stratified fluid particles continue falling where the turbulence intensity is low, but are trapped by a discoidal patch where the turbulence is high.

of the order of several cm/s) a concentration n of order $10 - 10^2$ cm^{-3}. This value is quite realistic for dust clouds.

In the course of the extension of a patch its thickness is reduced, and the shear intensity necessary for supporting turbulence in the patch increases. When the available shear becomes insufficient to support turbulence, the patch drops the particles and becomes mixed with the ambient air.

An analogous explanation can be given of 'turbidity clouds' in the ocean.

12

Scaling: miscellaneous special problems

12.1 Mandelbrot fractals and incomplete similarity

12.1.1 The concept of fractals. Fractal curves

In the scientific and even popular literature of recent time *fractals* have been widely discussed. By fractals are meant those geometric objects, curves, surfaces, volumes and higher-dimensional bodies having a rugged form and possessing certain special properties of homogeneity and self-similarity. Such geometric objects were studied intensively by mathematicians at the end of the last and the beginning of present century, in particular in connection with the construction of examples of continuous nowhere-differentiable functions. To most physicists and engineers they seemed mathematical monsters having no applications in the problems of natural science and technology. In fact, it is not so.

The revival of interest in such objects and the recognition of their fundamental role in natural science and engineering is due to a series of papers by B. Mandelbrot and especially to his remarkable monographs (1975, 1977, 1982). Mandelbrot coined the very term 'fractal' and introduced the general concept of fractality. In the monographs and subsequent papers Mandelbrot and his followers showed that, contrary to what was expected, this concept, enclosing many known special examples, appeared to be exceptionally fruitful in such diverse and important applications as polymer physics, geomorphology, the theory of Brownian motion, turbulence theory, astrophysics, fracture theory and many others. In the monographs of Mandelbrot are referenced from a unified viewpoint the preceding works of other authors that relate to these topics[†].

In this section we will demonstrate the concept of Mandelbrot fractals for the simplest example, fractal curves. We will discuss the properties of homogeneity and self-similarity that make a continuous curve fractal, and we will show that the very idea of fractals is closely related to the incomplete similarity concept. A non-trivial example related to the fractality of respiratory organs will be presented in conclusion.

We will start from an instructive example. The famous English physicist L.F. Richardson (see Richardson, 1961; Mandelbrot, 1975, 1977, 1982) attempted (as a commission from the British Admiralty for which he worked at that time) to determine the length of the West coast of Britain. Richardson chose the following way of solving this problem, quite natural for ordinary smooth curves. He approximated the coastline on the most detailed map of Britain by a broken line composed of segments of constant length η, all vertices of which were situated on the coastline. The length L_η of this broken line was taken as an approximate value of the coastline's length corresponding to a given value of η. Richardson assumed at first that, when reducing η, corresponding values of the length of approximating broken lines L_η will tend to a definite finite limit that should be considered as the coastline's length.

Naturally, this is found to be the case when this method is used for a circle (Figure 12.1(a)). However, the West coastline turned out to be so rugged even down to the smallest scales available on the map, that the value L_η did not tend to a finite limit as the segment length η of the approximating broken line was reduced. Just the opposite: the value L_η tended to infinity as η tended to zero; throughout the available range of η the growth in L_η followed the power law (Figure 12.1(b)):

$$L_\eta = \lambda \eta^{1-D} \qquad (12.1)$$

where $\lambda > 0$ and D, $2 > D > 1$, were certain constants. For approximate lengths of separate parts of the same coastline between certain points of it relations of the form (12.1) were again obtained, with the same D, but a different, smaller value of λ. When such processing was performed later for the coastline of Australia (Figure 12.1(c)), the power-type (scaling) law remained, but at this time both λ and D were found to be different.

[†] Mandelbrot's success was so complete that nowadays people try to find fractals everywhere. I have to emphasize therefore that fractals in their turn are very special objects.

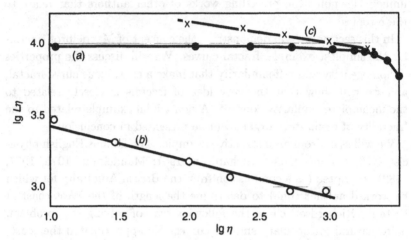

Figure 12.1. The dependence of the length L_η of a broken line, approximating the coastline, on the length η of its segment: (*a*) a circle, (*b*) the West coast of Britain, (*c*) the Australian coast. (After Mandelbrot, 1977).

As is seen, D is dimensionless; however, λ has the unusual dimension of length raised to a non-integer fractional power.

Formal passage to the limit $\eta \to 0$ in the relation (12.1) gives a result rather poor in its content: the length of the coastline determined by the method proposed, and even the length of each part of it, appeared to be infinite. The most essential point here is that if one tries to use a more detailed map in the hope that there the desired limit will appear, he or she will discover that such a map is somewhat meaningless because due to tides the very concept of the coastline is restricted by rather large scales.

It follows from (12.1) that the parts of the coastline can be compared by a certain measure of their extension, although not by their length. In fact, let us approximate two pieces of the coastline by broken lines with the same segment length η. In both cases relations of the form (8.1) are obtained:

$$L_\eta^{(1)} = \lambda^{(1)}\eta^{1-D}, \quad L_\eta^{(2)} = \lambda^{(2)}\eta^{1-D}. \tag{12.2}$$

As is seen the ratio $L_\eta^{(2)}/L_\eta^{(1)} = \lambda^{(2)}/\lambda^{(1)}$ does not depend on the segment length η. Therefore, the extent of certain parts of the coastline

can be compared, not, however, by their lengths but by the corresponding coefficients $\lambda^{\dagger}$. Thus, the very approach of measuring the extent of
the coastline by the same means as for smooth curves is found to be
inapplicable.

An adequate representation of the coastline appears to be not a smooth
curve of the type of a circle or ellipse, for which the limiting length of an
approximating broken line L_η is finite, but a curve of the von Koch triad
type, which was considered in the Introduction. Indeed, for the latter
curves the relation obtained by Richardson as an empirical equation,
(12.1), is also valid, if we denote $\lambda = 3d^{1+\alpha}$.

It follows from relation (12.1) that the number of segments of the
length η of the approximating broken line is

$$N_\eta = L_\eta/\eta = \lambda\eta^{-D}. \tag{12.3}$$

The quantity L_η, the length of the approximating broken line, tends
to infinity as $\eta \to 0$, because $D > 1$. Let us construct a square on
each segment of the approximating broken line. The total area of these
squares is equal to $N\eta^2 = \lambda\eta^{2-D}$. This quantity tends to zero as $\eta \to 0$,
because $D < 2$. Therefore, roughly speaking, the length of this curve
is infinite, and the area is equal to zero. However, a finite quantity,
different from zero, is obtained in the limit as $\eta \to 0$, if the number of
segments in the approximating broken line is multiplied by η raised to
a power D, intermediate between one and two:

$$N\eta^D = \lambda. \tag{12.4}$$

The constant D is called the *Hausdorff dimension* of the curve considered. For the Hausdorff dimension of the von Koch triad the double
inequality $1 < D < 2$ is valid. The same follows for the coastlines:
for the West coast of Britain $D \simeq 1.24$, and for the Australian coastline $D \simeq 1.13$ (Figure 12.1). Thus, for these curves also the Hausdorff
dimension lies between one and two. However, the length of the approximating broken line for ordinary smooth curves is bounded, so for
smooth curves $D = 1$. It is clear that the Hausdorff dimension is defined
not for all continuous curves, but only for those where the relation (12.1)
for the length of the approximating broken lines holds. Let us now give
a formal definition of fractal curves.

[†] As far as is known to the present author, L.F. Richardson did this measurement
by the commission of the British Admiralty, in connection with some argument
with the British Treasury concerning the necessary number of coastguards. The
regulation in force was that there should be one coastguard for a certain length
of coastline. Apparently, the answer did not completely satisfy the admirals, let
alone the Treasury.

The fractal curve is a continuous curve for which the Hausdorff dimension is strictly larger than unity:

$$D > 1. \tag{12.5}$$

From what was said it follows that the von Koch triad is a fractal curve. As was shown by Richardson's analysis, presented above, the coastlines of the British West coast and of Australia are also adequately approximated by fractal curves.

Note that the constancy of the Hausdorff dimension along the whole fractal curve is not necessary. To be fractal, a continuous curve should allow, in the vicinity of each point, a local approximation of the curve by broken lines whose lengths are represented by a relation of the type (12.1), where D is in general more than unity, but can be different for different points. A simple example of a fractal curve with varying Hausdorff dimension is obtained if we change at a certain step of the von Koch triad construction the elementary operation, making it different for various segments of the broken line. Thus, on the first segment we may, for example, leave the elementary operation as it was. On the second segment we may divide the segment into five equal parts and replace the second and fourth parts each by two sides of an equilateral triangle constructed on their base. On the third segment we may divide the segment by seven equal parts and replace the second, fourth, and sixth parts, etc. As a result on the first part of limiting curve the Hausdorff dimension will equal $\ln 4 / \ln 3$, as before, on the second part it will equal $\ln 7 / \ln 5$, on the third, $\ln 10 / \ln 7$, etc.

The consideration of fractals presented above for the example of fractal curves can be in principle extended very simply to surfaces, volumes, and, in general, to objects of arbitrary topological dimension. For instance, surfaces should be approximated by ones composed from tetrahedrons (see section 12.3).

12.1.2 Incomplete similarity of fractals

Let us explain the properties of fractal curves considered above which led to a scaling law of growth of the length of the approximating broken lines when reducing the segment length. Consider a continuous closed line, whose diameter (the distance between the furthest points) is equal to d. Approximate the curve considered by a broken line with constant segment length[†], its vertices being situated on the curve. It is clear

---

[†] Obviously, the last segment can have length less than η, but this does not matter for $\eta \to 0$.

that the number of segments N_η of the broken line depends on the dimensional parameters d and η. The quantity N_η is dimensionless; therefore, dimensional analysis gives in a standard way

$$N_\eta = f(d/\eta). \tag{12.6}$$

Let us take another approximating broken line with a lesser length of segment, $\xi < \eta$. Consider the portion of the basic curve between two neighbouring vertices of the first broken line and let us attempt to determine the number of vertices of the second curve contained in this portion. The von Koch triad has two very important properties. The first is *homogeneity*: all portions of the basic curve between neighbouring vertices of the first broken line generate equal numbers of segments of the second broken line. The second is *self-similarity* (the similarity of the curve to its part): the number of segments of the broken line with segment length ξ that are placed between neighbouring vertices of the broken line with segment length η depends only on the ratio η/ξ, not on η and ξ separately. We shall assume that the curve considered also possesses the properties of homogeneity and self-similarity. Now consider the broken line with segment length equal to the diameter of the curve. The number of segments of such a broken line is equal, according to (12.6), to $f(1)$. Thus, each segment of the broken line, equal to the diameter of the curve contains $f(d/\eta)/f(1)$ segments of the broken line with segment length η. According to the self-similarity property, the analogous expression with d replaced by η, and η by ξ holds also for the number $N_{\xi\eta}$ of segments of a second broken line, with segment length ξ, that are contained between two neighbouring vertices of the broken line with segment length η:

$$N_{\xi\eta} = f(\eta/\xi)/f(1). \tag{12.7}$$

However, due to the homogeneity of the curve the same relation holds for all segments of the broken line with segment length η, whose number is equal to $f(d/\eta)$. Therefore, on the one hand the total number of segments of the second broken line contained in the basic curve will be equal to

$$\frac{f(d/\eta)f(\eta/\xi)}{f(1)}. \tag{12.8}$$

On the other hand, owing to the same formula (12.6) the number of segments of the second broken line contained in the basic curve is equal to $f(d/\xi)$. Equating these two relations we obtain a functional equation for the function f:

$$f(x)f(y/x) = f(y)f(1) \tag{12.9}$$

where $x = d/\eta$ and $y = d/\xi$, so that $\eta/\xi = y/x$. We have met already a relevant functional equation, (1.6) in chapter 1, equation (1.6). Equation (12.9) is solved in an analogous way, and we obtain its solution in the form

$$f(x) = Cx^D \qquad (12.10)$$

where $C = f(1)$ and D are constants. Bearing in mind that $L_\eta = N_\eta \eta$, we obtain from (12.6) and (12.10)

$$L_\eta = \lambda \eta^{1-D} \qquad (12.11)$$

where $\lambda = CL^D$, i.e. the relation (12.1). For the von Koch triad, for instance, $C = 3$, $D = 1.2618\ldots$.

Thus, we have shown that for a continuous closed curve possessing the properties of homogeneity and self-similarity the scaling law (12.1) is valid, D having a constant value over the whole curve. If $D > 1$, the curve is fractal.

However, the requirements of homogeneity and self-similarity are very restrictive ones, so the set of curves exactly satisfying them is rather narrow. It is unlikely, for instance, that the curves representing the coastline, would satisfy this property exactly. We will show that the properties of homogeneity and self-similarity are not necessary for a curve to be fractal: the much weaker properties of *local homogeneity* and *local self-similarity* are sufficient.

The latter properties imply that for every point on such a curve a small vicinity Δ can be found where the curve has the following property. The leading term in the asymptotic representation of the number of vertices $N_{\xi\eta}$ of the approximating broken line with segment length ξ between two neighbouring vertices of the broken line with segment length η, depends, as $\eta/\xi \to \infty$ only on the ratio η/ξ. We may assume therefore that, with accuracy to small quantities, the number of vertices $N_{\xi\eta}$ of the broken line with segment length ξ inside a segment of the broken line with segment length η does not depend on the position of this latter segment within the vicinity Δ or on the values of η and ξ given that the ratio $\eta/\xi \gg 1$ is held fixed:

$$N_{\xi\eta} = f(\eta/\xi). \qquad (12.12)$$

Consider now a third broken line with still smaller segment length $\zeta \ll \xi$. Due to local homogeneity and self-similarity the number of its segments within one segment of length η positioned in the vicinity Δ is, on the one hand, equal, with accuracy to small quantities, to $f(\eta/\zeta)$. On the other hand it is equal to the product of the number $f(\eta/\xi)$ of segments of length ξ inside one segment of length η times the number of segments of length ζ inside one segment of length ξ. Equating the two expressions we obtain a functional equation for the function f,

$$f(x)f(y/x) = f(y), \qquad (12.13)$$

which coincides with equation (1.6). Here $x = \eta/\xi$ and $y = \eta/\zeta$. The solution to this equation is represented by $f(x) = x^D$. The value of D can be different for various parts of the basic curve. Going from the number of segments to the length of the broken line we obtain that for the lengths of approximating broken lines in the vicinity of each point of a continuous curve possessing the properties of local homogeneity and local self-similarity a scaling asymptotic relation is valid,

$$L_\xi = \eta^D \xi^{1-D} + \ldots \qquad (12.14)$$

where the dots refer to quantities small in comparison with the first term. If $D > 1$ this means that the curve considered is fractal.

We emphasize again that the set of curves having the properties of local homogeneity and local self-similarity is more rich than the set of curves of the van Koch triad type, which possess the very special properties of complete homogeneity and self-similarity.

Fractals reveal the properties of incomplete similarity. Let us show this for the same example, fractal curves. In fact, the length of a broken line of segment length ξ that approximates the continuous curve between two of its points a distance η apart depends on the dimensional parameters η and ξ. Dimensional analysis gives

$$L_\eta = \eta \Phi(\eta/\xi). \qquad (12.15)$$

For a smooth (or piecewise smooth) curve, as $\xi \to 0$, i.e. as $\eta/\xi \to \infty$, the function Φ tends to a finite limit, $\Phi(\infty)$. By definition the value

$$L_0 = \Phi(\infty)\eta \qquad (12.16)$$

is the length of a portion of a smooth curve between two of its points a the distance η apart. For instance, if the curve considered is a half-circle, having the segment η as its diameter, $\Phi(\infty) = \pi/2$. Therefore for smooth curves we have complete similarity in the parameter η/ξ at $\eta/\xi \to \infty$.

For fractal curves a finite limit of the function $\Phi(\eta/\xi)$ as $\eta/\xi \to \infty$ does not exist; the limit is equal to infinity. However, it follows from the relation (12.14) that as $\eta/\xi \to \infty$ the function $\Phi(\eta/\xi)$ has a scaling asymptotic representation,

$$\Phi(\eta/\xi) \simeq (\eta/\xi)^{D-1}, \qquad (12.17)$$

i.e., incomplete similarity occurs in the parameter η/ξ as $\eta/\xi \to \infty$. It is clear also that the Hausdorff dimension D depends on the geometric properties of the form of the curve and cannot be obtained from dimensional considerations.

We note in conclusion that, passing from geometric objects to the physical objects represented by them, we can simply identify fractality with incomplete similarity.

12.2 Example: scaling relationship between the breathing rate of animals and their mass. Fractality of respiratory organs

Every animal possesses a respiratory organ that absorbs oxygen from the environment. At first sight, the part of the organ that directly assimilates the oxygen may be schematically represented as a line (this will be the case if the respiratory organ consists of one or more whiskers), a surface, or some volume that, like a kidney, contains a multitude of small absorbent sacs separated by pores along which water, or air, containing oxygen moves. (As we shall see later, the actual situation is more complicated.) Thus, the respiratory organ of an animal can be characterized by some specific absorptive capacity β_n, i.e., the mass of oxygen absorbed per unit time per unit length $(n = 1)$, unit area $(n = 2)$, or unit volume $(n = 3)$ of the respiratory organ, respectively. Of course, the specific absorptive capacity β_n may depend on external conditions: the temperature, the composition of the environment, the time of day, the speed at which the animal is moving, etc.

Our basic assumption is that the breathing rate of the animal, i.e., the mass of oxygen it absorbs per unit time, R, is determined by the following quantities: the body mass of the animal, W, the density of its body ρ, and the specific absorptive capacity of its respiratory organ β_n. Thus, it is assumed that the external factors only exert an influence via this specific absorptive capacity of the respiratory organ. Hence

$$R = f(W, \rho, \beta_n). \tag{12.18}$$

We now note an important feature: the mass of oxygen absorbed and the body mass of the animal may be measured in independent units. This is natural, since the change in the body mass of the animal due to breathing in and out is small, and may be neglected. Thus, we choose the $LMTM_{O_2}$ class of systems of units, in which M_{O_2} is the dimension of the mass of oxygen absorbed, which is, according to what we have just said, independent of the dimension of the body mass of the animal, M.

The dimensions of the parameter R and of the governing parameters W, ρ, and β_n are, as may easily be seen, given by the following relations:

$$[R] = M_{O_2}T^{-1}, \quad [W] = M, \quad [\rho] = ML^{-3}, \quad [\beta_n] = M_{O_2}L^{-n}T^{-1}.$$
$$(12.19)$$

Thus, the number of governing parameters is equal to three; they all have independent dimensions, and, according to dimensional analysis, the relation (12.18) can be written in the following dimensionless form:

$$\Pi = \frac{R}{\beta_n(W/\rho)^{n/3}} = \text{const}. \tag{12.20}$$

Hence, we have

$$R = AW^\alpha, \quad A = \text{const}\,\beta_n\rho^{-\alpha}, \quad \alpha = n/3, \tag{12.21}$$

i.e., a scaling relationship between the breathing rate of an animal and its body mass. According to the foregoing, if the respiratory organ consists of whiskers, α should be equal to $1/3$; if the respiratory organ is a surface, α should be equal to $2/3$; finally, if the oxygen absorption occurs in a volume, α should be equal to unity.

Biological data (see Figure 12.2 for some instructive examples) indicate that a relationship of the form (12.21) is in good agreement with experiment. However, we can say that it is also fairly well-established that, as a rule, the exponent α lies between $2/3$ and unity, and very rarely takes on these extreme values.

We interpret this result in the following way. Respiratory organs do not have smooth surfaces like a sphere or an ellipsoid, but fractal surfaces, i.e., surfaces whose planar cross sections are fractal curves similar to the Koch curve discussed in the Introduction and in the previous sections of this chapter.

More precisely, we give the name fractal to a surface that, although continuous, has an extremely broken shape and possesses the property described as follows.

We inscribe polyhedra consisting of triangles with side length η within the surface, just as we inscribed broken lines within the Koch curve. Then, as η tends to zero, the total surface area of the polyhedron, S_η, does not approach a finite limit as it does for a smooth surface such as a sphere. On the contrary, S_η goes to infinity according to the scaling law

$$S_\eta = \sigma\eta^{2-D}, \tag{12.22}$$

where σ is some constant having dimension L^D, where D is a dimensionless constant greater than two, but less than three. The constant D is the *Hausdorff dimension* of the fractal surface.

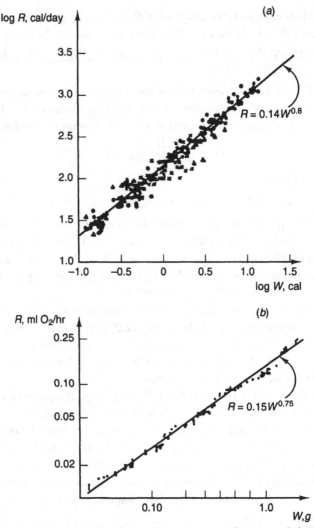

Figure 12.2. The rate R of oxygen absorption as a function of the body weight of sea animals. (*a*) Mysids: ▲, sea mysids; •, farm mysids; ×, laboratory mysids (Shushkina, Kus'micheva and Ostapenko, 1971). The rate of oxygen absorption and the body weight of the animals are expressed in energy units (which are convenient for biologists). (*b*) *Rhithropanopeus harrisii tredentatus* crabs (Nikolaeva *et al.*, 1975).

Clearly, the area of each face of the inscribed polyhedron is $(\sqrt{3}/4)\eta^2$. From this and (12.22), it follows that the number of faces in the inscribed polyhedron depends on η in the following way: $N_\eta = \text{const}\,\eta^{-D}$.

Thus, for fractal surfaces, the surface area of the inscribed polyhedron tends to infinity as the side length η tends to zero. At the same time, if a prism with altitude η is constructed on each face of the polyhedron, the total volume contained within all such prisms will be $V_\eta = N_\eta (\sqrt{3}/4) \eta^3$. It tends to zero as $\eta \to 0$, since $D < 3$. However, there is some measure of the fractal surface that is intermediate between area and volume; since the quantity $N_\eta \eta^D$ ($2 < D < 3$) approaches a finite limit as η goes to zero, this limit can be used as a measure of the surface mentioned above. Clearly, if the surface of the respiratory organ is a fractal, the specific absorptive capacity of this organ, β_n, should not be defined as the rate of absorption per unit area or volume but per unit of this intermediate dimension. Thus, β_n has dimension

$$[\beta_n] = [R]L^{-D} \, , \qquad (12.23)$$

where D is the fractal dimension of the respiratory organ; D is not restricted to integer values. A comparison of this result with the data presented above (Figure 12.2) and data obtained by other biologists indicates that self-consistency is obtained if one assumes that the respiratory organ is a fractal surface, with, for example, $D = 2.4$ for man and sturgeon, $D = 2.4$ for mysids (small sea animals) (Shushkina *et al.*, 1971), $D = 2.25$ for the *Rhithropanopeus harrisii tredentatus* crab (Nikolaeva *et al.*, 1975), etc. The idea that respiratory organs are fractals is also in qualitative agreement with the anatomical data. The analysis performed above was presented in the paper Barenblatt and Monin (1983).

12.3 The spreading of a ground-water mound

12.3.1 Mathematical model

This problem, briefly outlined in the Introduction, is of practical importance, in particular for ecology, and its analysis is instructive in many respects. Moreover, here every step can be justified rigorously. Therefore we will discuss it at some length. Consider a stratum consisting of a gas-filled porous medium (for example, sandstone) containing a ground-water (or liquid waste) mound on top of an underlying horizontal impermeable bed (Figure 12.3).

Under the influence of gravity, the mound spreads out and flows along the impermeable bed.

We shall discuss this problem using the following simplifying basic assumptions: (1) the mound is axially symmetric, and (2) the height of the mound $h(r,t)$ decreases with increasing radius from the very beginning of the motion.

The fluid motion in a porous medium is slow, so that the water pressure within the mound may be assumed to obey the hydrostatic law

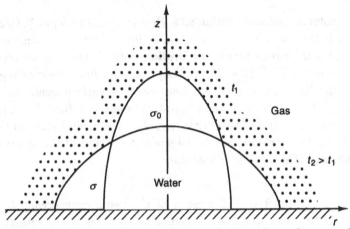

Figure 12.3. A schematic diagram showing the spreading of a ground-water mound.

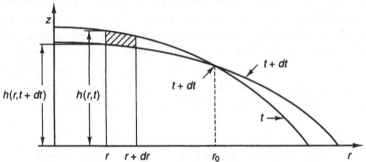

Figure 12.4. Derivation of the equation describing the spreading of a ground-water mound.

$p = \rho g(h - z)$, where ρ is the ground-water density and g is the gravitational acceleration (we shall neglect the gas pressure). Thus, the total head, i.e., the quantity $p + \rho g z$ well-known from hydraulics, remains constant throughout the height of the mound within the mound, and is equal to $\rho g h$. The fundamental law that is assumed to hold in the theory of filtration (fluid seepage in porous media) is the *Darcy law* (see Polubarinova–Kochina, 1962). According to this law, the flux of fluid (flow rate per unit area per unit time) is proportional to the gradient of the total head, i.e., in this case, the gradient of the quantity $\rho g h$. This implies that the radial component of the filtration flux is constant throughout the height of the mound (h is evidently independent of z). Thus, the total flux of water through a cylindrical surface of area $2\pi r h$ (Figure 12.4) is, according to Darcy's law,

$$q = -\frac{k}{\mu}(\partial_r p)2\pi r h = -\frac{k}{\mu}\rho g(\partial_r h)2\pi r h = -\frac{k\rho g\pi}{\mu}r\partial_r h^2 , \qquad (12.24)$$

where k is the permeability coefficient, a constant property of the porous medium with dimensions of area and of order 10^{-8} cm^2, and μ is the dynamic viscosity of the ground water; the coefficient of proportionality in the relation for Darcy's law is thus k/μ. Furthermore, we introduce the fractional volume of the stratum occupied by pores – the porosity of the medium – which we denote by m; m is generally of order 10^{-1}. It is important to note that when water enters an empty pore, it does not occupy its entire volume, but only some fraction σ; the remaining fraction is occupied by gas. At the same time, the ground water never completely flows out of an initially full pore; some fraction of the water is held back by capillary forces. We shall denote this fraction by σ_0 and assume that the quantities m, σ, and σ_0 are constants.

Thus the situation in those portions of the mound where the water is leaving previously filled pores ($r < r_0$, Figure 12.4) is different from that in those portions of the mound where the water is filling previously empty pores ($r > r_0$). The quantity $\partial_t h$ vanishes at the radius $r = r_0$; r_0 is obviously a function of time.

In principle, the basic equation for the mound-height h can be derived in the same way as was the equation of heat conduction in chapter 2. Namely, we derive the rate of change in the volume of water within a volume element of the mound lying between the cylinders with radii r and $r + dr$. The rate of change is due to the difference in the flux of water through these cylindrical surfaces. This quantity is then set equal to the rate of change in the volume of water due to the decrease in the water saturation from σ to σ_0 where the water is flowing out of the pores and the mound-height is decreasing ($r < r_0$ and $\partial_t h < 0$, Figure 12.4) and the increase in the water saturation from zero to σ where the water is flowing into the pores and the mound-height is increasing ($r > r_0$ and $\partial_t h > 0$).

As a result, the following equation for the mound-height is obtained:

$$\partial_t h = \begin{cases} (\kappa_1/r)\partial_r(r\partial_r h^2) & (\partial_t h \leq 0), \\ (\kappa/r)\partial_r(r\partial_r h^2) & (\partial_t h \geq 0), \end{cases} \tag{12.25}$$

where

$$\kappa_1 = \frac{k\rho g}{2m\mu(\sigma - \sigma_0)}, \quad \kappa = \frac{k\rho g}{2m\mu\sigma}. \tag{12.26}$$

The solution h must be continuous; the flux of water, which is proportional to the derivative $\partial_r h^2 = 2h\partial_r h$, must also be continuous. For non-zero h, the latter condition reduces to the requirement that the derivative $\partial_r h$ be continuous; at $h = 0$, the derivative $\partial_r h$ may have a discontinuity even though $\partial_r h^2$ remains continuous (cf. the discussion of very intense thermal waves in subsection 2.2.1).

Equation (12.25) must be supplemented by an initial condition. We assume that at the initial time the water is concentrated only within a certain finite region of the stratum, the water saturation in the mound is equal to σ and the total volume of water in the mound is equal to V. Without loss of generality, we may write the initial height distribution of the mound, $h(r, 0)$, in the form

$$h(r, 0) = \frac{V}{2\pi m \sigma r_*^2} h_0 \left(\frac{r}{r_*} \right), \qquad (12.27)$$

where r_* is the initial radius of the mound and $h_0(s)$, $s = r/r_*$, is a monotonic non-increasing dimensionless function such that $h_0(s)$ is equal to zero for $s \geq 1$ and the integral

$$\int_0^1 s h_0(s) ds \qquad (12.28)$$

is equal to unity.

Thus, we have a mathematical formulation of the problem at hand, and we seek a solution to (12.25) that is continuous, has a continuous derivative $\partial_r h^2$, and satisfies initial condition (12.27).

12.3.2 Dimensional analysis of the problem

The solution h depends on the following governing parameters: the independent variables r and t in (12.25), and the parameters κ_1, κ, $Q = V/2\pi m \sigma$, and r_*, which enter equation (12.25) and initial condition (12.27), so that

$$h = f(Q, \kappa_1, t, r, r_*, \kappa). \qquad (12.29)$$

The dimensions in the HLT class of the local ground-water mound-height h and of the governing parameters are as follows:

$$[h] = H, \quad [Q] = HL^2, \quad [t] = T,$$
$$[r] = [r_*] = L, \quad [\kappa_1] = [\kappa] = L^2 T^{-1} H^{-1}, \qquad (12.30)$$

where the dimension H of the mound-height may be assumed to be independent, since the ratio of the mound-height to the horizontal size of the mound does not appear explicitly among the governing parameters[‡].

Dimensional analysis yields

$$\Pi = \Phi(\Pi_1, \Pi_2, \Pi_3), \qquad (12.31)$$

[‡] Also, the mound-height can be replaced by the total head $\rho g h$ (which obviously has an independent dimension) without changing the subsequent relations. This we did in the Introduction. This is another good example of an additional group.

where

$$\Pi = \frac{h}{Q^{1/2}\kappa_1^{-1/2}t^{-1/2}}, \quad \Pi_1 = \frac{r}{Q^{1/4}\kappa_1^{1/4}t^{1/4}},$$

$$\Pi_2 = \frac{r_*}{Q^{1/4}\kappa_1^{1/4}t^{1/4}}, \quad \Pi_3 = \frac{\kappa}{\kappa_1}. \tag{12.32}$$

We are interested in long time scales, when the influence of the details of the initial conditions – the initial shape of the mound – has disappeared. It is therefore natural to discuss the limiting similarity relationships with respect to the parameters Π_1 and Π_2. Since the radius r can be chosen arbitrarily (in particular, it can be increased in such a way that Π_1 remains finite as t increases), we shall begin by assuming that the parameter Π_2 is small while the parameters Π_1 and Π_3 are finite and, in accordance with the recipe presented in chapter 5, we shall assume complete similarity in the parameter Π_2.

First of all, we note that if this assumption turns out to be valid, the solution obtained will correspond to that for an instantaneous, line-concentrated source. Thus, we seek a solution to (12.25) of the following form:

$$\Pi = \Phi(\Pi_1, 0, \Pi_3) = \Phi_1(\Pi_1, \Pi_3),$$

$$h = \frac{Q^{1/2}}{\kappa_1^{1/2}t^{1/2}}\Phi_1\left(\xi, \frac{\kappa}{\kappa_1}\right), \quad \xi = \Pi_1 = \frac{r}{Q^{1/4}\kappa_1^{1/4}t^{1/4}}. \tag{12.33}$$

Under the assumption of complete similarity in the parameter Π_2, dimensional analysis yields the following expression for the radius r_0 of the cylinder at which $\partial_t h$ vanishes:

$$r_0 = \xi_0\left(\frac{\kappa}{\kappa_1}\right)Q^{1/4}\kappa_1^{1/4}t^{1/4}, \tag{12.34}$$

where ξ_0 is a dimensionless quantity which might depend on κ/κ_1 but does not depend on time.

Substituting (12.33) and (12.34) into (12.25), we obtain for the function Φ_1 an ordinary differential equation with discontinuous coefficient:

$$\frac{d^2\Phi_1^2}{d\xi^2} + \frac{1}{\xi}\frac{d\Phi_1^2}{d\xi} + \frac{\xi}{4}\frac{d\Phi_1}{d\xi} + \frac{\Phi_1}{2} = 0, \quad (\xi \leq \xi_0),$$

$$\frac{\kappa}{\kappa_1}\left(\frac{d^2\Phi_1^2}{d\xi^2} + \frac{1}{\xi}\frac{d\Phi_1^2}{d\xi}\right) + \frac{\xi}{4}\frac{d\Phi_1}{d\xi} + \frac{\Phi_1}{2} = 0, \quad (\xi \geq \xi_0). \tag{12.35}$$

Equation (12.35) may easily be solved: we multiply both sides by ξ and obtain an equation in total differentials. Integrating, we have

$$\xi\frac{d\Phi_1^2}{d\xi} + \frac{1}{4}\xi^2\Phi_1 = 0, \quad (\xi \leq \xi_0),$$

$$\xi\frac{d\Phi_1^2}{d\xi} + \frac{\kappa_1}{4\kappa}\xi^2\Phi_1 = 0, \quad (\xi \geq \xi_0). \tag{12.36}$$

We have set the constants of integration equal to zero in both cases for the following reasons.

First, along the axis of symmetry, i.e., at $\xi = 0$, the dimensionless filtration flux, which is proportional to $\xi d\Phi_1^2/d\xi$ is equal to zero, and the dimensionless mound-height $\Phi_1(\xi, \kappa, \kappa_1)$ is finite; therefore, the constant in the first equation (12.36) is equal to zero.

Second, unlike the linear heat conduction equation, the nonlinear equation (12.25) possesses the following property: if the initial height distribution in the mound has a compact support, i.e., it is different from zero over only a finite region, then at all times t the height remains different from zero only over a finite region, i.e., for r less than some $r_1(t)$ (cf. the discussion of very intense thermal waves in chapter 2). Under the assumption of complete similarity with respect to the parameter Π_2, dimensional analysis yields

$$r_1 = \xi_1 \left(\frac{\kappa}{\kappa_1}\right) Q^{1/4}\kappa_1^{1/4}t^{1/4}. \tag{12.37}$$

From continuity of the mound-height and the filtration flux at $r = r_1$, it follows that the mound-height $h(r_1, t)$ and the filtration flux, i.e., $r_1\partial_r h^2(r_1, t)$, vanish simultaneously at $r = r_1$; thus, we find that Φ_1 and $\xi d\Phi_1^2/d\xi$ vanish simultaneously at $\xi = \xi_1$ and, consequently, that the constant in the second equation of (12.36) is also equal to zero.

For $\kappa_1 = \kappa$, i.e., zero residual water saturation $\sigma_0 = 0$, the assumption of complete similarity with respect to the parameter Π_2 is valid. The corresponding self-similar solution for an instantaneous concentrated source can be explicitly constructed in this case. We leave this to the reader; the resulting expression for the function Φ, which can be obtained exactly as in chapter 2 can be written in the form

$$\Phi_1 = \begin{cases} (8 - \xi^2)/16 & (\xi \le \xi_1 = \sqrt{8}), \\ 0 & (\xi \ge \xi_1), \end{cases}$$

so that for $\sigma_0 = 0$, i.e., $\kappa_1 = \kappa$, the limiting mound-height distribution for a concentrated, instantaneous source is given by the following relation (Barenblatt, 1952):

$$h = \frac{Q^{1/2}}{\kappa^{1/2}t^{1/2}}\Phi_1\left(\frac{r}{(Q\kappa t)^{1/4}}\right)$$
$$= \begin{cases} \frac{Q^{1/2}}{16\kappa^{1/2}t^{1/2}}\left(8 - \frac{r^2}{(Q\kappa t)^{1/2}}\right) & (r \le r_1 = \sqrt{8}(Q\kappa t)^{1/4}), \\ 0 & (r \ge r_1). \end{cases} \tag{12.38}$$

As is evident, the derivative $d\Phi_1^2/d\xi$ is continuous at $\xi = \xi_1$, although $d\Phi_1/d\xi$ has a finite discontinuity. This implies that the flux of liquid, which is proportional to $\partial_r h^2$, is also continuous.

For $\kappa_1 \ne \kappa$, i.e., $\sigma_0 \ne 0$, we reach a contradiction, since a change in sign of $\partial_t h$ should occur within the mound at a radius where the mound-height and the flux are non-zero. Thus, $r_0 < r_1$ and $\xi_0 < \xi_1$. However, setting $\xi = \xi_0$ in (12.36) and subtracting the second equation from the first, we find that $\Phi_1(\xi_0) = 0$ for $\kappa_1 \neq \kappa$, which is impossible. Thus, the assumption of complete self-similarity with respect to the small parameter Π_2 turns out to be incorrect for $\kappa_1 \neq \kappa$.

12.3.3 Intermediate asymptotics. Self-similar solution of the second kind

Following the general recipe presented in chapter 5, we shall now make the next-most-complicated assumption, incomplete similarity for $\kappa_1 \neq \kappa$, i.e., for $\sigma_0 \neq 0$. Since the two parameters Π_1 and Π_2 tend to zero with increasing time t, we shall assume that both these parameters are small. Under the assumption of incomplete similarity in the parameter Π_2, we obtain a scaling representation of the function Φ at Π_2 small:

$$\Phi(\Pi_1, \Pi_2, \Pi_3) = \Pi_2^\gamma \Phi_1 \left(\frac{\Pi_1}{\Pi_2^\delta}, \Pi_3 \right) ,$$

where γ and δ are constants.

From this expression and the general relation (12.31) for the solution, we find that

$$h = \frac{Q^{(2-\gamma)/4} r_*^\gamma}{(\kappa_1 t)^{(2+\gamma)/4}} \Phi_1 \left(\frac{r}{r_*^\delta (Q\kappa_1 t)^{(1-\delta)/4}}, \frac{\kappa}{\kappa_1} \right) ,$$

$$r_0 = \zeta_0 \left(\frac{\kappa}{\kappa_1} \right) r_*^\delta (Q\kappa_1 t)^{(1-\delta)/4}, \quad r_1 = \zeta_1 \left(\frac{\kappa}{\kappa_1} \right) r_*^\delta (Q\kappa_1 t)^{(1-\delta)/4} .$$

$$(12.39)$$

We now substitute (12.39) into the first equation of (12.25), and introduce the notations

$$\alpha = (2+\gamma)/4, \quad \beta = (1-\delta)/4,$$

$$A = Q^{(2-\gamma)/4} r_*^\gamma \kappa_1^{-\alpha}, \quad B = (Q\kappa_1)^\beta r_*^{1-4\beta}, \quad (12.40)$$

$$\zeta = r/Bt^\beta .$$

We thus have

$$h = \frac{A}{t^\alpha} \Phi_1 \left(\zeta, \frac{\kappa}{\kappa_1} \right) \quad r_0 = \zeta_0 B t^\beta, \quad r_1 = \zeta_1 B t^\beta, \quad (12.41)$$

$$\frac{d^2 \Phi_1^2}{d\zeta^2} + \frac{1}{\zeta} \frac{d\Phi_1^2}{d\zeta} + \frac{B^2}{A\kappa_1} t^{\alpha + 2\beta - 1} \left(\alpha \Phi_1 + \beta \zeta \frac{d\Phi_1}{d\zeta} \right) = 0, \quad (\zeta \leq \zeta_0).$$

$$(12.42)$$

However, since the function Φ_1 cannot depend directly on the time t, because it is a function of ζ only, the exponent $\alpha + 2\beta - 1$ must be equal to zero, so that $\alpha = 1 - 2\beta$. From this and (12.40), we find that $A = B^2/\kappa_1$. We finally find that the function Φ_1 satisfies the equation

$$\frac{d^2\Phi_1^2}{d\zeta^2} + \frac{1}{\zeta}\frac{d\Phi_1^2}{d\zeta} + (1 - 2\beta)\Phi_1 + \beta\zeta\frac{d\Phi_1}{d\zeta} = 0 \qquad (12.43a)$$

for $\zeta < \zeta_0$, i.e., by virtue of (12.41), for $(1 - 2\beta)\Phi_1 + \beta\zeta d\Phi/d\zeta > 0$, and the equation

$$\frac{d^2\Phi_1^2}{d\zeta^2} + \frac{1}{\zeta}\frac{d\Phi_1^2}{d\zeta} + \frac{\kappa_1}{\kappa}\left[(1 - 2\beta)\Phi_1 + \beta\zeta\frac{d\Phi_1}{d\zeta}\right] = 0 \qquad (12.43b)$$

for $(1 - 2\beta)\Phi_1 + \beta\zeta d\Phi_1/d\zeta < 0$, i.e., for $\zeta > \zeta_0$. From the requirement that the filtration flux be equal to zero at the axis of the mound, we obtain a boundary condition on the function Φ_1:

$$d\Phi_1/d\zeta = 0 \text{ for } \zeta = 0. \qquad (12.44)$$

The desired solution Φ_1 must be continuous and have a continuous derivative $d\Phi_1^2/d\zeta$.

Note that by renormalizing the constant B appropriately we can always ensure that the radius of the expanding mound corresponds to $\zeta = \zeta_1 = 1$. The quantities Φ_1 and $d\Phi_1^2/d\zeta$ must vanish at $\zeta = 1$, since the height of the mound and the filtration flux must be continuous at the edge of the mound and the function Φ_1 must be identically equal to zero for $\zeta \geq 1$. Consequently, approaching the point $\zeta = 1$ from $\zeta < 1$, we find from (12.43b) that, at $\zeta = 1$,

$$2\left(\frac{d\Phi_1}{d\zeta}\right)^2 + \frac{\kappa_1}{\kappa}\beta\frac{d\Phi_1}{d\zeta} = 0. \qquad (12.45)$$

From this, we obtain two conditions on the function Φ at $\zeta = 1$:

$$\Phi_1(1) = 0, \quad d\Phi_1/d\zeta = -\kappa_1\beta/2\kappa. \qquad (12.46)$$

For $\zeta > 1$, $\Phi \equiv 0$, so that the derivative $d\Phi_1/d\zeta$ undergoes a discontinuity at the point $\zeta = 1$, while $d\Phi_1^2/d\zeta$ is continuous.

However, the function Φ_1 must also satisfy condition (12.44):

$$d\Phi_1(0)/d\zeta = 0.$$

In general, the solution to a second-order equation cannot satisfy the three boundary conditions mentioned above for arbitrary β. However, there exist exceptional values of β – eigenvalues – for which all three boundary conditions are satisfied. Hence, the exponent β in the law describing the extension of the mound,

$$r_1 = Bt^\beta, \qquad (12.47)$$

is determined by solving an eigenvalue problem rather than via dimensional analysis; this is completely analogous to the situation for the examples in chapters 3 and 4. This eigenvalue problem may easily be

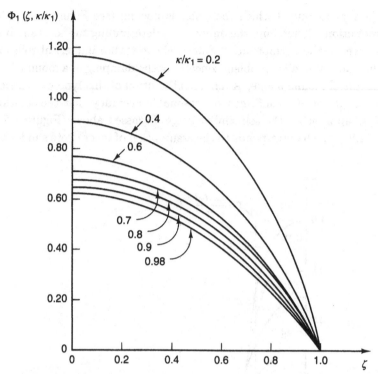

Figure 12.5. The function $\Phi_1(\zeta, \kappa/\kappa_1)$ obtained by solving the eigenvalue problem.

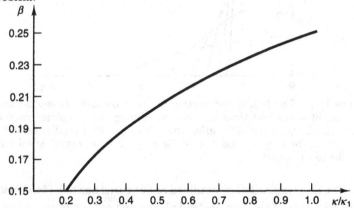

Figure 12.6. The relationship between β and κ/κ_1 obtained by solving the eigenvalue problem.

solved numerically; the results are shown in Figures 12.5 and 12.6. We find that the self-similar solution gives the form of the height distribution in the mound for large times up to a constant B. There is no integral mass-conservation law for this problem: some of the water is retained in the regions out of which the water is flowing (see Figure 0.7 in the Introduction). Therefore, the only way of determining the constant B is to match up the asymptotics obtained above with a numerical solution to the non-self-similar problem concerning the damping of a mound for a given initial mound shape. A numerical solution of this type was carried out, and its results confirmed to reasonable accuracy that the solution rapidly approaches the self-similar stage discussed above (Figures 12.7 and 12.8), which corresponds to the assumption of incomplete similarity.

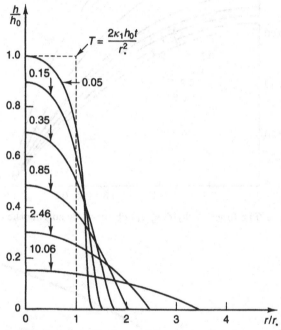

Figure 12.7. The height distribution in the ground-water mound as a function of space and time obtained by solving the complete non-self-similar problem numerically, using the following initial conditions at $t = 0$: $h \equiv h_0$ for $r \leq r_*$ and $h \equiv 0$ for $r \geq r_*$ (the mound is initially cylindrical in shape).

The limiting similarity law obtained above turns out to be informative. Indeed, by virtue of this law, instead of carrying out a numerical integration to large times one may terminate the numerical integration at the beginning of the self-similar stage and determine the constant B from the numerical calculation (at which point the calculation of the damping of the mound may be considered complete). However, for $\kappa_1 \neq \kappa$, i.e., $\sigma_0 \neq 0$, this self-similar solution no longer corresponds to an instantaneous source consisting of a finite mass of water concentrated at the axis

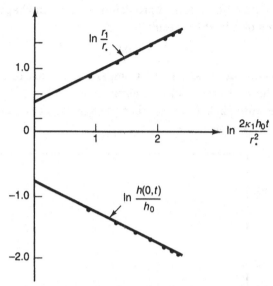

Figure 12.8. The maximum height and radius of the mound as functions of time obtained from a numerical solution to this problem rapidly approaches a scaling law with an exponent identical to that obtained by solving the eigenvalue problem.

(as for $\kappa_1 = \kappa$). Letting the initial radius of the mound, r_*, go to zero, we must (in accordance with (12.40), (12.41)) increase the initial mass of the fluid in the mound so that the product

$$Q r_*^{1/\beta - 4} \tag{12.48}$$

remains constant while the limiting regime remains invariant (i.e., B remains constant).

As one can see, the situation in the problem of the spreading of a ground-water mound is rather peculiar. For zero residual water saturation σ_0, i.e., $\kappa_1 = \kappa$, the exponent β is equal to $1/4$, and complete similarity holds in the limiting stage: this limiting stage corresponds to a concentrated, instantaneous source. For $\sigma_0 > 0$ and $\kappa_1 \neq \kappa$, the limiting stage possesses incomplete similarity and no longer corresponds to a concentrated, instantaneous source.

The solution presented above was obtained in the paper by Kochina, Mikhailov, and Filinov (1983) under the author's supervision.

12.3.4 Application of renormalization group and ϵ-expansion. Asymptotic conservation laws

In the paper by Chen, Goldenfeld and Oono (1991) the problem considered above was solved by the renormalization-group method in a way completely analogous to the Goldenfeld, Martin and Oono (1990) solution to the modified heat source problem presented in chapter 8.

The authors obtained the result

$$\frac{1}{4} - \beta = \frac{\epsilon}{16} + 0.048\epsilon^2 + O(\epsilon^3) \tag{12.49}$$

where we denote $\kappa_1 = \kappa(1 + \epsilon)$. Comparison of this result with the numerical solution to the nonlinear eigenvalue problem presented above is shown in Figure 12.9. It is seen that the agreement is very good even up to rather large values of ϵ.

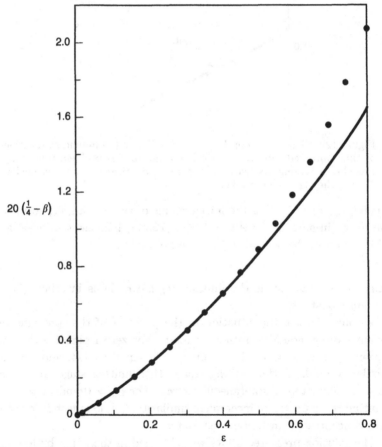

Figure 12.9. The eigenvalue β as a function of ϵ. The data points determined by numerically solving the nonlinear eigenvalue problem presented above are denoted by •. The continuous curve is the renormalization group calculation of equation (12.49) (from Chen, Goldenfeld and Oono, 1991).

We will obtain this result with accuracy to ϵ using a different technique, also demonstrated in chapter 8 for the problem of the modified heat source.

Indeed, the non-integrable conservation law can be easily obtained by integration of equation (12.25)

$$\frac{d}{dt}\int_0^{r_1} rh(r,t)dr = \kappa\epsilon \left(r\partial_r h^2\right)_{r=r_0(t)}. \tag{12.50}$$

Assuming again an asymptotic solution in a form invariant with respect to the renormalization group,

$$h = \frac{B^2}{\kappa_1 t^\alpha}\Phi(\zeta,\epsilon), \quad \zeta = \frac{r}{r_1(t)}, \quad r_1(t) = Bt^\beta,$$

$$r_0 = \zeta_0(\epsilon)Bt^\beta, \quad \alpha = 1 - 2\beta \tag{12.51}$$

we obtain

$$\int_0^{r_1} rh(r,t)dr = \frac{B^4}{\kappa_1 t^{\alpha-2\beta}}\int_0^1 \Phi(\zeta,\epsilon)\zeta d\zeta \tag{12.52}$$

and

$$\kappa\epsilon(r\partial_r h^2)_{r=r_0(t)} = \epsilon\frac{B^4}{\kappa_1 t^{2\alpha}}\left(\zeta\frac{d\Phi^2(\zeta,\epsilon)}{d\zeta}\right)_{\zeta=\zeta_0} \tag{12.53}$$

so that the conservation law (12.50) gives us, bearing in mind that $\alpha - 2\beta = 4(1/4 - \beta)$,

$$-\left(\frac{1}{4}-\beta\right)\int_0^1 \Phi(\zeta,\epsilon)\zeta d\zeta = \epsilon\left(\zeta\frac{d\Phi^2(\zeta,\epsilon)}{d\zeta}\right)_{\zeta=\zeta_0}. \tag{12.54}$$

For $\epsilon = 0$ we obtain easily from the solution (12.38)

$$\Phi(\zeta,0) = \frac{1}{16}(1-\zeta^2), \quad \zeta_0 = \frac{1}{\sqrt{2}}, \tag{12.55}$$

so that for small $\epsilon > 0$ the following relations hold:

$$\Phi(\zeta,\epsilon) = \frac{1}{16}(1-\zeta^2) + O(\epsilon), \quad \zeta_0 = \frac{1}{\sqrt{2}} + O(\epsilon). \tag{12.56}$$

Substituting (12.56) into (12.54) we obtain

$$\frac{1}{4} - \beta = \frac{\epsilon}{16}, \tag{12.57}$$

i.e., the result (12.49) with accuracy to $O(\epsilon^2)$.

It is interesting that for this problem there exists an invariant, asymptotic conservation law, similar to that noted by G.K. Batchelor and P.F. Linden in the problem of a turbulent burst (chapter 10). The current amount of water in an 'active' state, i.e., not retained by the porous medium, at the self-similar stage is equal, according to (12.52), to

$$Q(t) = 2\pi m\sigma \int\limits_0^{r_1} rh(r,t)dr = \text{const} \frac{B^4}{\kappa_1 t^{1-4\beta}}. \tag{12.58}$$

In the case of $\sigma_0 = 0$ it remains constant. If $\sigma_0 \neq 0$, as may be seen from (12.51) and (12.58), the quantity

$$Q(t)r_1^{1/\beta-4} = \text{const} \frac{B^{1/\beta}}{\kappa_1} = \text{const} \, Qr_*^{1/\beta-4}$$

remains constant with time.

The problem of ground-water mound extension considered in this section, and its generalization, has attracted the attention of mathematicians. From the mathematical viewpoint the basic problem is reduced to the initial-value problem (12.25)–(12.27) (the mathematicians have coined for this equation the term '*modified porous medium equation*'; it is of special interest because it combines two essentially different types of nonlinearity).

Hulshof and Vázquez (1993) performed a rigorous investigation of the family of self-similar solutions (12.41). They proved an important theorem according to which a solution belonging to this family is the asymptotics at $t \to \infty$ of the solution to the initial-value problem (12.25)–(12.27). And here again, as we have seen repeatedly in this book, the constant A is an integral of the solution. Indeed, replacing the initial condition (12.27) by the solution to (12.25)–(12.27), $h(r,t_1)$ for an arbitrary time t_1, does not change A.

Of special interest in this respect is the paper by Aronson and Vázquez (1993). Here a general method, called by the authors the implicit function theorem (IFT) method, was developed for determining the exponents in the self-similar variables (the anomalous exponents) as functions of the parameters. In the papers by Aronson and Graveleau (1993) and Angenent and Aronson (1993), the *focussing problem* was formulated and solved for the nonlinear heat conduction ('porous medium') equation

$$\partial_t \theta = \frac{\kappa}{r^\nu} \left(\partial_r r^\nu \partial_r \theta^{n+1} \right)$$

($\nu = 1$ for plane radial symmetry, $\nu = 2$ for spherical radial symmetry) repeatedly considered in this book. The focussing problem (see also Diez *et al.*, 1992) is a complete analogue for this equation of the Guderley very intense implosion problem in gas dynamics. The rigorous investigations performed in these papers are especially instructive because there is also non-uniform convergence to the self-similar asymptotics, as in the

Guderley and the von Weizsäcker–Zeldovich impulsive loading problems. Such rigorous investigation is inaccessible nowadays for the problems mentioned in gas dynamics.

Afterword

The technique of scaling, incomplete similarities, and intermediate asymptotics seems now to be a recognized well-established tool in frequent use. Any attempt to expose in this book, or even to reference here, every essential application of the technique would make the book a monster. I would like to make here only one exception, for a field from which I myself am far – particle physics. I was impressed indeed by the papers of A.M. Baldin and his associates (see the review Baldin and Didenko, 1990). In these papers perfect scaling laws were obtained embracing many orders of magnitude of the relevant quantities. I was delighted to see that the technique of intermediate asymptotics was substantially helpful here too.

However, there exist many problems of recognized importance where this technique has not yet been fully explained, but for which results of substantial value can be expected from its application. I will briefly mention two of them here.

The equation

$$\partial_t h + \kappa \partial_x (h^n \partial_{xxx}^3 h) = 0 \tag{1}$$

is basic in several models of special viscous flows that have attracted the attention of applied mathematicians and physicists (see the recent papers by L.P. Kadanoff and his colleagues and associates: Bertozzi, Brenner, Dupont and Kadanoff, 1993; Boatto, Kadanoff and Olla, 1993). Here h is a fluid-film thickness or relevant quantity, t is the time, x is the space coordinate, κ and n are positive constants. The case $n = 1$

is relevant to the flow in a Hele–Shaw cell, $n = 3$ corresponds to a surface-tension-driven viscous flow on a solid surface. There are some interpretations for other values of n too. However, for all n there remains an interesting problem that is not yet completely solved – the boundary conditions at the contact boundary line where $h = 0$ (for $n = 3$ see especially Dussan V. and Davis, 1986; Dussan V., Ramé and Garoff, 1991).

Equation (1) seems to be a genuine Pandora's box of incomplete similarities and self-similarities of the second kind, not less so, and perhaps even more so, than the 'porous medium' equation

$$\partial_t \theta = \kappa \partial_{xx}^2 \theta^{n+1} \tag{2}$$

and its modifications, considered repeatedly in the present book. Basic mathematical information concerning equation (1), including a definition of the generalized solutions and existence (but not uniqueness!) theorems, was obtained by Bernis and Friedman (1990). Some important results were announced recently by Beretta, Bertsch and Dal Passo (1995); Bertozzi and Pugh (1995). Bernis, Peletier and Williams (1992) performed a natural step: they constructed source-type self-similar solutions corresponding to a δ-function initial condition and possessing the property of continuity of h and of the flux $h^3 \partial_{xxx}^3 h$, which preserve the integral

$$\int_{-\infty}^{\infty} h(x,t)dx. \tag{3}$$

These solutions have finite support at any time, i.e., they are different from zero only at $|x| < a(t) < \infty$, and satisfy the conditions $\partial_x h(\pm a(t), t) = 0$. And here questions begin to arise. Firstly, it appears that such source-type solutions exist for $n < 3$ only. But what are the asymptotics as $t \to \infty$ of the solution to the non-self-similar initial-value problem if the fluid is concentrated initially ($h(x,0) \neq 0$) in a small but finite region, and n is greater than or equal to three? Furthermore, let us consider for $0 < n < 3$ the sequence of initial data

$$h(x,0) = \frac{3}{4a_0^3}(a_0^2 - x^2), \quad |x| \leq a_0; \quad h(x,0) \equiv 0, \quad |x| > a_0, \tag{4}$$

which are 'overturned parabolas' with the integral (3) equal to unity, converging as $a_0 \to 0$ to a δ-function. It seems natural to expect the source-type solutions to be asymptotics of the solutions with initial values such as (4) for $a(t) \gg a_0$. However, as is easily seen, (4) is a steady solution of (1) itself, possessing the property of continuity of h and of the flux $h^3 \partial_{xxx}^3 h$. So, what is the class of initial conditions for which the

Bernis–Peletier–Williams source-type solution is an intermediate asymptotics? And what are the asymptotics of the solutions for general initial conditions having a finite support?

Furthermore, let n again be in the interval $0 < n < 3$ where source-type solutions appear to exist. What is the asymptotics for 'dipole-type' initial and boundary conditions, similar to those considered in chapter 2? Indeed, in contrast to the porous medium equation (2), for equation (1) the dipole moment

$$\int_{-\infty}^{\infty} h(x, t)x\,dx$$

is not preserved. A self-similarity of the second kind can be expected here!

The following example belongs to an entirely different branch of applied mechanics: contact mechanics. Nearly a hundred years ago the great Swedish engineer J.A. Brinell proposed a test for estimating the quality and homogeneity of metals. The test is very simple and based on indentation of a metal surface with a small very hard indentor, often a ball. During the test the maximum load L and the diameter of the circular impression a are measured. (The details and references can be found in a comprehensive book, Johnson (1985).)

After the Brinell invention, activity began in areas closely related to the main subject of this book, scaling and intermediate asymptotics. This activity continues until the present time. Indeed, E. Meyer established, rather soon after Brinell's invention, that in an *intermediate range* of the ratio a/D, where D is the diameter of the indenting ball, something like

$$\sim 0.01 < \frac{a}{D} < \sim 0.5\,,$$

a *scaling law* is valid

$$\frac{L}{\pi a^2} = k\left(\frac{a}{D}\right)^{1/n}\,, \tag{5}$$

where k, n are material constants, k has the dimensions of pressure, and n is dimensionless. Much later H. O'Neill established that, in an ordinary uniaxial tensile test, the true (Cauchy) stress τ and the finite plastic strain (a logarithmic measure) γ, again in an intermediate interval, are also related by a scaling law

$$\tau = \kappa\gamma^{1/n} \tag{6}$$

where n is *the same dimensionless material constant* as in the Meyer scaling law (5), and κ is another material constant with the dimension of pressure, like k in (5). Thus, we have two scaling laws with the same powers. And a real sensation was the discovery by Tabor (1951) of a fundamental scaling-law relation between k and κ:

$$\frac{k}{\kappa} = \alpha\beta^{(1/n)} \tag{7}$$

where α and β are *universal constants*: $\alpha \simeq 2.8$, $\beta \simeq 0.4$. Universal means the same for all metals! The law (7) was obtained by careful processing of experimental data obtained for many metals. Apparently, similar scaling relations are valid for creep. Laws like the Tabor scaling law (7) seem to be very simple after the main guess; such laws always, in fact, mark something very fundamental.

Recently R. Hill, B. Storåkers, and their associates performed a fundamental analysis of the indentation problem using the classical similarity approach and extensive numerical analysis (see Hill, Storåkers and Zdunek, 1989; Hill, 1992; Storåkers, and Larsson, 1994). Their results are very instructive, but I still have a definite feeling that here is a fascinating self-similarity of the second kind. Its discovery will be interesting and important and will clarify the internal nature of these scaling laws.

A different, practically important, and challenging (for mathematicians) branch of solid mechanics – the theory of shells – should also be mentioned. The instructive recent paper by Andrianov and Kholod (1993) showed the perspectives of intermediate asymptotics in this theory.

The self-similarities and travelling waves mainly considered in this book form an important class of intermediate asymptotics. However, it should be emphasized that this class is only a very special one.

In general form the latter statement is trivial, but I want to mention here an important and natural complication which is by no means so. Apparently one of the first observations of this complication goes back to a rather early paper by Shkadinsky, Khaikin and Merzhanov (1971). In this paper a model of gasless combustion of solid fuel was studied numerically. (Later physical experiments were reported confirming the mathematical model.) At certain critical values of the parameters a conventional intermediate asymptotics of travelling-wave type became unstable, and periodic fluctuations began. Further departure from the stability threshold led to period doubling and the classical transition to chaos.

In a recent paper by G.I. Sivashinsky and his colleagues (Frankel, Roytburd and Sivashinsky, 1994) the reader can find an instructive example and a short but comprehensive review. More recently still

Brailovsky and Sivashinsky (1994) performed numerical experiments using the system

$$\partial_t\theta = \partial_{xx}^2\theta + CH(\theta - 1)$$
$$\partial_t C = -\mu CH(\theta - 1)$$

(8)

(H is the Heaviside step-function) under the boundary and initial conditions

$$Vt \leq x < \infty, \quad \theta(Vt, t) = 1 + \Delta, \quad \theta(\infty, t) \equiv 0,$$
$$C(\infty, t) \equiv 1, \quad C(x, 0) \equiv 1, \quad \theta(x, 0) \equiv 0$$

(9)

(the parameters $\mu > 0$ and $\Delta > 0$ were fixed.) It was found that for large V a travelling wave is the stable intermediate asymptotics. At a certain critical value $V = V_{cr}$ periodic oscillations begin, and with further reduction in V new frequencies appear. The nonlinear system (8) is convenient: the equations in fact turn out to be linear, and this simplifies the stability analysis. So, intermediate asymptotics makes contact with chaos and the transition to chaos, so fashionable now. How should we deal practically with periodic, quasi-periodic, or stochastic intermediate asymptotics? Naturally, the very concept of intermediate asymptotics should be modified, and nowadays it is clear how such modification should be done. By the way, a natural question arises – what is the counterpart of this behaviour for self-similarities? Periodicity in t for travelling waves should correspond to periodicity in log t for self-similarities. Natural, non-artificial, examples of such phenomena are clearly of interest. Apparently, one such example is known: the Forsyth effect, the jump-like extension of a fatigue crack (Forsyth, 1976; see also Barenblatt and Botvina, 1993).

Let us note another important point. We understand by intermediate asymptotics the behaviour of the solution in the space and time regions where the solutions no longer depend on the details of the initial and boundary conditions, yet the system is still far from being in its ultimate equilibrium state. It is natural to inquire further whether this intermediate-asymptotic regime is independent, in a certain well-defined sense, of the details of the model, i.e., of the equations themselves? In other words, whether this intermediate asymptotics is in a certain sense also *structurally stable*? This problem statement was advanced recently by N. Goldenfeld, Y. Oono and their students (Paquette, Chen, Goldenfeld and Oono, 1994; Paquette and Oono, 1994). They considered several instructive examples. These examples showed that the renormalization group may be viewed also as a means of extracting structurally stable predictions from model equations: i.e., that a class of models all give rise to the same behaviour at large times (and/or at large distances). Indeed, this notion is analogous to the situation familiar in critical phenomena, where a class of microscopic statistical mechanical models all give rise to the same thermodynamic behaviour close to the critical temperature. If the intermediate asymptotics happens to be *structurally unstable*, then, generally speaking, it ceases to be invariant, and, in particular may lose self-similarity or translational invariance. A variety of new phenomena may appear, including bifurcations, boundary layers, limit cycles, etc. A good example is the periodic phenomena in combustion mentioned before. Nevertheless, as is demonstrated by Chen, Goldenfeld and Oono (1994), the renormalization group remains an advantageous method of global asymptotic analysis.

These are only a few examples of current problems involving scaling, self-similarities, and intermediate asymptotics. When solved, they will produce new knowledge, but also new numerous challenges.

The best way I can see, under the circumstances, of finishing the book is with lines from Rudyard Kipling's *The Palace*:

'They sent me a Word from the Darkness. They whispered and called me aside.

They said 'The end is forbidden'. They said 'Thy use is fulfilled.

Thy Palace shall stand as that other's – the spoil of a King who shall build.'

I called my men from my trenches, my quarries, my wharves, and my sheers.

All I had wrought I abandoned to the faith of the faithless years.

Only I cut on the timber – only I carved on the stone:

'After me cometh a Builder. Tell him, I too have known!'

References

Ablowitz, M.J. & Clarkson, P.A. (1991). *Solitons, Nonlinear Evolution Equations and Inverse Scattering.* Cambridge University Press.

Abramowitz, M. & Stegun, I.A., eds. (1970). *Handbook of Mathematical Functions.* Dover Publications, New York.

Adamsky, V.B. (1956). Integration of a system of autosimulating equations for the problem of a short-duration shock in a cold gas. *Soviet Phys. Acoustics* **2** (1), 1–7.

Aldushin, A.P., Zeldovich, Ya.B. & Khudyaev, S.I. (1979). *Flame propagation in a reacting gas mixture.* Preprint, Institute of Chemical Physics, Chernogolovka.

Alexandrov, S.E. & Goldstein, R.V. (1993a). On the separated flows in the theory of plasticity. *Izvestiya, Russian Ac. Sci. Mech. Solids* **4**, 144–149.

Alexandrov, S.E. & Goldstein, R.V. (1993b). The flow of plastic mass in a converging channel: the singularities of a solution. *Doklady, Russian Ac. Sci.* **332** (3), 314–316.

Amit, D. (1989). *Field Theory, the Renormalization Group and Critical Phenomena,* 2nd edition, World Scientific, Singapore etc.

Andrade, E.N. da C. (1910). On the viscous flow of metals and allied phenomena. *Proc. Roy. Soc.* **A84**, 1–12.

Anderson, D.M. & Davis, S.H. (1993). Two-fluid viscous flow in a corner. *J. Fluid Mech.* **257**, 1–31.

Andrianov, I.V. & Kholod, E.G. (1993). Intermediate asymptotics in nonlinear dynamics of shells. *Izvestiya, Russian Ac. Sci., Mech. Solids* **2**, 172–7.

Andrushchenko, V.A., Barenblatt, G.I. & Chudov, L.A. (1975). Self-similar propagation of strong blast waves in the presence of radiation or energy release at the wave front. In *Progress in the Mechanics of Deformable Media, collection of papers dedicated to the 100th anniversary of B.G. Galerkin,* Shapiro, G.S. (ed.), 35–44, (in Russian) Nauka, Moscow.

Angenent, S.B. & Aronson, D.G. (1993). The focusing problem for the radially symmetric porous medium equation. *Euro. J. Appl. Math.*, (to appear).

Aronson, D.G. & Graveleau, J. (1993). A self-similar solution to the focusing problem for the porous medium equation. *Euro. J. Appl. Math.* **4**, 65–81.

Aronson, D.G. & Vázquez, J.L. (1993). Anomalous exponents in nonlinear diffusion. IMA Preprint No. 1165, University of Minnesota.

Bailey, R.W. (1929). *Transactions of Tokyo Sect. Meeting of the World Power Conference*, Tokyo.

Baldin, A.M. & Didenko, L.A. (1990). Asymptotic properties of hadron matter in relative four-velocity space. *Fortschritte der Physik* **38** (4), 261–332.

Barenblatt, G.I. (1952). On some unsteady motions of fluids and gases in a porous medium, *Prikl. Mat. Mekh.* **16** (1), 67–78.

Barenblatt, G.I. (1953). On the motion of suspended particles in a turbulent flow, *Prikl. Mat. Mekh.* **17** (3), 261–274.

Barenblatt, G.I. (1954). On limiting self-similar motions in the theory of unsteady filtration of gas in a porous medium and the theory of the boundary layer, *Prikl. Mat. Mekh.* **18** (4), 409–414.

Barenblatt, G.I. (1955). On the motion of suspended particles in a turbulent flow, occupying a half-space or a plane open channel of finite depth, *Prikl. Mat. Mekh.* **19** (1), 61–88.

Barenblatt, G.I. (1956). On certain problems of the theory of elasticity, which arise in the theory of the hydraulic fracture of the oil stratum. *Appl. Math. Mech. (PMM)* **20** (4), 475–486.

Barenblatt, G.I. (1959a). On the equilibrium cracks formed in brittle fracture. *Appl. Math. Mech. (PMM)* **23**: (3), 434–44; (4), 706–21; (5), 893–900.

Barenblatt, G.I. (1959b). The problem of thermal self-ignition. In: Gelfand, I.M. *Some Problems of the Theory of Quasi-Linear Equations,*, Russian Mathematical Surveys Vol. **14** (2), 137–42.

Barenblatt, G.I. (1962). Mathematical theory of equilibrium cracks in brittle fracture. *Adv. Appl. Mech.* **7**, 55–129.

Barenblatt, G.I. (1964). On certain general concepts of the mathematical theory of brittle fracture. *Appl. Math. Mech. (PMM)* **28** (4), 630–43.

Barenblatt, G.I. (1977). Strong interaction of gravity waves and turbulence. *Izvestiya, USSR Ac. Sci., Atmos. Oceanic Phys.* **13** (8), 581–83.

Barenblatt, G.I. (1978a). Dynamics of turbulent spots and intrusions in a stably stratified fluid. *Izvestiya, USSR Ac. Sci., Atmos. Oceanic Phys.* **14** (2), 139–45.

Barenblatt, G.I. (1978b). Self-similarity of temperature and salinity distributions in the upper thermocline. *Izvestiya, USSR Acad. Sci., Atmos. Oceanic Phys.* **14** (11), 820–23.

Barenblatt, G.I. (1979). *Similarity, Self-similarity, and Intermediate Asymptotics* (1st Russian edition Gidrometeoizdat, Leningrad, 1978; 2nd Russian edition, Gidrometeoizdat, Leningrad, 1982). Plenum, New York, London.

Barenblatt, G.I. (1983). Self-similar turbulence propagation from an instantaneous plane source. In *Non-linear dynamics and turbulence*, Barenblatt, G.I., Iooss, G. & Joseph, D.D. (eds.), 48–60, Pitman, Boston.

Barenblatt, G.I. (1987). *Dimensional Analysis*, Gordon and Breach, New York, London.

Barenblatt, G.I. (1991). On the scaling laws (incomplete self-similarity with respect to Reynolds number) for the developed turbulent flows in tubes. *C.R. Acad. Sci. Paris,* **313**, Sér. II, 107–12.

Barenblatt, G.I. (1993a). Scaling laws for fully developed turbulent shear flows. Part 1. Basic hypotheses and analysis. *J. Fluid Mech.* **248**, 513–20.

Barenblatt, G.I. (1993b). Intermediate asymptotics, scaling laws and renormalization group in continuum mechanics. *Meccanica* **28**, 177–83.

Barenblatt, G.I. (1993c). Some general aspects of fracture mechanics. In *Modelling of Defects and Fracture Mechanics,* Herrmann, G. (ed.), pp. 29–50. Springer-Verlag, Vienna, New York.

Barenblatt, G.I. (1994). *Scaling Phenomena in Fluid Mechanics.* Cambridge University Press.

Barenblatt, G.I. & Botvina, L.R. (1981). Incomplete self-similarity of fatigue in the linear range of crack growth. *Fatigue of Engineering Materials and Structures* **3**, 193–212.

Barenblatt, G.I. & Botvina, L.R. (1982). A note concerning power-type constitutive equations of deformation and fracture of solids. *Int. J. Eng. Sci.,* **20** (2), 187–91.

Barenblatt, G.I. & Botvina, L.R. (1983). The self-similarity of fatigue fracture. The damage accumulation. *Izvestiya, USSR Ac. Sci., Mech. Solids* **44**, 161–5.

Barenblatt, G.I. & Botvina, L.R. (1986). Similarity methods in mechanics and physics of fracture. *Physical and Chemical Mechanics of Materials* (1), 57–62.

Barenblatt, G.I. & Botvina, L.R. (1993). Self-oscillatory modes of fatigue fracture and the formation of self-similar structures at the fracture surface. *Proc. Roy. Soc. London* **A442**, 489–94.

Barenblatt, G.I. & Christianovich, S.A. (1955). On the failure of the roof in mineworkings. *Izvestiya, USSR Ac. Sci., Techn. Sci.* **11**, 73–86.

Barenblatt, G.I., Entov, V.M. & Ryzhik, V.M. (1990). *Theory of Fluid Flows Through Natural Rocks.* Kluwer Academic Publishers, Dordrecht, Boston, London.

Barenblatt, G.I., Galerkina, N.L. & Lebedev, I.A. (1992). Mathematical model of lower quasi-homogeneous oceanic layer: general concepts and sealing-off model. *Izvestiya, Russian Ac. Sci., Atmos. Oceanic Phys.* **28** (1), 68–74.

Barenblatt, G.I., Galerkina, N.L. & Lebedev, I.A. (1993). Mathematical model of lower quasi-homogeneous oceanic layer: effects of temperature and salinity stratification and tidal oscillations. *Izvestiya, Russian Ac. Sci., Atmos. Oceanic Phys.* **29** (4), 537–42.

Barenblatt, G.I., Galerkina, N.L. & Luneva, M.V. (1987). Evolution of turbulent burst. *Inzhenerno-Fizichesky Zh. (Zh. Eng. Phys.)* **53**, 733–40.

Barenblatt, G.I. & Gavrilov, A.A. (1974). On the theory of self-similar degeneracy of homogeneous isotropic turbulence. *Sov. Phys. JETP* **38** (2), 399–402.

Barenblatt, G.I. & Goldenfeld, N.D. (1995). Does fully developed turbulence exist? Reynolds number independence versus asymptotic covariance. *Phys. Fluids* **7** (12), 3078–3082.

Barenblatt, G.I. & Golitsyn, G.S. (1974). Local structure of mature dust storms. *J. Atmos. Sci.* **31**, 1917–33.

Barenblatt, G.I., Guirguis, R.H., Kamel, M.M., Kuhl, A.L., Oppenheim, A.K. & Zeldovich, Ya.B. (1980). Self-similar explosion waves of variable energy at the front. *J. Fluid Mech.* **99** (4), 811–58.

Barenblatt, G.I. & Krylov, A.P. (1955). On elasto-plastic regime of filtration. *Izvestiya, USSR Ac. Sci., Tech. Sci.* **2**, 14–26.

Barenblatt, G.I. & Monin, A.S. (1976). *Similarity Laws for Stratified Turbulent Shear Flows. Report of the Fourth All-Union Congress on Theoretical and Applied Mechanics*, 41, Naukova Dumka. Kiev.

Barenblatt, G.I. & Monin, A.S. (1979a). Similarity laws for turbulent stratified shear flows. *Arch. Rat. Mech. Anal.* **70** (4), 307–17.

Barenblatt, G.I. & Monin, A.S. (1979b). On a plausible mechanism of the phenomenon of discoidal formations in the atmosphere. *Doklady, USSR Ac. Sci.,* **246** (4) 834–837.

Barenblatt, G.I. & Monin, A.S. (1983). Similarity principles for the biology of pelagic animals. *Proc. Natl. Acad. Sci. USA* **80** (6), 3540–42.

Barenblatt, G.I. & Prostokishin, V.M. (1993). Scaling laws for fully developed turbulent shear flows. Part 2. Processing of experimental data. *J. Fluid Mech.* **248**, 521–9.

Barenblatt, G.I. & Sivashinsky, G.I. (1969). Self-similar solutions of the second kind in nonlinear filtration. *Appl. Math. Mech. (PMM)* **33** (5), 836–45.

Barenblatt, G.I. & Sivashinsky, G.I. (1970). Self-similar solutions of the second kind in the problem of propagation of intense shock waves. *Appl. Math. Mech. (PMM)* **34** (4), 655–62.

Barenblatt, G.I. & Vishik, M.I. (1956). On the finite speed of propagation in the problems of unsteady filtration of fluid and gas in a porous medium. *Appl. Math. Mech. (PMM)* **20** (4), 411–17.

Barenblatt, G.I. & Zeldovich, Ya.B. (1957a). On the dipole-type solution in the problem of a polytropic gas flow in a porous medium. *Appl. Math. Mech. (PMM),* **21** (5), 718–20.

Barenblatt, G.I. & Zeldovich, Ya.B. (1957b). On the stability of flame propagation. *Appl. Math. Mech. (PMM)* **21** (6), 856–9.

Barenblatt, G.I. & Zeldovich, Ya.B. (1971). Intermediate asymptotics in mathematical physics. *Russian Math. Surveys* **26** (2), 45–61.

Barenblatt, G.I. & Zeldovich, Ya.B. (1972). Self-similar solutions as intermediate asymptotics. *Ann. Rev. Fluid Mech.* **4**, 285–312.

Batchelor, G.K. (1967). *An Introduction to Fluid Dynamics*. Cambridge University Press.

Batchelor, G.K. & Linden, P.F. (1992). Discussion at the Fluid Mechanics Seminar, DAMTP, University of Cambridge.

Bechert, K. (1941). Differentialgleichungen der Wellenausbreitung in Gasen. *Ann. Phys.* **39** (5), 357–72.

Belyaev, V.S. & Gesentzwei, A.N. (1978). Shear instabilities of internal waves in the ocean. *Izvestiya, USSR Ac. Sci., Atmos. Oceanic Phys.* **14** (6), 459–63.

Belyaev, V.S., Losovatsky, I.D. & Ozmidov, R.V. (1975). Relationships between small-scale turbulence parameters and local stratification conditions in the ocean. *Izvestiya, USSR Ac. Sci., Atmos. Oceanic Phys.* **11** (7), 448–52.

Benbow, J.J. (1960). Cone cracks in fused silica. *Proc. Phys. Soc.* **B75**, 697–99.

Benilov, A.Yu. (1973). Generation of ocean turbulence by surface waves. *Izvestiya, USSR Ac. Sci., Atmos. Oceanic Phys.* **9** (3), 160–4.

Beretta, E., Bertsch, M. & Dal Passo, R. (1995). Non-negative solutions of a fourth-order nonlinear degenerate parabolic equation. *Arch. Rat. Mech. Anal.* **129** (2), 175–200.

Bernis, F. & Friedman, A. (1990). Higher order nonlinear degenerate parabolic equations. *J. Diff. Equations* **83** (1), 179–206.

Bernis, F., Peletier, L.A. & Williams, S.M. (1992). Source type solutions of a fourth order nonlinear degenerate parabolic equation. *Nonlinear Anal., Theory, Meth. Applic.* **18** (3), 217–34.

Bertozzi, A.L., Brenner, M.P., Dupont, T.F. & Kadanoff, L.P. (1993). Singularities and similarities in Interface Flows. Preprint, Ryerson Laboratory, University of Chicago.

Bertozzi, A.L. & Pugh, M. (1995). The lubrication approximation for thin viscous films: regularity and long time behavour of weak solutions. *Comm. Pure Appl. Math.* (in press).

Bertsch, M., Dal Passo, R. & Kersner, R. (1994). The evolution of turbulent bursts: the $b - \epsilon$ model. *Euro. J. Appl. Math.* **5** (4), 537–557.

Birkhoff, G. (1960). *Hydrodynamics, a Study in Logic, Fact, and Similitude*, 2nd edition. Princeton University Press.

Bluman, G.W. & Cole, J.D. (1974). *Similarity Methods for Differential Equations*, Springer-Verlag, New York, Heidelberg, Berlin.

Boatto, S., Kadanoff, L.P. & Olla, P. (1993). Travelling wave solutions to thin film equations. Preprint, Ryerson Laboratory, University of Chicago.

Bogolyubov, N.N. & Shirkov, D.V. (1955). On the renormalization group in quantum electrodynamics. *Doklady, USSR Ac. Sci.*, **103** (2), 203–6.

Bogolubov, N.N. & Shirkov, D.V. (1959). *Introduction to the Theory of Quantized Fields*. Wiley Interscience, New York, London.

Bose, E. & Bose, M. (1911). Über die Turbulenzreibung verschiedener Flüssigkeiten. *Physikalische Zeitschrift* **12** (4), 126–35.

Bose, E. & Rauert, D. (1909). Experimentalbeitrag zur Kenntnis der turbulenten Flüssigkeitsreibung. *Physikalische Zeitschrift* **10** (12), 406–9.

Botvina, L.R. (1989). *The Kinetics of Fracture of Structural Materials*. Nauka, Moscow.

Brailovsky, I. & Sivashinsky, G.I. (1994). Oscillatory propagation of reaction waves sustained by external sources of energy (to appear).

Bricmont, J. & Kupiainen, A. (1992). Renormalization group and the Ginzburg-Landau equation. *Comm. Math. Phys.* **150**, 193–208.

Bridgman, P.W. (1931). *Dimensional Analysis*. Yale University Press, New Haven.

Brushlinsky, K.V. & Kazhdan, Ya.M. (1963). On auto-models in the solution of certain problems of gas dynamics. *Russian Math. Surveys* **18** (2), 1-22.

Budiansky, B. & Carrier, G.F. (1973). The pointless wedge. *SIAM J. Appl. Math.* **25** (3), 378–87.

Bui, H.D. (1977). *Mécanique de la Rupture Fragile*. Masson, Paris.

Cane, B.J. & Greenwood, G.W. (1975). The nucleation and growth of cavities in iron during deformation at elevated temperatures. *Metal Sci.* **9** (2), 55–60.

Carothers, S.D. (1912). Plane strain in a wedge. *Proc. Roy. Soc. Edinburgh* **23**, 292–306.

Carrier, G.F. & Pearson, C.E. (1976). *Partial Differential Equations, Theory and Technique*. Academic Press, New York, San Francisco, London.

Carslaw, H.W. & Jaeger, J.C. (1960). *Conduction of Heat in Solids*, 2nd edition. Clarendon, Oxford.

Castaing, B., Gagne, Y. & Hopfinger, E.J. (1990). Velocity probability density functions of high Reynolds number turbulence. *Physica D.* **46** 177–200.

Chen, L.-Y. & Goldenfeld, N. (1992). Renormalization-group theory for the propagation of a turbulent burst. *Phys. Rev.* **A45** (8), 5572–4.

Chen, L.-Y., Goldenfeld, N. & Oono, Y. (1991). Renormalization-group theory for the modified porous-medium equation. *Phys. Rev.* **A44** (10), 6544–50.

Chen, L.-Y., Goldenfeld, N. & Oono, Y. (1994). Renormalization group theory for global asymptotic analysis. *Phys. Rev. Lett.* (submitted).

Chernyi, G.G. (1961). *Introduction to Hypersonic Flow* (trans. R.F. Probstein). Academic Press, New York.

Cole, J.D. (1968). *Perturbation Methods in Applied Mathematics*. Blaisdell, Toronto, London.

Cole, J.D. & Wagner, B.A. (1995) On self-similar solutions of Barenblatt's non-linear filtration equation. *Euro. J. Appl. Math* (in press).

Collins, R.E. (1961). *Flow of Fluids through Porous Materials*. Reinhold, New York.

Corino, E.R. & Brodkey, R.S. (1969). A visual investigation of the wall region in turbulent flow. *J. Fluid Mech.* **37** (1), 1-30.

Daniell, P.J. (1930). The theory of flame motion. *Proc. Roy. Soc.* **A126**, 393–402.

Dempsey, J.P. (1981). The wedge subjected to tractions: a paradox resolved. *J. Elasticity* **11**, 1–10.

Diez, J.A., Gratton, R. & Gratton, J. (1992). Self-similar solution of the second kind for a convergent viscous gravity current. *Phys. Fluids* **A4** (6), 1148–55.

Drazin, P.G. & Johnson, R.S. (1989). *Solitons: An Introduction*. Cambridge University Press.

Dryden, H.L. (1943). A review of the statistical theory of turbulence. *Quart. J. Appl. Math.* **1**, 7–42.

Dundurs, J. & Markenscoff, X. (1989). The Sternberg-Koifer conclusion and other anomalies of the concentrated couple. *ASME J. Appl. Mech.* **56**, 240–5.

Dussan, V., E.B. & Davis, S.H. (1986). Stability in systems with moving contact lines. *J. Fluid Mech.* **173**, 115–30.

Dussan, V., E.B., Ramé E. & Garoff, S. (1991). On identifying the appropriate boundary conditions at a moving contact line: an experimental investigation. *J. Fluid Mech.* **230**, 97–116.

Efimov, S.S. & Tsarenko, V.M. (1980). Self-similarity of the temperature distribution in the upper thermocline. *Izvestiya, USSR Ac. Sci., Atmos. Oceanic Phys.* **16** (6), 429–33.

Eilenberger, G. (1981). *Solitons. Mathematical Methods for Physicists*. Springer-Verlag, Berlin, Heidelberg, New York.

Einstein, H.A. & Ning Chen (1955). *Effects of Heavy Sediment Concentration Near the Bed on the Velocity and Sediment Distribution*. University of California MRD Series Report No. 8.

Entov, V.M. (1994). Private communication.

Fedorov, K.N. (1976). *Fine Thermohaline Structure of Ocean Water*. Gidrometeoizdat, Leningrad.

Fisher, R.A. (1937). The wave of advance of advantageous genes. *Ann. Eugenics*, **7**, 355–69.

Fordy, A.P. (ed.) (1990). *Soliton Theory: A Survey of Results*. Manchester University Press, Manchester, New York.

Forsyth, P.J.E. (1976). Some observations and measurements on mixed fatigue tensile crack growth in aluminium alloys. *Scripta Metall.* **10**, 383–6.

Fourier, J. (1822). *Théorie analytique de la chaleur*. Firmin Didot, Paris.

Frankel, M., Roytburd, V. & Sivashinsky, G. (1994). A sequence of period doubling and chaotic pulsations in a free-boundary problem modelling thermal instabilities. *SIAM J. Appl. Math.* (to appear).

Gad-el-Hak, M. & Corrsin, S. (1974). Measurements of the nearly isotropic turbulence behind a uniform jet grid. *J. Fluid Mech.* **62** (1), 115–43.

Gardner, C.S.J., Greene, J.M., Kruskal, M.D. & Miura, R.M. (1967). A method for solving the Korteweg-de-Vries equation. *Phys. Rev. Lett.* **19**, 1095–97.

Gell-Mann, M. & Low, F.E. (1954). Quantum electrodynamics at small distances. *Phys. Rev.* **95**, 1300–12.

Germain, P. (1973). Méthodes asymptotiques en mécanique des fluids. In *Dynamics of Fluids*, R. Balian & J.L. Peube (eds.), 7–147. Gordon and Breach, London, etc.

Germain, P. (1986a). Mécanique, tome I. Ecole Polytechnique, Ellipses, Paris.

Germain, P. (1986b). Mécanique, tome II. Ecole Polytechnique, Ellipses, Paris.

Ginzburg, I.S., Entov, V.M. & Theodorovich, E.V. (1992). Renormalization group method for the problem of convective diffusion with irreversible sorption. *Appl. Math. Mech. (PMM)* **56** (1), 59–96.

Goldenfeld, N. (1989). The approach to equilibrium: scaling and the renormalization group. Invited lecture at the Conference on Non-linear Phenomena, Moscow, USSR Ac. Sci., 19–22 September.

Goldenfeld, N. (1992). *Lectures on Phase Transitions and the Renormalization Group*. Addison-Wesley.

Goldenfeld, N., Martin, O. & Oono, Y. (1989). Intermediate asymptotics and renormalization group theory. *J. Scient. Comput.* **4**, 355–72.

Goldenfeld, N., Martin, O. & Oono, Y. (1991). Asymptotics of partial differential equations and the renormalization group. In *Proc. NATO Advanced Research Workshop on Asymptotics Beyond all Orders, La Jolla*, S. Tanvera (ed.). Plenum Press.

Goldenfeld, N., Martin, O., Oono, Y. & Liu, F. (1990). Anomalous dimensions and the renormalization group in a non-linear diffusion process. *Phys. Rev. Lett.* **65** (12), 1361–64.

Goldenfeld, N. & Oono, Y. (1991). Renormalization group theory for two problems in linear continuum mechanics. *Physica A.* **177**, 213–19.

Goldstein, S. (1939). A note on the boundary layer equations. *Proc. Camb. Phil. Soc.* **35**, 338–40.

Goldstein, R.V. & Vainshelbaum, V.M. (1978). Material scale length as a measure of fracture toughness in fracture mechanics of plastic materials. *Int. J. Fracture*, **14** (2), 185–201.

Golitsyn, G.S. (1973). *Introduction to the Dynamics of Planetary Atmospheres*. Gidrometeoizdat, Leningrad.

Gossard, E.E. & Hooke, W.H. (1975). *Waves in the Atmosphere*. Elsevier, New York.

Grebenev, V.N. (1992). The dynamic system that arises in the problem of the evolution of a turbulent layer of a homogeneous fluid. *Comput. Math. and Math. Phys.* **32** (1), 103–13.

Griffith, A.A. (1920). The phenomenon of rupture and flow in solids. *Phil. Trans. Roy. Soc. London* **A221**, 163–98.

Guderley, K.G. (1942). Starke kugelige und zylindrische Verdichtungsstösse in der Nähe des Kugelmittelpunktes bzw. der Zylinderachse. *Luftfahrtforschung* **19** (9), 302–12.

Häfele, W. (1955). Zur analytischen Behandlung ebener, starker, instationärer Stosswellen. *Z. Naturforschung* **10a** (9/10), 693–7.

Hahn, H.G. (1976). *Bruchmechanik*. Teubner, Stuttgart.

Hain, K. & Hörner, S.V. (1954). Instationäre starke Stossfronten. *Z. Naturforschung* **9a** (12), 993–1004.

Hanjalic, K. & Launder, B.E. (1972). A Reynolds stress model of turbulence, and its application to thin shear flows. *J. Fluid Mech.* **52**, 609–38.

Harmon, L.D. (1973). Recognition of faces. *Scientific American* **229** (5), 70–82.

Hastings, S.P. & Peletier, L.A. (1992). On a self-similar solution for the decay of turbulent bursts. *Euro. J. Appl. Math.* **3**, 319–41.

Heiser, F.A. & Mortimer, W. (1972). Effects of thickness and orientation on fatigue crack growth rate in 4340 steel. *Met. Trans.* **3**, 2119–23.

Hill, R. (1992). Similarity analysis of creep indentation tests. *Proc. Roy. Soc. London* **A436**, 617–30.

Hill, R., Storåkers, B. & Zdunek, A.B. (1989). A theoretical study of the Brinell hardness test. *Proc. Roy. Soc. London* **A423**, 301–30.

Hinch, E.J. (1991). *Perturbation Methods*. Cambridge University Press.

Hinze, J.O. (1959). *Turbulence. An Introduction to its Mechanism and Theory*. McGraw-Hill, New York, Toronto, London.

Hinze, J.O. (1962). Turbulent pipe-flow, in *Mécanique de la turbulence*, 63–76. Edition du Centre Nat. Rech. Sci. Paris.

Hulshof, J. (1993). Self-similar solutions of the $k - \epsilon$ model for turbulence. Report No. W93-11, Mathematical Institute, University of Leiden.

Hulshof, J. & Vázquez, J.L. (1993). Self-similar solutions of the second kind for the modified porous medium equation. Report No. W93-04, Mathematical Institute, University of Leiden.

Huppert, H.E. (1982). The propagation of two-dimensional and axisymmetric viscous gravity currents over a rigid horizontal surface. *J. Fluid Mech.* **121**, 43–58.

Inglis, C.E. (1922). Some special cases of two-dimensional stress and strain. *Trans. Inst. Naval Arch.* **64**, 253–8.

Irwin, G.R. (1949). Fracture dynamics, in *Fracturing of Metals*, 147–66. ASM, Cleveland, OH.

Irwin, G.R. (1957). Analysis of stresses and strains near the end of a crack traversing a plate. *J. Appl. Mech.* **24**, 361–4.

Irwin, G.R. (1958). Fracture, in *Handbuch der Physik, Bd VI*, pp. 551-90. Springer, Berlin.

Irwin, G.R. (1960). Fracture mode transition for a crack traversing a plate. *Trans. ASME, Ser. D* **82**, 417–25.

Jeffrey, A. & Kakutani, T. (1972). Weak non-linear dispersive waves: a discussion centered around the Korteweg-de-Vries equation. *SIAM Review* **14** (4), 582–643.

Johnson, K.L. (1985). *Contact Mechanics.* Cambridge University Press.

Kadanoff, L.P. (1966). Scaling laws for Ising model near T_c. *Physics* **2** (6), 263–72.

Kadanoff, L.P., Götze, W., Hamblen, D., Hecht, R., Lewis, E.A.S., Paleiauskas, V.V.I., Rayl, M., Swift, J., Aspnes, D. & Kane, J. (1967). Static phenomena near critical points: theory and experiment. *Rev. Mod. Phys.* **39** (2), 395–431.

Kalashnikov, A.S. (1987). Some problems of qualitative theory of non-linear second-order parabolic equations. *Russian Math. Surveys* **42**, 169–222.

Kamenomostskaya, S.L. (Kamin) (1957). On a problem of the theory of filtration. *Doklady, USSR Ac. Sci.*, **116** (1), 18–20.

Kamin, S., Peletier, L.A. & Vázquez, J.-L. (1991). On the Barenblatt equation of elasto-plastic filtration. *Indiana Univ. Math. J.* **40** (4), 1333–62.

Kamin, S. & Vázquez, J.L. (1992). The propagation of turbulent bursts. *Euro. J. Appl. Math.* **3**, 263–72.

Kanel', Ya.I. (1962). On the stabilization of solutions of Cauchy problems met within the theory of combustion. *Matem. Sb.* **59** (101), 245–88.

Kao, T.W. (1976). Principal stage of wake collapse in a stratified fluid: two-dimensional theory. *Phys. Fluids* **19** (8), 1071–4.

Kapitza, S.P., (1966). A natural system of units in classical electrodynamics and electronics. *Sov. Phys. Uspekhi* **9**, 184.

Karpman, V.I. (1975). *Non-linear Waves in Dispersive Media.* Pergamon, Oxford.

Keller, L.V. & Friedmann, A.A. (1924). Differentialgleichungen für die turbulente Bewegung einer kompressiblen Flüssigkeit. In *Proc. First Int. Congress Appl. Mech.*, pp. 395–405. J. Waltman Jr, Delft.

Kerchman, V.I. (1971). On self-similar solutions of the second kind in the theory of unsteady filtration. *Appl. Math. Mech. (PMM)* **35** (1), 158–62.

Kestin, J. & Richardson, P.D. (1963). Heat transfer across turbulent incompressible boundary layers. *Int. J. Heat. Mass Transfer* **6** (2), 147–89.

Kevorkian, J. & Cole, J.D. (1980). *Perturbation Methods in Applied Mathematics.* Springer-Verlag, New York, Heidelberg, Berlin.

Kim, H.T., Kline, S.J. & Reynolds, W.C. (1971). The production of turbulence near a smooth wall in a turbulent boundary layer. *J. Fluid Mech.* **50** (1), 133–60.

Kistler, A.L. & Vrebalovich, T. (1966). Grid turbulence at large Reynolds numbers. *J. Fluid Mech.* **26** (1), 37–47.

Kitaigorodsky, S.A. & Miropolsky, Yu.Z. (1970). *Izvestiya, USSR Ac. Sci., Atmos. Oceanic Phys.* **6** (2), 97–102.

Kline, S.J., Reynolds, W.C., Schraub, F.A. & Runstadler, P.W. (1967). The structure of turbulent boundary layers. *J. Fluid Mech.* **30** (4), 741–74.

Kochin, N.E., Kibel', I.A. & Roze, N.V. (1964). *Theoretical Hydromechanics,* Vol. 1. Interscience, New York. Vol. 2 available from ASTIA as AD129210.

Kochina, I.N., Mikhailov, N.N. & Filinov, M.V. (1983). Groundwater mound damping. *Int. J. Eng. Sci.* **21** (4), 413–21.

Kolmogorov, A.N. (1941). The local structure of turbulence in incompressible fluids at very high Reynolds numbers. *Doklady, USSR Ac. Sci.* **30** (4), 299–303.

Kolmogorov, A.N. (1942). The equations of turbulent motion of incompressible fluids. *Izvestiya, USSR Ac. Sci., Phys.* **6** (1-2), 56–8.

Kolmogorov, A.N. (1954). On a new variant of the gravitational theory of motion of suspended sediment. *Vestn. MGU* **3**, 41–5.

Kolmogorov, A.N. (1962). A refinement of previous hypotheses concerning the local structure of turbulence in a viscous incompressible fluid at high Reynolds number. *J. Fluid Mech.* **13** (1), 82–5.

Kolmogorov, A.N., Petrovsky, I.G. & Piskunov, N.S. (1937). Investigation of the diffusion equation connected with an increasing amount of matter and its application to a biological problem. *Bull. MGU* **A1** (6), 1–26.

Korotaev, G.K. & Panteleev, N.A. (1977). Experimental investigations of hydrodynamic instability in the oceans. *Oceanology USSR* **17** (6), 914–53.

Kovasznay, L.S.G., Kilens, V. & Blackwelder, R.F. (1970). Large-scale motion in the intermittent region of a turbulent boundary layer. *J. Fluid Mech.* **41** (2), 283–325.

Kulikovsky, A.G. & Lyubimov, G.A. (1965). *Magnetohydrodynamics.* Addison-Wesley, Reading MA.

Lagerstrom, P.A. & Casten, R.J. (1972). Some basic concepts underlying singular perturbation techniques. *SIAM Review* **14** (1), 63–120.

Landau, L.D. & Lifshitz, E.M. (1986). *Theory of Elasticity*, 2nd edition. Pergamon Press, London.

Landau, L.D. & Lifshitz, E.M. (1987). *Fluid Mechanics*, 2nd edition. Pergamon Press, London.

Launder, B.E. & Spalding, D.B. (1972). *Mathematical Models of Turbulence.* Academic Press, London.

Launder, B.E., Morse, A.P., Rodi, W. & Spalding, D.B. (1972). Prediction of free shear flows – a comparison of six turbulence models. NASA Report SP 321.

Launder, B.E. & Spalding, D.B. (1974). The numerical computation of turbulent flows. *Comp. Math. Appl. Mech. Eng.* **3**, 269–89.

Lax, P.D. (1968). Integrals of nonlinear equations of evolution and solitary waves. *Comm. Pure Appl. Math.* **21** (5), 467–90.

Liebowitz, H. (ed.) (1968a). *Fracture. An Advanced Treatise, Vol. I.* Academic Press, New York, London.

Liebowitz, H. (ed). (1968b). *Fracture. An Advanced Treatise, Vol II.* Academic Press, New York, London.

Lighthill, J. (1978). *Waves in Fluids.* Cambridge University Press.

Linden, P.F. (1975). The deepening of a mixed layer in a stratified fluid. *J. Fluid Mech.* **71** (2), 385–405.

Ling, S.C. & Huang, T.T. (1970). Decay of weak turbulence. *Phys. Fluids* **13** (12), 2912–20.

Ling, S.C. & Wan, C.A. (1972). Decay of isotropic turbulence generated by a mechanically agitated grid. *Phys. Fluids* **15** (8), 1363–9.

Loitsiansky, L.G. (1939). Some basic laws of isotropic turbulent flow. *Proc. Central Aero-Hydrodynamic Institute, Moscow* **440**, 3–23. (In Russian.) Translated as Loitsiansky, L.G. (1945). Some basic laws of isotropic turbulent flow. NACA Technical Memo. No. 1079.

Ma, S.-K. (1976). *Modern Theory of Critical Phenomena.* Benjamin/Cummings, Reading MA.

Mandelbrot, B. (1975). *Les objects fractals: forme, hasard et dimension.* Flammarion, Paris.

Mandelbrot, B. (1977). *Fractals, Form, Chance and Dimension.* W.H. Freeman and Co., San Francisco.

Mandelbrot, B. (1982). *The Fractal Geometry of Nature.* W.H. Freeman and Co., San Francisco.

Maxworthy, T. (1973). Experimental and theoretical studies of horizontal jets in a stratified fluid. In *Proc. Int. Symposium on Stratified Flows, Novosibirsk, 1972,* 611–18. Am. Soc. Civ. Eng., New York.

McMahon, T.A. (1971). Rowing: a similarity analysis. *Science* **173**, 23 July 1971, 349–51.

Meyer, F. (1955). Zur Darstellung starker Stossfronten durch Homologie-Lösungen. *Z. Naturforschung* **10a** (9/10), 693–7.

Migdal, A.B. (1977). *Qualitative Methods in Quantum Theory.* W.A. Benjamin, Reading, MA.

Miles, J.W. (1961). On the stability of heterogeneous shear flow. *J. Fluid Mech.* **10** (4), 496–508.

Miles, J.W. (1963). On the stability of heterogeneous shear flow. *J. Fluid Mech.* **16** (2), 209–27.

Millionshchikov, M.D. (1939). Decay of homogeneous turbulence in a viscous incompressible fluid. *Doklady, USSR Ac. Sci.* **22** (5), 236–40.

Moffatt, H.K. (1964). Viscous and resistive eddies near a sharp corner. *J. Fluid Mech.* **18** (1), 1–18.

Moffatt, H.K. & Duffy, B.R. (1980). Local similarity solutions and their limitations. *J. Fluid Mech.* **96** (2), 299–313.

Monin, A.S. (1950). Turbulence in the atmospheric surface layer. *Coll. Sci. Inform. Hydromet. Science USSR, Moscow* (1), 13–27.

Monin, A.S. & Obukhov, A.M. (1953). Dimensionless characteristics of turbulence in the surface layer of the atmosphere. *Doklady, USSR Ac. Sci.* **93** (2), 223–6.

Monin, A.S. & Obukhov, A.M. (1954). Basic relationships for turbulent mixing in the surface layer of the atmosphere. *Proc. Inst. Theor. Geophys., USSR Ac. Sci.* **24** (151), 163–87.

Monin, A.S. & Ozmidov, R.V. (1981). *Oceanic Turbulence.* Gidrometeoizdat, Leningrad.

Monin, A.S. & Yaglom, A.M. (1971). *Statistical Fluid Mechanics. Mechanics of Turbulence,* Vol. 1. MIT Press, Cambridge, London.

Monin, A.S. & Yaglom, A.M. (1975), *Statistical Fluid Mechanics. Mechanics of Turbulence,* Vol. 2. MIT Press, Cambridge, London.

Monin, A.S. & Yaglom, A.M. (1992). *Statistical Fluid Mechanics: Theory of Turbulence,* Vol. 1, 2nd Russian edition. Gidrometeoizdat, St Petersburg.

Munk, W. (1966). Abyssal recipes. *Deep Sea Research* **13**, 707–30.

Murray, J.D. (1977). *Lectures on Non-linear Differential Equation Models in Biology.* Clarendon Press, Oxford.

Muskhelishvili, N.I. (1963). *Some Basic Problems of the Mathematical Theory of Elasticity,* 2nd English edition. P. Noordhoff, Groningen.

Muskhelishvili, N.I. (1966). *Some Basic Problems of Mathematical Theory of Elasticity,* 5th Russian edition. Nauka, Moscow.

Nigmatulin, R.I. (1965). A plane strong explosion on a boundary of two ideal, calorically perfect gases. *Bulletin MGU, Ser. Matem. Mekh.* **1**, 83–7.

Nikuradse, J. (1932). Gesetzmässigkeiten der turbulenten Strömung in glatten Röhren. VDI Forschungscheft No. 356.

Nikolaeva, G.G. *et al.* (1975). Metabolism rate and size-weight characteristics of the *Rhithropanopeus harrisii tredantatus* crab from the Caspian Sea. *Oceanology USSR* **15**, 99–100.

Norton, F.H. (1929). *Creep of Steel at High Temperatures.* McGraw Hill, New York.

Novikov, S., Manakov, S.V., Pitaevsky, L.P. & Zakharov, V.E. (1984). *Theory of Solitions: The Inverse Scattering Method.* Consultants Bureau, New York, London.

Nowell, A.R.M. & Hollister, C.D. (1985). The objectives and rationale of HEBBLE. *Marine Geology* **66**, 1–12.

Obukhov, A.M. (1941). On the distribution of energy in the spectrum of a turbulent flow. *Doklady, USSR Ac. Sci.* **32** (1), 22–4.

Obukhov, A.M. (1946). Turbulence in thermally inhomogeneous atmosphere. *Proc. Inst. Theor. Geophys., USSR Ac. Sci.* **1**, 95–115.

Obukhov, A.M. (1962). Some specific features of atmospheric turbulence. *J. Fluid Mech.* **13** (1), 77–81.

Offen, G.R. & Kline, S.J. (1975). A proposed model of the bursting process in turbulent boundary layers. *J. Fluid Mech.* **70** (2), 209–28.

Oleinik, O.A. (1957). Discontinuous solutions of nonlinear differential equations. *Uspekhi Mat. Nauk* **12**, 3(75), 3–73.

Oleynik, O.A., Kalashnikov, A.S. & Chzhou Yui-lin (1958). The Cauchy problem and boundary problems for equations of the type of unsteady filtration. *Izvestiya, USSR Ac. Sci., Ser. Mat.* **22**, 667–704.

Oppenheim, A.K., Kuhl, A.C. & Kamel, M.M. (1972). On self-similar blast waves headed by the Chapman–Jouguet detonation. *J. Fluid Mech.* **55** (2), 257–70.

Oppenheim, A.K., Kuhl, A.L., Lundström, E.A. & Kamel, M.M. (1971). A systematic exposition of the conservation equations for blast waves. *J. Appl. Mech.* **38** (4), 783–94.

Oppenheim, A.K., Lundström, E.A., Kuhl, A.C. & Kamel, M.M. (1972). A parametric study of self-similar blast waves. *J. Fluid Mech.* **52** (4), 657–82.

Orowan, E. (1949). Fracture and strength of solids. *Rep. Progr. Phys. Soc. London* **12**, 185–232.

Ovsyannikov, L.V. (1978). *Group Analysis of Differential Equations.* Nauka, Moscow.

Panasyuk, V.V. (1968). *Limiting Equilibrium of Brittle Bodies with Cracks.* Naukova Dumka, Kiev.

Paquette, G.C., Chen, L.-Y., Goldenfeld, N. & Oono, Y. (1994). Structural stability and renormalization group for propagating fronts. *Phys. Rev. Lett.* **72**, 76–9.

Paquette, G.C. & Oono, Y. (1994). Structural stability and selection of propagation fronts in semilinear partial diffferential equations. *Phys. Rev. E* **49**, 2368–88.

Paris, P.C. & Erdogan, F. (1963). A critical analysis of crack propagation laws. *J. Basic Eng. Trans. ASME, Ser. D.* **85**, 528–34.

Parkhomenko, V.P., Popov, S.P. & Ryzhov, O.S. (1977a). On the influence of the initial velocity of particles on the unsteady axisymmetric gas motions. *Uchenye Zapiski (Research Notes) TSAGI,* **8** (3), 32–8.

Parkhomenko, V.P., Popov, S.P. & Ryzhov, O.S. (1977b). On the influence of the initial velocity of particles on the unsteady spherically symmetric gas motions. *Comput. Math. and Math. Phys.* **15** (5), 1325–9.

Parvin, M. & Williams, J.G. (1975). The effect of temperature on the fracture of polycarbonate. *J. Mater. Sci.* **10** (11), 1883–6.

Patashinsky, A.Z. & Pokrovsky, V.L. (1966). On the behaviour of ordering systems near the phase transition point. *J. Exp. Theor. Phys.* **50** (2), 439–47.

Pattle, R.E. (1959). Diffusion from an instantaneous point source with a concentration-dependent coefficient. *Quart. J. Mech. Appl. Math.* **12**, 407–9.

Pedlosky, J. (1979). *Geophysical Fluid Dynamics.* Springer-Verlag, New York, Heidelberg, Berlin.

Petrovsky, I.G. (1967). *Lectures on Partial Differential Equations.* Saunders, Philadelphia.

Phillips, O.M. (1967). The generation of clear-air turbulence by the degradation of internal waves. In *Atmospheric Turbulence and the Propagation of Radio Waves*, 130–8. Nauka, Moscow.

Phillips, O.M. (1976). Energy loss mechanisms from low-mode waves. Report on the Soviet-American Conference on Internal Waves, Novobirsk, December 1976.

Phillips, O.M. (1977). *The Dynamics of the Upper Ocean*, 2nd edition. Cambridge University Press, Princeton.

Polubarinova-Kochina, P.Ya. (1962). *Theory of Groundwater Movement.* Princeton University Press.

Prandtl, L. (1932a). Meteorologische Anwendungen der Strömungslehre. *Beiträge Phys. Atmos.* **19** (3), 188–202.

Prandtl, L. (1932b). Zur turbulenten Strömung in Röhren und längs Platten. *Ergebn. Aerodyn. Versuchsanstalt, Göttingen* **B4**, 18–29.

Prandtl, L. (1945). Ueber ein neues Formelsystem für die ausgebildete Turbulenz. *Nach. Ges. Wiss. Göttingen, Math.-Phys. Kl.*, 6–18.

Praskovsky, A. & Oncley, S. (1994). Measurement of Kolmogorov constant and intermittency exponent at very high Reynolds numbers. *Physics of Fluids* **6** (9), 2886–2889.

Prostokishin, V.M. (1994). Private communication.

Raizer, Yu.P. (1968). A high-frequency high-pressure gas flow discharge as a slow combustion process. *J. Appl. Mech. Tech. Phys.* **9** (3), 239–43.

Raizer, Yu.P. (1970). Physical foundations of the theory of cracks in brittle fracture. *Soviet Phys. Uspekhi* **100** (2), 329–47.

Raizer, Yu.P. (1977). *Laser-induced Discharge Phenomena.* Consultants Bureau, New York.

Rao, K.N., Narasimha, R. & Badri Narayanan, M.A. (1971). The bursting phenomenon in a turbulent boundary layer. *J. Fluid Mech.* **48** (2), 339–52.

Reynolds, O. (1895). On the dynamical theory of incompressible viscous fluids and the determination of the criterion. *Phil. Trans. Roy. Soc. London* **186**, 123–64.

Reynolds, W.C. (1976). Computation of turbulent flows. *An. Rev. Fluid Mech.* **8**, 183–208.

Richardson, L.F. (1922). *Weather Prediction by Numerical Process.* Cambridge University Press.

Richardson, L.F. (1961). The problem of contiguity: an appendix of statistics of deadly quarrels. *General Systems Year Book* **6**, 139–87.

Roesler, F. (1956). Brittle fracture near equilibrium. *Proc. Phys. Soc.* **B69**, 981–92.

Rosen, J.B. (1954). Theory of laminar flame stability, I, II. *J. Chem. Phys.* **22** (4), 733–48.

Samarsky, A.A. & Sobol', I.M. (1963). Examples of numerical computation of temperature waves. *Comput. Math. and Math. Phys.* **3** (4), 702–16.

Sapunkow, Ia.G. (1967). Convergent detonation waves under Chapman-Jouguet conditions in media with variable and constant initial densities. *Appl. Math. Mech. (PMM)* **31** (5), 945–8.

Schlichting, H. (1968). *Boundary Layer Theory*, 6th edition. McGraw-Hill, New York.

Sedov, L.I. (1944). Decay of isotropic turbulent motions of an incompressible fluid. *Doklady, USSR Ac. Sci.* **42** (3), 121–4.

Sedov, L.I. (1945). On some unsteady motions of compressible fluids. *Prikl. Mat. Mekh.* **9** (4), 293–311.

Sedov, L.I. (1946). Propagation of strong shock waves. *Prikl. Mat. Mekh.* **10**, 241–50, (Pergamon Translations, No. 1223).

Sedov, L.I. (1959). *Similarity and Dimensional Methods in Mechanics*. Academic Press, New York.

Sedov, L.I. (1971). *A Course in Continuum Mechanics*. Wolters-Noordhoff, Groningen.

SethuRaman, S. (1980). A case of persistent breaking of internal gravity waves in the atmospheric gravity waves in the atmospheric surface layer over the ocean. *Boundary-layer Meteorology* **19** (1), 67–80.

Shchelkachev, V.N. (1959). *Development of Oil-water Strata Under Elastic Drive*. Gostoptekhizdat, Moscow.

Shkadinsky, K.G., Khaikin, B.I. & Merzhanov, A.G. (1971). Propagation of a pulsating exothermic reaction front in the condensed phase. *Comb. Expl. Shock Waves* **7**, 15–22.

Shushkina, E.A., Kus'micheva, V.I. & Ostapenko, L.A. (1971). Energy equivalent of body mass, respiration, and calorific value of mysids from the Sea of Japan. *Oceanology USSR* **11** (6), 880–3.

Sobolev, S.L. (1954). On a new problem of mathematical physics. *Izvestiya, USSR Ac. Sci., ser. mat.* **18** (1), 3–50.

Sneddon, I.N. (1951). *Fourier Transforms*. McGraw-Hill, New York.

Staniukovich, K.P. (1960). *Unsteady Motion of Continuous Media*. Pergamon Press, New York.

Sternberg, E. & Koiter, W.T. (1958). The wedge under a concentrated couple: a paradox in the two-dimensional theory of elasticity. *J. Appl. Mech.* **25** (4), 575–81.

Stewart, R.W. (1951). Triple velocity correlations in isotropic turbulence. *Proc. Camb. Phil. Soc.* **47**, 146–57.

Stommel, H. (1958). The abyssal circulation. *Deep Sea Research* **5**, 80–2.

Storåkers, B. & Larsson, P.L. (1994). On Brinell and Boussinesq indentation of creeping solids. *J. Mech. Phys. Solids* (in press).

Stückelberg, E.C.G. & Peterman, A. (1953). La normalisation des constantes dans la théorie des quanta. *Helvetica Physica Acta* **XXVI**, 499–520.

Swift, J. (1992) *Gulliver's Travels*. Wordsworth Classics (see p. 124).

Tabor, D. (1951). *Hardness of Metals*. Clarendon Press, Oxford.

Taffanel, M. (1913). Sur la combustion des mélanges gazeux et les vitesses de réaction. *C. R. Ac. Sci. Paris* **157**, 714–7.

Taffanel, M. (1914). Sur la combustion des mélanges gazeux et les vitesses de réaction. *C. R. Ac. Sci. Paris* **158**, 42–5.

Taylor, G.I. (1910). The conditions necessary for discontinuous motion in gases. *Proc. Roy. Soc.* **A84**, 371–7.

Taylor, G.I. (1935). Statistical theory of turbulence, I–IV. *Proc. Roy. Soc.* **A151**, 421–78.

Taylor, G.I. (1941). The formation of a blast wave by a very intense explosion. Report RC-210, 27 June 1941, Civil Defence Research Committee.

Taylor, G.I. (1950a). The formation of a blast wave by a very intense explosion. I, Theoretical discussion. *Proc. Roy. Soc.* **A201**, 159–74.

Taylor, G.I. (1950b). The formation of a blast wave by a very intense explosion. II. The atomic explosion of 1945. *Proc. Roy. Soc.* **A201**, 175–86.

Taylor, G.I. (1963). *Scientific Papers*, G.K. Batchelor (ed.), Vol. 3, *Aerodynamics and the Mechanics of Projectiles and Explosions*. Cambridge University Press.

Thompson, S.M. & Turner, J.S. (1975). Mixing across an interface due to turbulence generated by an oscillating grid. *J. Fluid Mech.* **67** (2), 349–68.

Ting, T.C.T. (1984). The wedge subjected to tractions: a paradox reexamined. *J. Elasticity* **14**, 235–47.

Townsend, A.A. (1976). *Structure of Turbulent Shear Flow*, 2nd edition. Cambridge University Press.

Turner, J.S. (1968). The influence of molecular diffusivity on turbulent entrainment across a density interface. *J. Fluid Mech.* **33** (4), 639–6.

Turner, J.S. (1973). *Buoyancy Effects in Fluids*. Cambridge University Press.

Turner, J.S. (1978). The temperature profile below the surface mixed layer. *Ocean Modelling* **11**, 6–8.

Van den Booghaart, A. (1966). Crazing and characterisation of brittle fracture in polymers. In *Proc. Conf. Phys. Basis of Yield and Fracture*, Oxford University Press.

Van Dyke, M. (1975). *Perturbation Methods in Fluid Mechanics*, 2nd edition. Parabolic Press, Stanford.

Van Dyke, M. (1982). *An Album of Fluid Motions*. Parabolic Press, Stanford.

Vanoni, V. (1946). Transportation of suspended sediment by water. *Trans. Am. Soc. Civil Eng.* **111**, 67–133.

Vavakin, A.S. & Salganik, R.L. (1975). On experimental determination of rate dependence of fracture toughness. *Izvestiya, USSR Ac. Sci., Mech. Solids* **5**, 127–33.

Vlasov, I.O., Derzhavina, A.I. & Ryzhov, O.S. (1974). On an explosion on the boundary of two media. *Comput. Math. and Math. Phys.* **14** (6), 1544–52.

von Kármán, Th. (1911). Über die Turbulenzreibung verschiedener Flüssigkeiten. *Phys. Zeit.* **12** (8), 1071–4.

von Kármán, Th. (1930). Mechanische Ähnlichkeit und Turbulenz. *Nachrichten Ges. Wiss. Göttingen, Math-Phys. Kl.*, 58–76.

von Kármán, Th. (1957). *Aerodynamics*. Cornell University Press, Ithaca.

von Kármán, Th. & Howarth, L. (1938). On the statistical theory of isotropic turbulence. *Proc. Roy. Soc. London* **A164** (917), 192–215.

von Koch, H. (1904). Sur une courbe continue sans tangente obtenue par une construction géometrique élémentaire. *Arkiv Mat. Astron. Fys.* **2**, 681–704.

von Neumann, J. (1941). The point source solution. National Defence Research Committee, Div. B, Report AM-9, June 30, 1941.

von Neumann, J. (1963). The point source solution, in *Collected Works*, Vol. VI, 219–37. Pergamon Press, Oxford, New York, London, Paris.

von Weizsäcker, C.F. (1954). Genäherte Darstellung starker instationärer Stosswellen durch Homologie-Lösungen. *Z. Naturforschung* **9a**, 269–75.

Weatherly, G.L. & Kelly, E.A. (1982). 'Too cold' bottom layer at the bottom of Scotia Rise. *J. Marine Res.* **40**, 985–1012.

Whitham, G.B. (1974). *Linear and Nonlinear Waves.* Wiley, New York.

Williams, M.L. (1952). Stress singularities resulting from various boundary conditions in angular corners of plates in extension. *J. Appl. Mech.* **19** (4), 526–8.

Wilson, K. (1971). Renormalization group and critical phenomena, I, II. *Phys. Rev.* **B4** (9), 3174–83, 3184–205.

Woods, J.D. (1968). Wave-induced shear instability in the summer thermocline. *J. Fluid Mech.* **32** (4), 792–800.

Wu, J. (1969). Mixed region collapse with internal wave generation in a density stratified medium. *J. Fluid Mech.* **35** (3), 531–44.

Yaglom, A.M. (1974). Data on the characteristics of turbulence in the surface layer of the atmosphere. *Izvestiya, USSR Ac. Sci., Atmos. Oceanic Phys.* **10** (6), 566–86.

Zatsepin, A.G., Fedorov, K.N., Voropayev, S.I. & Pavlov, A.M. (1978). Experimental study of the spreading of a mixed region in a stratified fluid. *Izvestiya, USSR Ac. Sci., Atmos. Oceanic Phys.* **14** (2), 170–3.

Zeldovich, Ya.B. (1942). On the distribution of pressure and velocity in products of detonation blasts, in particular for spherically propagating detonation waves. *Zhurn. Eksper. Teor. Fiz.* **12** (9), 389–406.

Zeldovich, Ya.B. (1948). On the theory of flame propagation. *Zhurn. Fiz. Khimii* **22** (1), 27–48.

Zeldovich, Ya.B. (1956). The motion of a gas under the action of a short term pressure shock. *Akust. Zh.* **2** (1), 28–38, (*Sov. Phys. Acoustics* **2**, 25–35).

Zeldovich, Ya.B. (1978). The flame propagation in a mixture reacting at the initial temperature. Preprint, Institute of Chemical Physics, Chernogolovka.

Zeldovich, Ya.B. (1992). *Selected Works. Volume 1, Chemical Physics and Hydrodynamics.* Princeton University Press, Princeton.

Zeldovich, Ya.B. & Barenblatt, G.I. (1958). The asymptotic properties of self-modeling solutions of the nonstationary gas filtration equations. *Soviet Phys. Doklady* **3** (1), 44–7.

Zeldovich, Ya.B., Barenblatt, G.I., Librovich, V.B. & Makhviladze, G.M. (1985). *The Mathematical Theory of Combustion and Explosions.* Consultants Bureau, New York, London.

Zeldovich, Ya.B. & Frank-Kamenetsky, D.A. (1938a). Theory of uniform propagation of flames. *Doklady, USSR Ac. Sci.* **19** (2), 693–7.

Zeldovich, Ya.B. & Frank-Kamenetsky, D.A. (1938b). Theory of uniform propagation of flames. *Zhurn. Fiz. Khimii* **12** (1), 100–5.

Zeldovich, Ya.B. & Kompaneets, A.S. (1950). On the theory of propagation of heat with thermal conductivity depending on temperature. In *Collection of Papers Dedicated to the 70th Birthday of A.F. Ioffe*, 61–71. Izd. Akad. Nauk USSR, Moscow.

Zeldovich, Ya.B. & Kompaneets, A.S. (1960). *Theory of Detonation.* Academic Press, New York.

Zeldovich, Ya.B. & Raizer, Yu.P. (1966). *Physics of Shock Waves and High Temperature Hydrodynamic Phenomena,* Vol. I. Academic Press, New York, London.

Zeldovich, Ya.B. & Raizer, Yu.P. (1967). *Physics of Shock Waves and High Temperature Hydrodynamic Phenomena,* Vol. II, Academic Press, New York, London.

Zheltov, Yu.P. & Christianovich, S.A. (1955). On the hydraulic fracture of the oil stratum. *Izvestiya, USSR Ac. Sci. Techn. Sci.* **5**, 3–41.

Zhukov, A.I. & Kazhdan, Ia.M. (1956). Motion of a gas due to the effect of a brief impulse. *Soviet Phys. Acoustics* **2** (4), 375–381.

Index

Printed in the United States
by Baker & Taylor Publisher Services

Printed in the United States
By Bookmasters